Industrial
Solid-State
Electronics

Devices and Systems

Industrial Solid-State Electronics

Devices and Systems

Timothy J. Maloney

Monroe County Community College
Monroe, Michigan

Prentice-Hall, Inc., Englewood Cliffs, New Jersey   07632

Library of Congress Cataloging in Publication Data

MALONEY, TIMOTHY.
   Industrial solid-state electronics.

   Includes index.
   1. Industrial electronics.   2. Electronic control.
3. Semiconductors.   I. Title.
TK7881.M34 1979   621.38′5   79-12098
ISBN   0-13-463406-3
ISBN   0-13-463398-9 (limited paperback edition)

Printed in the United States of America

10   9   8

PRENTICE-HALL INTERNATIONAL, INC.,   *London*
PRENTICE-HALL OF AUSTRALIA PTY. LIMITED,   *Sydney*
PRENTICE-HALL OF CANADA, LTD.,   *Toronto*
PRENTICE-HALL OF INDIA PRIVATE LIMITED,   *New Delhi*
PRENTICE-HALL OF JAPAN, INC.,   *Tokyo*
PRENTICE-HALL OF SOUTHEAST ASIA PTE. LTD.,   *Singapore*
WHITEHALL BOOKS LIMITED,   *Wellington, New Zealand*

Contents

3   **TRANSISTOR SWITCHES FOR MEMORY**
AND COUNTING   **73**

4 SCRs 132

5 UJTs 153

8 **OPERATIONAL AMPLIFIERS**   **261**

9 **FEEDBACK SYSTEMS AND**
SERVOMECHANISMS   **286**

Preface

This book is intended for two groups of people. The first group consists of students enrolled in electronic programs in community colleges, in two-year technical colleges and institutions, and in four-year engineering technology programs. The second group consists of industrial maintenance personnel and technicians who deal with solid-state systems.

The devices and systems which are presented have been chosen to represent a broad range of industrial applications. The overall aim has been to show how individual devices and circuits interrelate to form useful systems. The precise objectives of each chapter are stated at the beginning so they can be kept constantly in mind as the chapter material is studied.

Whenever a new device is introduced, first its *chief* characteristics are described, and then its uses in industrial systems. The more esoteric aspects of the various devices are left for another book which is not committed to a study of applications. For example, when IC op amps are discussed, the offset problem is identified, and the techniques used to correct it are explained. However, the precise internal *cause* of op amp offset is not covered, since that subject is not really the concern of an applications book.

If two devices tend to overlap each other's area of use, then the finer distinctions between them are made. Relative advantages and disadvantages of the different possible approaches to industrial problems are pointed out whenever necessary. For example, magnetic relays historically have handled almost all of industry's needs for control logic. In the last decade, solid-state logic has taken over many of the jobs previously done by relays, but relay control logic still has some very attractive features. Those good features are clearly explained, and weighed against the benefits of solid-state control logic.

The circuits and systems which are used to illustrate the ideas under discussion are invariably taken from real-life industrial situations. The circuits are usually not exact duplicates of actual circuits that would be used; they have been simplified so the reader can concentrate on the main ideas without getting bogged down in non-essential details. For example, if the subject being discussed is a power supply decoupling network between stages in a servo amplifier, then naturally those circuit components will be drawn. But if we wish to concentrate on other aspects of a servo amplifier, the decoupling components will be deleted from the circuit drawing, to keep it simple.

Knowledge of trigonometry or calculus is not required. However, formula manipulation and power-of-ten notation are used in the solution of example problems, and the reader should be familiar with the prefixes and symbols for engineering notation (milli, micro, Kilo, etc.). No special knowledge of the binary number system is prerequisite. What is necessary to know about binary numbers is covered in this book. It is assumed, however, that the reader is comfortable with the fundamental ideas of electricity (Ohm's law, Kirchhoff's laws), and that he has some familiarity with electronic power supplies and amplifiers, and with the basic ideas of ac circuits (leading and lagging phase relationships, for example). In short, the material is suitable for a second-year electronics course that seeks to stress applications of the ideas learned in first-year courses. Electronic devices that are not usually covered during the first year of study are introduced fresh. For instance, no prior exposure is expected to SCRs, UJTs, breakover thyristors, or op amps.

For readers already working in the field of industrial electronics, this book should bridge the gap between a strictly academic treatment of electronic ideas and the "cook book" treatment provided in most maintenance instruction manuals.

My thanks and appreciation go to Jeanne Deinzer for her superb work in typing the final manuscript. Thanks are also due to Dan Metzger for his astute proofreading, which has eliminated numerous defects from the manuscript, and to Frank McElhannon for his help in preparing some of the drawings.

TIMOTHY J. MALONEY

Monroe, Michigan

1

Transistors as

ON-OFF Devices

OBJECTIVES

After completing this chapter, you will be able to:

1. Explain the operation of a transistor switch and how it differs from a linear transistor amplifier
2. Calculate resistor sizes for a transistor switch
3. Explain the advantages of some common variations of the basic transistor switch
4. Discuss the uses of transistor switches as indicators and as interfacers

1-1 THE TRANSISTOR SWITCH

Besides their uses as amplifiers of ac signals, transistors can also be used as switches. Transistors cannot exactly duplicate the action of normal switches having mechanical contacts, but they do have certain advantages over mechanical switches. The schematic diagram of a transistor switch is shown in Fig. 1-1.

Figure 1-1. Elementary transistor switch.

Figure 1-1 shows the load resistor placed in the collector lead of the transistor, in series with the transistor's main current path. Input voltage V_{in} determines whether the transistor switch will be *open*, preventing the flow of current through the load, or *closed*, allowing current to flow. When V_{in} is a low voltage, there can be no current flow into the base. With no base current, there can be no current through the collector, and since the load is in the collector lead, it too carries no current. Under this condition the transistor is behaving like an open switch in series with the load. When a transistor is operating like this, it is said to be cut OFF or turned OFF.

1-1-1 Calculations involving transistor switches

When a transistor switch is supposed to be cut OFF, V_{in} must be below 0.6 V if the transistor is a silicon type. This ensures that no base current flows into the transistor, because it takes 0.6 V to forward-bias the base-emitter junction of a silicon transistor. To guarantee sure turn-OFF, switching circuits are designed so that V_{in} is less than 0.3 V when the transistor is supposed to be OFF. The closer to ground the input voltage is held, the more certain is the turn-OFF of the transistor switch.

To energize (deliver current to) the load, the transistor must act like a closed switch. This is accomplished by raising V_{in} high enough to drive the transistor into *saturation*. Saturation is the condition in which the collector current is so high that the entire supply voltage, V_{CC}, is dropped across the load resistance. Ideally, the collector current to cause this is given by

$$I_{C(sat)} = \frac{V_{CC}}{R_{LD}},$$

which is Ohm's law applied to the collector circuit. Base current is given by

$$I_{B(sat)} = \frac{I_{C(sat)}}{\beta} = \frac{V_{CC}}{\beta R_{LD}}, \tag{1-1}$$

which expresses the relation between dc collector current and dc base current. The term β in Eq. (1-1) is taken to mean the dc beta (dc current gain) of the transistor, as opposed to the ac beta. Dc beta and ac beta may be quite different for some transistors.

Therefore, to close the switch, V_{in} must be high enough to deliver an amount of current into the base given by Eq. (1-1). Since the base circuit is simply a resistor in series with the base-emitter junction, V_{in} can be calculated by

$$V_{in} = I_{B(sat)} R_B + 0.6 \text{ V},$$

$$V_{in} = \frac{V_{CC} R_B}{\beta R_{LD}} + 0.6 \text{ V}. \tag{1-2}$$

If the base voltage is equal to or greater than the value given by Eq. (1-2), the transistor becomes a closed switch and the load has the entire supply voltage impressed across it.

To summarize, a transistor can act like a mechanical switch in series with a load; the means of opening and closing the switch is the input voltage rather than a mechanical actuator, solenoid plunger, or relay armature, which are the normal methods of operating a mechanical switch.

To avoid confusion, all transistor switches discussed in this chapter will use *npn* transistors. Transistor switches can certainly be built using *pnp* transistors, but they are far less common than *npn* types.

Example 1-1

Refer to Fig. 1-2. What input voltage is required to close the switch (saturate the transistor)? How much load current flows when this happens? How much current must the input source deliver to the base?

Figure 1-2. Transistor switch for turning a light bulb on and off.

Solution. From Eq. (1-2), at saturation, the load resistor will drop the entire supply voltage, so

$$I_{C(sat)} = \frac{V_{CC}}{R_{LD}} = \frac{24 \text{ V}}{16 \, \Omega} = 1.5 \text{ A}.$$

From Eq. (1-1),

$$I_{B(\text{sat})} = \frac{V_{CC}}{\beta R_{LD}} = \frac{24 \text{ V}}{(150)(16 \ \Omega)} = 10 \text{ mA}.$$

The input voltage is given by

$$V_{\text{in}} = I_{B(\text{sat})} R_B + 0.6 \text{ V} = (10 \text{ mA})(1 \text{ K}) + 0.6 \text{ V} = 10.6 \text{ V}$$

Example 1-1 shows that a rather large load current, 1.5 A, can be switched by a small input voltage and small input current. Contrary to what might be expected, the transistor does not necessarily have to be a power transistor with elaborate heat sinking, because the large current flow is accompanied by zero or very low collector-to-emitter voltage. Therefore power, the product of current and voltage, is small.

1-1-2 Comparison of transistor switches versus mechanical switches

So far, the transistor switch has been discussed under the assumption that it becomes a dead short circuit upon closing. This is not quite true. No transistor can saturate so hard that the collector-to-emitter voltage is absolutely zero. Most small signal silicon transistors have a collector-to-emitter voltage at saturation, $V_{CE(\text{sat})}$, of about 0.2 V. A special-purpose *switching transistor* has a lower $V_{CE(\text{sat})}$, usually about 0.1 V at reasonable load currents. At high load currents, $V_{CE(\text{sat})}$ tends to rise somewhat. In most calculations, it is not necessary to take $V_{CE(\text{sat})}$ into account, but it is important to be aware of it when taking measurements in switching circuits. This slight voltage drop is the chief disadvantage of the transistor compared to the mechanical switch. A small voltage drop is seldom a major problem in itself, but because of it, transistor switches cannot be stacked in series as mechanical switches often are. For example, relay contacts are commonly wired in series with each other and with other switches, as shown in Fig. 1-3(a). The equivalent circuit with transistors, Fig. 1-3(b), is not good practice, because the small voltages across each transistor add together to produce a large total voltage drop.

Transistor switches are perfectly capable of operating in parallel, however, as shown in Fig. 1-4.

Transistors have several advantages over conventional mechanical switches:

Figure 1-3. (a) Various mechanical switches connected in series. All the switches must be closed to complete the circuit. (b) Transistor switches connected in series. This is usually inadvisable.

(a)   (b)

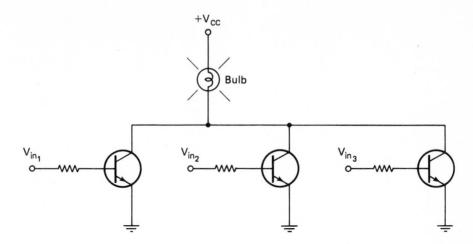

Figure 1-4. Transistor switches in parallel. Closing any one of the switches will turn on the light bulb.

a. They have no moving parts, so they are free from wear and tear and can operate an unlimited number of times. Conventional switches suffer contact wear and cannot be depended on for more than a few million operations. Also, because transistors have no exposed contacts, it is impossible for foreign matter to collect on the contacts and prevent good closure. This problem plagues conventional switches in dirty environments.

b. Transistor switches are much faster acting than conventional switches, which have closing times measured in milliseconds. Transistor switches have closing times (turn-ON times) measured in microseconds.

c. Transistor switches have no inherent *bounce*, whereas mechanical switches do. Bounce is the problem of switch contacts closing and opening many times in quick succession before they close for good. Figure 1-5(a) shows an exaggerated waveform

Figure 1-5. (a) Voltage waveform showing contact bounce in a mechanical switch. The switch bounces four times before final closure. (b) Voltage waveform of a transistor switch showing no bounce. In both cases the switch is actuated at time t_1.

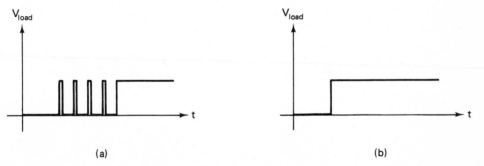

(a)   (b)

of load voltage versus time when a mechanical switch energizes a load. Figure 1-5(b) shows the waveform of load voltage when a transistor switch energizes it.

d. When a transistor switch powers an inductive load, there is no spark when it opens. When a mechanical switch opens to stop current through an inductive load, the inductor's kickback often causes an arc across the contacts. This arc not only pits the contact surfaces, but it may be hazardous in some circumstances.

1-1-3 Testing transistor switches

Unfortunately, it is not possible to tell by visual inspection if a transistor switch is closed or open, as it is for most mechanical switches. A meter is needed to detect switch closure. Figure 1-6 will be used to explain the procedure for checking a transistor switching circuit. In general, when the switch is closed, V_{CE} will read near 0 V, and when it is open, V_{CE} will equal V_{CC}.

Figure 1-6. The voltages which should be measured when troubleshooting a transistor switch.

The V_{CC} reading in the open condition is due to the fact that the load is carrying no current and therefore is dropping no voltage. The entire supply voltage is dropped across the switch, just as in a conventional switching circuit.

If the load fails to energize when expected, it is first necessary to check V_{in}. This voltage must be high enough to guarantee transistor turn-ON. If it isn't, the problem is in the signal source, not the transistor itself.

If V_{in} is sufficiently high to turn ON the switch but the load still fails to energize, the supply voltage should be checked where it connects to the top of the load.

When turned ON, a silicon transistor should have a V_{BE} of about 0.6 V. If V_{in} is sufficiently high for turn-ON but V_{BE} is much higher or lower than 0.6 V, for instance, 1.5 or 0.2 V, the base-emitter junction is bad and the transistor must be replaced. Large power transistors carrying heavy load currents sometimes have a V_{BE} in excess of 1 V. Therefore in such cases a V_{BE} reading of 1.5 V would not necessarily indicate a bad junction. In these situations the manufacturer's specifications should be consulted.

If V_{BE} is normal and base current is certainly flowing, then V_{CE} should be inspected. If V_{CE} is close to V_{CC}, the transistor has a bad collector-base junction and should be replaced.

If V_{CE} is close to 0 V but the load stays deenergized, the load itself is probably open-circuited. In that case the load must be replaced.

When V_{in} goes to a low voltage, the transistor is supposed to turn OFF and deenergize the load. If the load fails to deenergize when expected, there is probably a short circuit from collector to base or from collector to emitter, and the transistor must be replaced.

1-2 VARIATIONS OF THE BASIC TRANSISTOR SWITCH

Sometimes it is not certain that a supposedly low input voltage will be low enough to turn OFF a transistor switch. This happens whenever the input voltage is close to 0.6 V, say 0.5 V. In these marginal situations, steps must be taken to guarantee that the transistor turns OFF. Figure 1-7 shows the two most common schemes for this purpose.

In Fig. 1-7(a), a diode is inserted in the base lead. This causes an increase in the input voltage necessary to start current flow in the base circuit. The 0.6-V turn-ON voltage of the diode is now added to the 0.6-V turn-ON voltage of the base-emitter junction. Therefore, if V_{in} is close to 0.6 V because of some malfunction of the source, the transistor remains securely OFF, and the switch remains open.

Figure 1-7. Variations of the basic transistor switch to ensure turn-OFF. (a) A diode placed in the base lead. (b) A hold-OFF resistor for pulling the base voltage below 0.6 V.

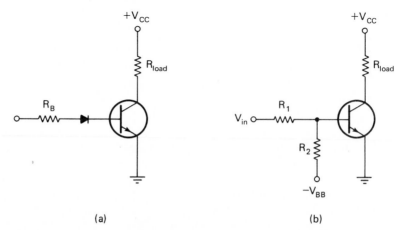

(a)   (b)

Figure 1-7(b) shows a hold-OFF resistor, R_2, added to the circuit. This resistor does what its name suggests, namely holds the transistor OFF when V_{in} is in the marginal range. Inspection of Fig. 1-7(b) shows that R_1 and R_2 form a series voltage divider as long as the base-emitter junction is not forward biased. The voltage at the base is always more negative (less positive) than V_{in}. Therefore, if V_{in} is marginal, the voltage at the base is pulled below 0.6 V by the hold-OFF resistor connected to the negative power·supply. When V_{in} is in the high range, however, the base voltage is sufficient to turn the transistor ON even with the hold-OFF resistor installed.

Example 1-2

For the circuit of Fig. 1-8, imagine three separate input voltages: (a) 0.1 V, (b) 0.6 V, and (c) 3.5 V. For each input voltage, calculate the base voltage and state whether the transistor is ON or OFF.

Figure 1-8. Transistor switch with specific component values, for use in Example 1-2.

Solution. a. Begin by assuming that the base-emitter is reverse biased, so the resistors are effectively in series. Then, for $V_{in} = 0.1$ V,

$$V_{R_2} = \frac{(6.8 \text{ K})(2.1 \text{ V})}{6.8 \text{ K} + 2.2 \text{ K}} = 1.6 \text{ V},$$

where V_{R_2} is the voltage across R_2. The 2.1 V is the total potential difference across the series combination, or the voltage difference between $+0.1$ and -2.0 V. The base voltage relative to ground can be found by

$$V_B = -2.0 \text{ V} + 1.6 \text{ V} = -0.4 \text{ V}.$$

According to the voltage divider, V_B will be -0.4 V. This would indeed reverse-bias the base-emitter junction; therefore the original assumption was valid that R_1 and R_2 could be treated in series. Because the base is negative with respect to the emitter, the transistor is certainly OFF, and the switch is open.

b. For the normally marginal input voltage of 0.6 V, the voltage across R_2 can be found by

$$V_{R_2} = \frac{(6.8 \text{ K})(2.6 \text{ V})}{6.8 \text{ K} + 2.2 \text{ K}} = 1.9 \text{ V}$$

Therefore $V_B = -2.0 \text{ V} + 1.9 \text{ V} = -0.1 \text{ V}$. Again, because the base-emitter junction is not forward biased, the assumption to treat the resistors as a series circuit is valid and the transistor is OFF.

c. If the input voltage is $+3.5$ V, the calculation gives

$$V_{R_2} = \frac{(6.8 \text{ K})(5.5 \text{ V})}{6.8 \text{ K} + 2.2 \text{ K}} = 4.1 \text{ V}.$$

The 5.5 V in the above equation is the total voltage drop across the resistor combination, from $+3.5$ to -2.0 V. The calculation for base voltage yields

$$V_B = -2.0 \text{ V} + 4.1 \text{ V} = +2.1 \text{ V},$$

which is impossible. The base-emitter junction will be forward biased and will not allow V_B to rise much above 0.6 V. In this case the resistors cannot be viewed as a series circuit. The calculation for base voltage gives an impossible value because the initial assumption was false. The transistor will be turned ON, and the switch will be closed.

Speedup Capacitors. To increase the speed of a transistor switch in those applications where very fast switching is needed, the approach in Fig. 1-9 is often used. The capacitor across R_B is usually referred to as a *speedup capacitor*, because it speeds up the turning ON and the turning OFF of the transistor. When V_{in} rises above ground level and starts delivering base current to the transistor, the capacitor at first acts like a short circuit, since it cannot charge up instantly. Therefore the current into the base during the first instant of turning ON is larger than normal because R_B is effectively short-circuited. This larger than normal current serves to drive the transistor into saturation in a shorter time. Later, after the capacitor is charged up, it becomes an open circuit and does not affect the operation of the transistor.

When the input voltage falls back to ground, turning the transistor switch OFF, the capacitor tends to reverse-bias the base-emitter junction for a short time, because it is charged $+$ on the left, as shown in Fig. 1-9, at the instant the input voltage starts to fall. When V_{in} reaches zero, effectively connecting the input terminal to the emitter, the negative voltage on the left side of the capacitor

Figure 1-9. A speedup capacitor in parallel with the base resistor. C is usually between 100 and 1000 pF.

reverse-biases the base-emitter junction, driving the transistor into cut-OFF sooner than would otherwise occur.

With proper selection of a speedup capacitor, the turn-ON and turn-OFF times of switching transistors can be reduced to a few tenths of a microsecond or less. Most speedup capacitors have a capacitance of several hundred picofarads.

Sometimes the load in a transistor switching circuit is not inserted directly in the collector lead but as shown in Fig. 1-10. This is similar to the method of connecting the load of a small-signal ac amplifier, except that in this case there is no output coupling capacitor. Note that the load is energized when the transistor turns OFF and is deenergized when the transistor turns ON. This behavior is the opposite of the previous situation, in which the load was in series with the collector lead. It is important to make a clear distinction in your mind between these two situations. Both are commonly encountered.

Figure 1-10. Transistor switching circuit with the load connected between collector and ground instead of in series with the collector lead.

Totem-pole Switches. In applications where the load has some associated capacitance (considered to be in parallel with R_{LD}), the method of Fig. 1-10 results in a slow rise in load voltage after the transistor turns OFF. This is due to the necessity of charging the load capacitance through R_C, the collector resistor. The larger the load capacitance and/or the larger the value of R_C, the greater is the time constant (R times C), and the slower the voltage rises across the load. Sometimes a slow buildup of voltage cannot be tolerated, so the circuit of Fig. 1-11 is used.

The totem-pole circuit is so named because one transistor is stacked on top of another. To energize the load, T_1 is turned ON and T_2 is turned OFF; this connects the top of the load to V_{CC}. To deenergize the load, T_1 is turned OFF and T_2 is turned ON; this connects the top of the load to ground potential. There is either no resistance (as shown in Fig. 1-11) or very low resistance in the collector lead of T_1, so the slow buildup of load voltage is eliminated. Instead, the top of the load is

Figure 1-11. Totem-pole transistor switch. At all times, one transistor is ON and the other transistor is OFF.

connected directly to the positive supply when T_1 turns ON. T_1 is said to "pull the load up" and is called a *pull-up transistor*. T_2 is called a *pull-down transistor*.

The input circuit of Fig. 1-11 is responsible for turning T_1 and T_2 ON and OFF and for ensuring that they are never both ON at the same time. Having T_1 and T_2 both ON at the same time would constitute a short circuit and would cause destruction of at least one transistor. Also, for this reason, totem-pole switches cannot be paralleled as shown in Fig. 1-4. T_1 of one switch and T_2 of another switch might turn ON at the same time and short-circuit the voltage supply. You may want to draw a pair of totem-pole switches with the outputs tied together to see how this can happen.

1-3 APPLICATIONS OF THE TRANSISTOR SWITCH

1-3-1 Indicating lamps

One common use of transistor switches is to drive indicating lamps. An indicating lamp is one which displays the condition at a certain point in a circuit. It may be used to indicate that a certain motor starter is energized, or that a certain limit switch is closed, or that a certain line in a digital circuit is at a HI* voltage.

For example, Fig. 1-12(a) shows a transistor switch indicating the output from a flip-flop, a digital memory device which we will study later. When the output of the flip-flop goes to a HI level, approximately 5 V for many flip-flops, the transistor turns ON and the lamp glows. An observer could tell the state of the flip-flop at a glance, without taking the trouble to actually hook up a meter to measure it.

*The meaning of the term *HI* in a digital circuit will be fully explained in Chapter 2.

Figure 1-12. Lamp drivers. (a) Straightforward transistor switch. (b) With an emitter follower added to raise the input impedance.

Sometimes a source has very low current-delivering capability, so that even the base current drawn by a transistor switch is an unacceptable load. In such cases, another variation on the basic switch is used, as shown in Fig. 1-12(b). When the output goes to a HI level, it turns on T_1, which is wired as an emitter follower. Because of the high input impedance of an emitter follower, the current drawn from the flip-flop is very low. The current originating in the collector lead of T_1 flows into the base of T_2, turning on T_2 and lighting the lamp.

Digital Readout. The circuit of Fig. 1-12(a) is used repetitively in *digital readout* applications. Digital readout means to take a binary number contained in a digital circuit and display it so that it is easily understood by a human being. As an example, a *decade counter* is a digital circuit capable of counting to 9; we will study decade counters in detail in Chapter 3. The important thing to realize here is that the number contained in a decade counter is not easily understood by a human. This is because a human would have to take four separate voltage measurements on a decade counter to find out what number it contains; even then the number would be in binary code and would have to be converted to decimal. This process is inconvenient and likely to produce mistakes. To avoid these difficulties electronic manufacturers make packaged circuits which can convert the voltages in a decade counter into a decimal numeral and switch on indicating lights to display that numeral. Circuits which perform this function are called *decoder/drivers* or simply *decoders*. The final output stages of decoders are transistor switches.

Figure 1-13 shows the output stages of a decoder/driver which would drive

Figure 1-13. Output transistor switches of a decimal decoder/driver. The 10 electrodes of the readout tube are wired to the collectors of the 10 transistor switches.

a 1-of-10 decimal indicating device, such as a Nixie* tube. A Nixie tube has 10 electrodes which are shaped like the numerals 0 through 9. The decoder/driver displays the numeral which it senses by switching on the appropriate electrode through a transistor switch (the decoder's *sensing* circuitry is not shown in Fig. 1-13, only its output circuitry). The electrical connection between the positive supply bus and the electrodes is through ionized gas. A 15-K current-limiting resistor is installed in series with the positive supply bus.

Other decoder/drivers have only 7 transistor switches instead of 10. These are designed for use with a different type of readout device, called a seven-segment readout. A numeral is formed by lighting up a particular combination of the seven segments. The geometrical layout of the segments is shown in Fig. 1-14.

For example, the numeral 7 would be formed by lighting up segments *a*, *b*, and *c*. The numeral 6 would be formed by lighting up segments *f*, *e*, *d*, *c*, and *g*. As far as the output switches are concerned, the only difference between seven-

*Trademark of Burroughs Corporation.

Figure 1-14. Layout of the segments in a seven-segment readout device.

segment decoders and decimal decoders (sometimes called 1-of-10 decoders) is that seven-segment decoders must have more than one transistor switch turned ON. This can be understood by referring to Fig. 1-15, which shows the seven output transistors and how they are connected to the segments. The 220-Ω resistors are present to limit current flow.

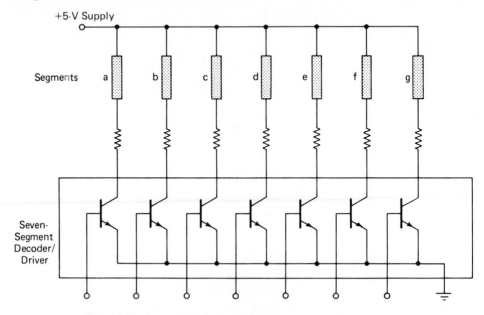

Figure 1-15. Output transistor switches of a seven-segment decoder/ driver. The seven segments are connected through resistors to the collectors of the seven transistor switches.

The segments themselves take a variety of forms in modern display devices. They may be incandescent filaments, light-emitting diodes (LEDs), or liquid crystals.

Figure 1-15 shows a 5-V bus line, whereas Fig. 1-13 shows a 180-V bus line. Readout devices operate on a wide variety of supply voltages. Of course, the transistors used to drive high-voltage readouts must themselves have a high collector-emitter breakdown voltage rating.

It should be noted that Figs. 1-13 and 1-15 show the positive ends of the indicating devices tied together and the negative ends connected to separate transistor switches. This configuration is called *common anode* and is the most frequently seen in industrial applications. However, there are some readout devices which are designed so that the negative ends are all tied together and the positive ends go to separate transistor switches. These are called *common cathode* devices. The terms anode and cathode come from the fact that the segments are often diodes (LEDs) which have an anode and a cathode. If the segments are incandescent filaments, the terms are still used even though there is no anode or cathode.

Example 1-3

In Fig. 1-15, which transistors would be turned ON to make the numeral 2? Which ones would be switched ON to make the numeral 8?

Solution. For 2, the transistors driving segments *a*, *b*, *g*, *e*, and *d* would be turned ON. For 8, all transistors would be turned ON.

1-3-2 Interfacing between different voltage
levels

Solid-state logic circuits in industrial settings will be taken up in detail in Chapter 2, but let it be stated right now by way of introduction that all such circuits can be broken up into three parts. These parts are referred to as the input section, the logic section, and the output section. To secure reliable operation, the input section and the output section must be operated at a high voltage, usually 115 V ac. Exactly why this is necessary will be discussed in Chapter 2. Since the logic section is comprised of low-voltage solid-state circuits, there must be some method of making the high-voltage input and output sections compatible with the low-voltage logic section. This problem of matching different sections with each other is called the *interface* problem, and the circuits which do the matching are called the *interface circuits*. Transistor switches find frequent use in interface circuits.

The most obvious interface application of a transistor switch is as a relay energizer. This application, shown in Fig. 1-16, is a simple and inexpensive way of matching the logic and output sections. A low-voltage signal from a circuit in the logic section supplies the base current to the transistor switch, which energizes a relay coil. The relay contact, which is electrically isolated from the logic circuitry, can then switch the high voltage to the output device, a solenoid valve in Fig. 1-16.

Figure 1-16. Transistor switch being used to interface between the logic section and the output section of a control circuit. When the transistor switch closes, it energizes the relay; the relay contact then causes the solenoid valve to energize.

The diode in parallel with the relay coil is a *counter-EMF diode*. Its purpose is to short out the inductive counter-EMF of the relay coil upon deenergization. Otherwise, the counter-EMF might take the form of high-voltage transient surges, which could damage the transistor switch.

To interface between input and logic sections, the transistor switch might be used as shown in Fig. 1-17. When the mechanical limit switch closes in Fig. 1-17, the step-down transformer is energized, causing the full-wave rectifier to deliver dc voltage to the base of the transistor switch. The transistor turns ON, and the collector voltage falls to zero. This 0-V signal is fed to the logic circuits to indicate that the mechanical limit switch has closed.

Figure 1-17. Transistor switch being used as part of the interface circuit between input and logic sections.

If the limit switch were open, the transformer would be deenergized, the transistor would turn OFF, and the collector voltage would equal V_{CC}. The logic circuits would be designed to interpret this V_{CC} signal to mean that the limit switch was open.

In Fig. 1-17, the load (the logic circuit) is not inserted in the collector lead but is connected between the collector and ground as previously introduced in Fig. 1-10. The circuit is still a switching circuit and not a conventional amplifier because the transistor is always operated either cut OFF or saturated. It is never biased in its linear operating region. R_1, R_2, and R_C in Fig. 1-17 would be properly sized to cause this to occur.

QUESTIONS AND PROBLEMS

1. If the load in Fig. 1-2 were two 20-Ω bulbs in parallel, how large an input voltage would be necessary to turn the transistor switch ON?
2. In Fig. 1-4, if V_{in_1} is holding the first transistor ON, will raising V_{in_2} have any effect on the bulb?

3. In a transistor switch using a silicon transistor, what particular problem is indicated if you measure 2.5 V between base and emitter?

4. If you measure a voltage equal to V_{CC} between the collector and emitter while base current is flowing, what particular problem is indicated?

5. In a transistor switch, if V_{in} equals 0 V but the load remains energized, what particular problem is indicated?

6. In what ways do specially designed switching transistors differ from standard small-signal transistors?

7. Give three practical reasons why specially designed switching transistors would be preferred over standard small-signal transistors for use in a transistor switch.

8. Roughly speaking, what is the life expectancy of a mechanical switch?

9. What is the life expectancy of a properly used transistor switch?

10. In Fig. 1-1, $V_{in} = V_{CC} = 10$ V, $R_C = 500$ Ω, and $R_B = 70$ K. What is the minimum dc beta which will produce saturation?

11. In Fig. 1-1, the resistance of the load is selectable. If $V_{in} = 3$ V, $V_{CC} = 12$ V, $R_B = 10$ K, and $\beta_{dc} = 80$, what load resistance should be selected to produce saturation? Is this value the maximum resistance or the minimum resistance which can be selected? Explain.

12. In Fig. 1-8, assume that $R_1 = 10$ K, $R_2 = 56$ K, and the negative supply is -4 V. The transistor is silicon.

   a. If $V_{in} = +2.0$ V, is the transistor switch open or closed?

   b. Repeat for $V_{in} = +0.8$ V.

   c. What value of V_{in} would cause doubt about the outcome?

SUGGESTED LABORATORY PROJECTS

PROJECT 1-1

Purpose

a. To build a transistor switch and measure its input and output voltage levels for both conditions

b. To observe the unpredictable behavior of a transistor switch when the input voltage is marginal

c. To alter the basic transistor switch to eliminate unpredictable behavior

Procedure

a. Construct the circuit of Fig. 1-1, with $R_C = 220$ Ω, $R_B = 10$ K, $V_{CC} = 10$ V, and V_{in} coming from a variable dc supply. The dc beta of the transistor should be accurately known.

   1. Increase V_{in} until the transistor saturates and measure the collector-to-emitter voltage for this particular collector current (measure $V_{CE(sat)}$).

   2. What input voltage is required to cause saturation? Does this agree with the results obtained from Eq. (1-2)?

    b. Gradually reduce V_{in} until the transistor is cut OFF.

      1. Measure V_{in}. Does this corroborate what you know about silicon transistors?

      2. With V_{in} set at this value, the input is said to be marginal. Verify this by inserting several different transistors of the same type. Why is the circuit behavior not consistent as transistors are changed?

      3. Insert a diode in the base lead as shown in Fig. 1-6(a). Repeat step 2. Is circuit behavior consistent? Why?

      4. Remove the diode and alter the circuit as shown in Fig. 1-6(b). Use a negative supply of -3 V and a hold-OFF resistor of 15 K. Repeat step 2.

      5. With the hold-OFF circuit installed, measure the new V_{in} which just barely causes cut-OFF. Does this agree with the calculated value? Refer to Example 1-2 to do the calculation.

PROJECT 1-2

The effect of a speedup capacitor can be demonstrated using the circuit of Project 1-1. A fast-rise pulse generator is substituted for the variable supply in Fig. 1-1. V_{CE} must be measured with a high-frequency oscilloscope in order to detect the improvement in turn-ON and turn-OFF times when a speedup capacitor is added to the circuit.

In general, the rise time of an oscilloscope must be at least five times as fast as the rise time of the pulse being measured in order to get an accurate measurement. Therefore if the transistor has a turn-ON time of roughly 200 nsec when the speedup capacitor is installed, which can reasonably be expected, the oscilloscope should have a rise time of about 40 nsec or less. An oscilloscope with a bandwidth of 10 MHz has a rise time of 33 nsec, so such an oscilloscope will be adequate. The pulse generator must also have a rise time of 40 nsec or less.

The best results will be seen for a pulse frequency of about 1 MHz and an oscilloscope sweep speed of about 0.1 μsec/cm.

By substituting various sizes of speedup capacitor and measuring the rise and fall times of V_{CE}, you can observe the capacitor's effects on the turn-ON and turn-OFF times of the transistor switch.

2

The Transistor Switch

as a Decision-Maker

In any industrial system, the control circuits constantly receive and process information about the conditions in the system. This information represents such things as the mechanical positions of movable parts; temperatures at various locations; pressures existing in pipes, ducts, and chambers; fluid flow rates; forces exerted on various detecting devices; speeds of movement; etc. The control circuitry must take all this empirical information and combine it with input from the human operators. Human operator input usually has the form of selector switch settings and/or potentiometer dial settings. Such operator input represents the desired system response or, in other words, the production results expected from the system.

Based on the comparison between system information and human input, the control circuitry *makes decisions*. These decisions concern the next action of the system itself, such as whether to start or stop a motor, whether to speed up or slow down a mechanical motion, whether to open or close a control valve, or even whether to shut down the system entirely because of an unsafe condition.

Obviously no real thought goes into the decision making done by the control circuits. Control circuits merely reflect the wishes of the circuit designer who foresaw all the possible input conditions and designed in the appropriate circuit responses. However, because the control circuits mimic the thoughts of the circuit designer, they are often called *decision-making circuits* or, more commonly, *logic circuits*.

OBJECTIVES

After completing this chapter, you will be able to:

1. Name the three parts of an industrial control circuit and describe the general function of each part
2. Describe how relays can be used to make decisions
3. Distinguish between normally open and normally closed relay contacts
4. Draw relay logic schematics showing series-connected contacts and parallel-connected contacts
5. Describe in detail the operation of a part-classifying system using relay logic
6. Explain the operation of the five most common solid-state logic gates: AND, OR, NAND, NOR, and NOT
7. Describe in detail the operation of a part-classifying system using solid-state logic
8. Name and explain the operation of the various circuits used for input signal conditioning in solid-state logic
9. Explain the purpose and operation of output amplifiers used with solid-state logic
10. Discuss the relative advantages and disadvantages of solid-state logic and relay logic
11. Describe in detail the operation of three real-life solid-state logic systems: a machine tool routing system, a first fault annunciator, and a machine tool drilling system
12. Discuss the characteristics of the five most common industrial logic families: RTL, DTL, HTL, TTL, and CMOS

2-1 SYSTEMS CONTAINING LOGIC CIRCUITS

An electrical control circuit for controlling an industrial system can be broken down into three distinct parts. These parts or sections are (a) input, (b) logic, and (c) output.

The *input section*, sometimes called the *information-gathering section* in this book, consists of all the devices which supply system information and human operator settings to the circuits. Some of the common input devices are push buttons, mechanical limit switches, pressure switches, and photocells.

The *logic section*, sometimes called the *decision-making section* in this book, is that part of the circuit which acts upon the information provided by the input section. It makes decisions based on the information received and sends orders to the output section. The logic section's circuits are usually built with magnetic relays, discrete transistor circuits, or integrated transistor circuits.* Fluidic devices can also be used for logic, but they are much less common than electromagnetic and electronic methods. We will not discuss fluidic devices. The essential ideas of logic circuits are universal, no matter what actual devices are used to build them.

The *output section*, sometimes called the *actuating device section* in this book, consists of the devices which take the output signals from the logic section and convert or amplify these signals into useful form. The most common actuating devices are motor starters and contactors, solenoid coils, and indicating lamps.

The relationship among these three parts of the control circuit is illustrated in Fig. 2-1.

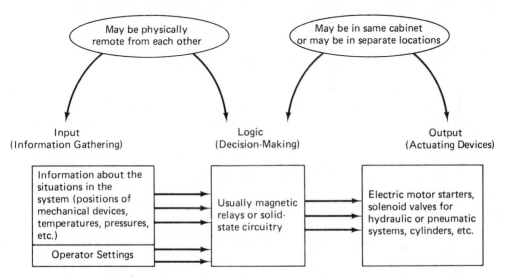

Figure 2-1. The relationship among the three parts of an industrial control system.

2-2 LOGIC CIRCUITS USING MAGNETIC RELAYS

For many years, industrial logic functions were performed almost exclusively with magnetically operated relays, and relay logic still enjoys wide popularity today. In this method of construction, a relay coil is energized when the circuit leading up to the coil is completed by closing certain switches or contacts. Figure 2-2 shows that relay *A* (R*A*) is energized if limit switch 1 (LS1) and pressure switch 4 (PS4) are closed.

*An integrated circuit (IC) is a circuit in which all the components (transistors, diodes, resistors, etc.) are formed within a single piece of silicon crystal. We will talk more about ICs as we go along.

Figure 2-2. A relay logic circuit in which the relay coil is controlled by input devices, namely a limit switch and a pressure switch.

The design of the circuit in Fig. 2-2 calls for relay *A* to be energized if a certain combination of events occurs in the system. The necessary combination is the closing of LS1 by whatever apparatus operates LS1 and, at the same time, the closing of PS4 by whatever liquid or gas affects PS4. If both of these things occur at the same time, relay *A* will energize. The terms *picked up* or just *picked* are often used to mean energized, and these terms will sometimes be employed in this book.

If either or both of the switches is open, then R*A* will be deenergized. The terms *dropped out* or *dropped* are often used to mean deenergized, and these terms will also be employed occasionally in this book.

If R*A* is deenergized, the contacts controlled by R*A* revert to their normal state. That is, the *normally closed* (N.C.) contacts are closed and the *normally open* (N.O.) contacts are open. On the other hand, if R*A* is energized, all the contacts associated with R*A* change state. The N.C. contacts go open, and the N.O. contacts go closed. Figure 2-2 shows only one of each kind of contact. Real industrial relays often have several contacts of each kind (several N.C. and several N.O. contacts).

Although this circuit is quite simple, it illustrates the two foundation ideas of relay logic circuits, and for that matter of all logic circuits:

a. A positive result (in this case picking up the relay) is conditional on several other events. The exact conditions needed depend on how the feed-in switch contacts are connected. In Fig. 2-2, *both* LS1 and PS4 must be closed because the contacts are connected in series. If the contacts were connected in parallel, *either* switch being closed would pick up the relay.

b. Once a positive result occurs, the result can branch out to many other circuit locations. It can thereby make its effects felt in several different places throughout the control circuits. Figure 2-2 shows R*A* having one N.O. contact and one N.C.

contact with each contact leading off to some other destination in the overall circuit. Therefore the action of R*A* would be felt at both of these other circuit destinations.

Expanding on these ideas concerning circuit logic, Fig. 2-3 shows how contacts which feed in to a relay coil sometimes are controlled by *other relays* instead of mechanical limit switches and other independent switches.

Figure 2-3. A relay logic circuit in which relay coils are controlled by the contacts of other relays.

In Fig. 2-3, the limit switch is mechanically actuated when hydraulic cylinder #3 is fully extended. Hydraulic cylinder #3 would be located somewhere in the mechanical apparatus of the industrial system and would have some sort of cam attached to it for contacting LS3. When the N.O. contact of LS3 goes closed, R*B* picks up. The "branching-out" idea is illustrated here because R*B* has three contacts, each one leading to a separate circuit location. The action of R*B* therefore affects the results at three other circuit locations, in this case R*F*, R*G*, and R*H*. This branching-out idea is often referred to as *fan-out*.

To appreciate the decision-making ability of such circuits, consider R*G* carefully. Imagine that R*G* has control over a solenoid valve which can pass or block the flow of water through a certain pipe. Therefore water will flow if the following conditions are met:

a. R*B* is picked, and
b. R*D* is picked, and
c. R*E* is dropped.

We have already seen that R*B* is controlled by hydraulic cylinder #3 through LS3. Relays R*D* and R*E*, though not described in Fig. 2-3, represent either conditions in the system, or human input, or a combination of both. For concreteness, imagine that R*D* will be picked if adequate water pressure is available and that R*E* will be picked if a certain type of contamination is detected in the water.

What happens here is that R*G* *makes a decision* on whether or not to permit water to flow. It makes its decision by consideration of three conditions:

a. R*B* (N.O.): Hydraulic cylinder #3 must be extended.
b. R*D* (N.O.): Adequate pressure must be available in the system.
c. R*E* (N.C.): The water must not be contaminated.

This therefore is a simple example of how relays are used to build a logic circuit.

2-3 A RELAY LOGIC CIRCUIT FOR A CONVEYOR/CLASSIFYING SYSTEM

To cement what we have learned about general logic systems, let us consider the logic for a specific system. The system layout is drawn schematically in Fig. 2-4(a).

Manufactured parts of varying height and weight come down the conveyor, moving to the right. A height detector measures the height of each part and classifies it as either short or tall, depending on whether the part is below or above a certain prescribed height. Likewise, a weighing device classifies it as either light or heavy depending on whether it is below or above a certain prescribed weight. Each part can therefore be put into one of four overall classifications. It is either (a) short/light, (b) short/heavy, (c) tall/light, or (d) tall/heavy.

The system then color-codes each part by spraying on a paint stripe of the proper color. After it has been painted, the part is sorted into the proper chute depending on its classification. There are four chutes, one for each classification. This sorting is done by having a diverter gate swing out to direct the part off the conveyor into the proper chute. Each chute has its own gate.

Referring to Fig. 2-4(a), we see that the system is divided into three zones.

A part is measured for height and weight in the testing zone. As it leaves the testing zone and enters the painting zone, the part actuates LS1. LS1 is a limit switch with a cat-whisker extension. Such switches are used when the actuating body does not have an exact repeatable position; parts moving on a conveyor belt are a good example. The part may be offset to either the left or the right side of the conveyor. To detect the passage of a part, the detecting switch must be able to respond to a body located anywhere on a line across the width of the conveyor.

As the part enters the painting zone, one of the four paint solenoid valves is opened up, applying a strip of paint as the part moves underneath it. As the part leaves the painting zone and enters the diverter zone, it strikes LS2, another cat-whisker limit switch. At this time the paint valve closes, and one of the four diverter gates swings out. When the part hits the diverter gate it is guided off the belt and into the proper chute. Figure 2-4(b) indicates how the gate swings out to block the path of the moving part. As the part slides down one of the chutes, it strikes the limit switch mounted in that chute, either LS3, LS4, LS5, or LS6. At this time, the diverter gate returns to its normal position, and the system is ready to receive another part in the testing zone.

Figure 2-4. (a) Physical layout of a conveyor/classifying system. (b) Top view of the diverting zone, showing the positions of the four diverting gates and the four chute limit switches.

The parts must be handled in such a way that a new part cannot enter the testing zone until the previous part has cleared one of the chute limit switches. This is because the system must retain the height/weight classification of a part until that part has completely cleared the system. It must retain the classification because it must hold the proper diverter gate out until the part has left the belt.

The relay logic to accomplish the above operation is shown in Fig. 2-5. The operation of the logic circuitry will now be explained. An equivalent solid-state logic circuit will be presented and explained in Sec. 2-6. This way, you can become acquainted with a complete practical logic circuit using relay construction. After

Figure 2-5. Control circuitry for the conveyor/classifying system, with the logic performed by magnetic relays.

an understanding of the system itself is gained, we will go on to study the same system using a more modern method of construction.

Let us start on line 9 of Fig. 2-5. The N.C. RCLR* contact is closed at the time a part enters the testing zone. While a part is in the testing zone, the height detector closes its contact if the part is tall but leaves the contact open if the part is short. This will pick RTAL if the part is tall or leave RTAL dropped if the part is short. If RTAL picks, it seals itself in with the N.O. RTAL contact on line 10. This is necessary because the height detector contact will return to the N.O. condition after the part has left the testing zone, but the system must retain the information about its tallness or shortness until the part has completely gone.

The actual operation of the height detector is of no concern to us at this time since we are concentrating on the system logic.

The weight detector on line 11 does the same thing. If the weight of the part is above the preset weight, the contact closes and picks RHVY, which then seals up with the N.O. contact on line 12. If the part is below the preset weight, the weight detector contact remains open, and RHVY remains dropped.

The circuitry between lines 13 and 16 picks the proper relay to indicate the classification of the part. If the part is short, the N.C. RTAL contact on line 13 will remain closed, applying power to the left side of the two RHVY contacts on lines 13 and 14. Then, depending on whether the part is light or heavy, either RSL (short/light) will pick or RSH (short/heavy) will pick.

The same circuit configuration is repeated on lines 15 and 16 through an N.O. contact of RTAL. If the part is tall, the N.O. RTAL contact will go closed, causing RTL (tall/light) to pick if RHVY is dropped or RTH (tall/heavy) to pick if RHVY is picked. Note that only one of the four relays, RSL, RSH, RTL, or RTH, can be picked for any part that is tested.

As the part leaves the testing zone and passes under the four paint nozzles, it strikes LS1. This momentarily closes the LS1 N.O. contact on line 1, causing RPZ to pick and seal up through its own N.O. contact on line 2. RPZ will remain sealed up until the N.C. RDZ contact on line 2 goes open. The part is now in the painting zone, and the RPZ N.O. contact on line 17 is closed. Therefore one of the paint solenoids will energize, causing the proper color of paint to flow onto the moving part. The paint solenoids and their controlling contacts appear on lines 17–20.

As the part leaves the painting zone, it strikes LS2 and momentarily closes the LS2 contact on line 3. This picks RDZ, which seals up through the N.O. RDZ contact on line 4. RDZ also breaks the seal on RPZ when the N.C. contact on line 2 opens, as mentioned above. Down on line 21, the N.O. RDZ contact goes closed,

---

*Relays are often named in accordance with the function they perform in the logic circuit. The name of a relay represents an abbreviation of its function. An example is RCLR, where the letters CLR are an abbreviation of the word *cleared*. The preceding R used in all relay names stands for the word *relay*. A more complete description of the relay's function is usually written alongside the coil, as an aid to understanding circuit operation. This helpful practice is followed in Fig. 2-5.

thereby causing one of the four diverter gates to swing out over the conveyor. The four solenoids which operate the four diverter gates are shown on lines 21–24.

When the part has been guided off the conveyor and down one of the chutes, one of the four chute limit switches will close momentarily. These switches are LS3, LS4, LS5, and LS6, and they are all wired in parallel on line 5–8. Therefore when any one of them closes, RCLR picks momentarily. The RCLR N.C. contact on line 4 breaks the seal on RDZ, indicating that the part has left the diverting zone. Also, the RCLR N.C. contact on line 9 breaks seal on RTAL and RHVY if either one was sealed up. The operation sequence of the system is now complete, and it is ready to receive a new part into the testing zone.

It was stated in Sec. 2-1 that control circuits are divided into three parts: input, logic, and output. The devices in Fig. 2-5 would be categorized as shown in Table 2-1.

TABLE 2-1 Categories of devices in Fig. 2-5

Information Gathering (input)	Decision Making (logic)	Actuating Devices (output)
LS1, LS2, LS3, LS4, LS5, LS6, height detector, weighing device	Relays RPZ, RDZ, RCLR, RTAL, RHVY, RSL, RST, RTL, RTH, and their associated contacts	Blue, yellow, red, and green solenoids; S/L, S/H, T/L, and T/H solenoids

2-4 LOGIC PERFORMED BY TRANSISTORS

We can see from the foregoing discussion how relay circuits make decisions. In simple terms, when two contacts are wired in series, the circuit function is called an *AND* function because the first contact *and* the second contact must be closed to energize the load (pick the relay). When two contacts are wired in parallel, the circuit function is an *OR* function because *either* the first *or* the second contact must be closed to energize the load. These two basic relay circuit configurations are illustrated in Fig. 2-6, along with two solid-state circuits for implementing the same functions.

In solid-state logic, instead of contacts being open or closed, input lines are LO or HI. Therefore in the solid-state circuits of Fig. 2-6, the X line going HI (going to $+5$ V) is equivalent to closing the RX contact in the relay circuit. The X line being LO (being at 0 V or ground potential) is equivalent to having the RX contact open. The same holds true for the Y and Z lines.

As for the circuit *result*, in a relay circuit the result is considered to be the energizing of a relay coil and the consequent switching of the contacts controlled by that relay. In a solid-state circuit the result is simply the output line going to a HI state.

Figure 2-6. (a) The AND logic function performed by relay circuitry and by solid state circuitry. (b) The OR function performed by relay circuitry and by solid-state circuitry.

With these equivalencies in mind, study the circuit of Fig. 2-6(a). If any one of the inputs is LO (ground voltage), the diode connected to that input will be forward biased. The bias current will come from the +5-V supply, through R_1, through the diode, out the cathode lead of the diode into ground. If a diode is forward biased, its anode can be no more than 0.6 V above the cathode potential. Therefore the anode tie point in Fig. 2-6(a) will be at +0.6 V relative to ground if any one of the X, Y, or Z inputs is LO. With only +0.6 V at the tie point, Q_1* will be cut OFF as explained in Chapter 1. Therefore the collector of Q_1 will deliver current into the base of Q_2, turning it ON. With Q_2 saturated, its collector is at approximately 0 V, so the output of the circuit is LO.

On the other hand, if all the inputs X, Y, and Z are HI (+5 V), then the anode tie point will not be pulled down to 0.6 V. Therefore there will be a current flow

---

*Transistors in electronic schematics can be identified by the letter Q or by the letter T. We will use Q in most situations. The letter T is preferred when Q is used for other purposes in the schematic drawing.

path through R_1 and into the base of Q_1. Q_1 will saturate, shutting OFF Q_2 and allowing the output to go to $+5$ V, a HI level. The solid-state circuit action is equivalent to the action of the relay circuit above it. All inputs must be present to get an output.

Figure 2-6(b) shows the OR function. In the solid-state transistor circuit, if any one of the inputs goes HI, Q_1 will turn ON (the resistors are sized to allow this), and its collector will be pulled down to ground. Therefore no base current will flow in Q_2, and it will turn OFF, allowing the output, W, to go to the HI level. Again, the action of the solid-state transistor circuit duplicates that of the relay circuit above it. If any one of the inputs is present, an output will be produced.

For both the relay OR circuit and the solid-state OR circuit, if all inputs are removed (all contacts open in the relay circuit, all inputs LO in the solid-state circuit), the circuit will not produce an output. That is, the relay circuit will fail to energize relay W, and the solid-state circuit will cause a LO signal to appear at output W.

2-5 LOGIC GATES—THE BUILDING BLOCKS OF SOLID-STATE LOGIC

In Sec. 2-4 we showed that solid-state circuits can perform logic functions. It would be very cumbersome and confusing to show every transistor, diode, and resistor in a solid-state logic diagram. Instead, we have invented symbols which stand for the logic function being performed by individual circuits. We then construct complex logic circuits by connecting together many simple individual logic circuits, such as the AND circuit of Fig. 2-6(a).

The simple individual logic circuits thus constitute the *building blocks* of an extensive logic circuit, with each building block indicated by a special symbol. These building blocks are popularly called *logic gates*, or just *gates*.

2-5-1 Noninverting gates: AND, OR

We have already encountered AND gates and OR gates in Sec. 2-4. In words, an AND gate is a circuit whose output goes HI only when *all* its inputs are HI. The schematic symbol for an AND gate is shown in Fig. 2-7(a), along with its truth table and an English-language explanation of its behavior.

Figure 2-7(a) shows an AND gate with three inputs, but any number of inputs is possible. The most common numbers are two, three, four, and eight inputs.

In the truth table of Fig. 2-7(a), the binary digit 1 represents a logical HI voltage, and the binary digit 0 represents a logical LO voltage. Truth tables are not difficult to interpret. They show the output of a digital circuit for every possible combination of inputs.

For example, the AND gate truth table in Fig. 2-7(a) shows all eight possible combinations of the three inputs A, B, and C. For each combination, the output level of the AND gate is specified. As a specific example, look at the second row

Gate	Schematic Symbol	Truth Table	Action

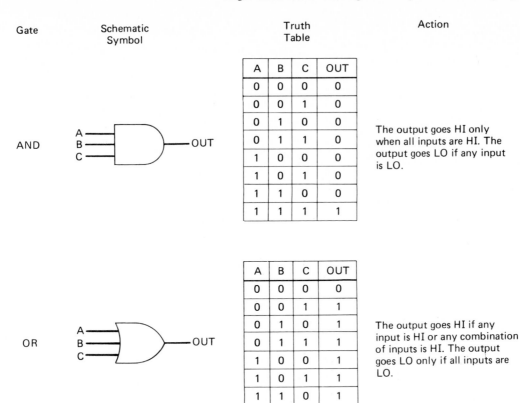

AND gate truth table:

A	B	C	OUT
0	0	0	0
0	0	1	0
0	1	0	0
0	1	1	0
1	0	0	0
1	0	1	0
1	1	0	0
1	1	1	1

The output goes HI only when all inputs are HI. The output goes LO if any input is LO.

OR gate truth table:

A	B	C	OUT
0	0	0	0
0	0	1	1
0	1	0	1
0	1	1	1
1	0	0	1
1	0	1	1
1	1	0	1
1	1	1	1

The output goes HI if any input is HI or any combination of inputs is HI. The output goes LO only if all inputs are LO.

Figure 2-7. (a) Behavior of an AND logic gate. (b) Behavior of an OR logic gate.

of the truth table. It says that if the *ABC* inputs are 001, respectively, the output will be a 0. The third row says that for *ABC* inputs of 010 the output is 0. The eighth row says that for *ABC* inputs of 111 the output is 1.

Truth tables are useful because they eliminate any ambiguity that might result from trying to explain the circuit action in words. A truth table may seem worthless when applied to a simple AND gate because the table expresses in a ponderous way something that is easy to understand in words. However, in more elaborate digital circuits it is difficult to express the circuit action in words.

It is good to learn to interpret logic gates intuitively, to know almost without thinking what the output will be for a given set of inputs. This can be done by reading and understanding the English-language description of the gate and mentally checking the truth table row by row against that description.

The symbol for an OR gate is shown in Fig. 2-7(b), along with its truth table and word description. The OR gate is so named because the output goes HI if *A or B or C* is HI. The truth table shows this clearly.

The electronic schematic drawings of Fig. 2-6 show some possible methods of constructing an AND gate and an OR gate. These are certainly not the only methods, nor are they necessarily the best methods. Although it is instructive and helpful to see how a gate can be constructed, once the gate is correctly wired up in a logic circuit, the internal construction is irrelevant. The input-output relationship from the black-box point of view is the only thing that is important.

2-5-2 Inverting gates: NAND, NOR, NOT

The inverting gates are those for which the output goes LO when the proper combination of inputs go HI. The symbols, truth tables, and word descriptions of the inverting gates are given in Fig. 2-8.

The simplest inverting gate is the NOT gate, often called an *inverter*. The output goes LO when the input is HI and goes HI when the input is LO. NOT gates are useful whenever you wish to use the digital level at a certain point to accomplish some circuit action but the level happens to be the wrong level. A NOT gate will change a HI level to a LO level and vice versa. The electronic construction of a NOT is just a transistor switch, as shown in Fig. 1-7(a).

The NAND gate is a gate whose output goes LO when all its inputs are HI. Another way of describing the behavior of a NAND is to say that the output goes HI if any input goes LO. Think about these two statements and convince yourself that they mean the same thing.

The NAND gate is equivalent to an AND gate followed by a NOT gate, as suggested in Fig. 2-9(a). A NAND gate might better be called an AND-NOT, but popular usage has moved the N to the front of the name.

The NOR gate is a gate whose output goes LO if any one of its inputs goes HI. Another way to describe a NOR is to say that the output goes HI only if all inputs are LO. Convince yourself that these statements mean the same thing. The truth table in Fig. 2-8(b) reflects this behavior.

The NOR is the logical equivalent of an OR gate followed by a NOT. Again the better name might be OR-NOT, but the name NOR is here to stay. These ideas are illustrated in Fig. 2-9(b).

You may wonder why NANDs and NORs are built and used at all, since they are certainly harder to think of intuitively. It is true that tracing through a multi-level logic circuit with NANDs and NORs is a lot more confusing than tracing through a circuit with ANDs and ORs. The more complex the logic circuit, the more confusing inverting logic becomes; this is true whether or not mathematical techniques are used to express what's happening in the circuit.

The reasons for the popularity of inverting gates are simple:

a. They are cheaper.
b. They are faster.
c. They burn less power.

To explain reason **a**, a NOR gate can be constructed in exactly the same way as the OR gate of Fig. 2-6(b) except that Q_2 is not needed. The only reason for attaching Q_2 at all is to make the output "follow" the inputs. If we are willing to

Gate	Schematic Symbol	Truth Table	Action

NOT (Inverter)

A	OUT
0	1
1	0

The output goes HI if the input is LO. The output goes LO if the input is HI.

NAND

A	B	C	OUT
0	0	0	1
0	0	1	1
0	1	0	1
0	1	1	1
1	0	0	1
1	0	1	1
1	1	0	1
1	1	1	0

The output goes LO only if all inputs are HI. The output goes HI if any input is LO.

NOR

A	B	C	OUT
0	0	0	1
0	0	1	0
0	1	0	0
0	1	1	0
1	0	0	0
1	0	1	0
1	1	0	0
1	1	1	0

The output goes LO if any input is HI or if any combination of inputs is HI. The output goes HI only if all inputs are LO.

Figure 2-8. (a) Behavior of a NOT gate. (b) Behavior of a NAND gate. (c) Behavior of a NOR gate.

live with gates in which the output doesn't follow the inputs, we can eliminate a transistor and simplify construction. To check this, mentally remove Q_2 from Fig. 2-6(b) and let the collector of Q_1 be the output. The resulting electronic circuit is a NOR gate.

In commercial gates, more than one transistor can usually be eliminated by changing from noninverting to inverting. Of course the resistors associated with these transistors are also removed. This results in a savings in components and cost.

Figure 2-9. (a) Equivalence of a NAND gate to an AND-NOT pair.
(b) Equivalence of a NOR gate to an OR-NOT pair.

Regarding reason **b**, if fewer transistors have to be switched, the gate can respond more quickly. When talking about gate response speed, the most useful term is *propagation delay time.* This is the time that elapses between the instant when the input changes level and the instant when the output changes level in response. Propagation delay times can be reduced by as much as 40% by using an inverting gate. In some applications such fast response is of primary importance.

Third, if there are fewer transistors in the gate, the gate draws less supply current and thereby dissipates less power. This can be important if the gates are powered by batteries or by any limited energy source.

2-5-3 Positive versus negative logic

In a *positive logic* system, the binary 1 or HI level is the more positive of the two voltage levels. Thus a system which has $+5$ and 0 V as the two possible voltage levels would have HI (1) represented by $+5$ V and LO (0) represented by ground potential. This is the natural way to define the HI and LO logic levels. Positive logic is assumed throughout this book.

However, some logic manufacturers use a *negative logic* system. In a negative logic system the binary 1 is indicated by the more *negative* of the two possible voltages. A negative logic system usually goes hand in hand with a negative supply voltage to the collectors of the transistors and the use of *pnp* transistors.

For example, one very popular line of industrial logic gates uses -20 and 0 V as the two possible voltage levels. The transistors are wired in common-emitter form as shown in Fig. 2-6, but they are *pnp* transistors with -20 V as the collector supply voltage. The logical 1 is considered -20 V, and the logical 0 is 0 V. The terms HI and LO are misleading in negative logic and are avoided. Instead, levels are always identified by the symbols 1 and 0.

As stated above, we will assume we are dealing with positive logic; but because negative logic is also popularly used, the student of industrial systems should be aware of it and should understand how it differs from positive logic.

2-6 A SOLID-STATE LOGIC CIRCUIT FOR THE CONVEYOR/CLASSIFYING SYSTEM

A solid-state version of the logic for controlling the classifying system of Fig. 2-4 will now be presented and explained.

In Fig. 2-10 the HI logic level is +15 V. As the part passes through the testing zone, the height and weight detectors close their contacts if the height and/or weight are above preset values. Concentrating on the height detector, if the contact closes, a HI will be applied to input 1 of OR3. This causes a HI at the output of OR3, which is then fed back to input 1 or AND3. This causes OR3 to seal up, just like a relay seals up. This happens because input 2 of AND3 is also HI at this time, causing AND3 to be enabled (the output goes HI), which puts a HI on input 2 of OR3. In this circuit arrangement, OR3 is sealed up even after the part leaves the testing zone and the height contact goes back to being open. The only way to break the seal of OR3 is to remove the HI at input 2 of AND3.

It was stated above that input 2 of AND3 is HI while the part is in the testing zone. This is so because of the situation at I2, which feeds input 2 of AND3. The input of I2 receives information from limit switches LS3–LS6, all wired in parallel. Since all four of these limit switches are released while the part is in the testing zone, there is no +15-V input applied to I2 at this time. Neither is there a 0-V signal applied to the I2 input. However, the presence of the 1-K resistor connected between the input and ground causes the inverter to treat this situation as if it were a LO input.

Therefore, with the I2 input LO, the output is inverted to a HI, which applies the HI to AND3. The output of OR3 will be maintained HI until the part actuates one of the chute limit switches. At that instant the I2 output will go LO, disabling AND3 and taking away the HI input to OR3. This will break the seal and allow the OR3 output to return to its LO state.

All the foregoing discussion assumed that the height detector contact actually did close, which indicated that the part was tall. Naturally, if the part were short, the contact would fail to close, and OR3 would remain off all through the cycle.

The note on the output line of OR3 describes the meaning of that line going HI. Thus if the output of OR3 is HI, we can conclude that the part is tall. On the other hand, if the output of OR3 is LO, the output of I3 will go HI, meaning that the part is short. The note on the output line of I3 conveys this meaning.

The weighing circuitry, comprised of the weight detector, OR4, AND4, and I4, is an exact duplicate of the height circuitry. Trace through the operation of these gates to be sure that you understand how they work.

AND gates 5–8 can be considered the classification gates. The input signals

Figure 2-10. Control circuitry for the conveyor/classifying system of Fig. 2-4, with the logic performed by solid-state logic gates.

to this group of gates come from the outputs of the height and weight detection circuitry. Each one of the AND gates has two inputs, representing a certain combination of height and weight outcomes. For example, the two input lines to AND5 are the two lines which indicate (a) the part is short and (b) the part is light. Therefore if the part is short and light, AND5 will be enabled. If the part is short and heavy, AND6 will be enabled, and so on.

The outputs of the classification AND gates feed into two other groups of AND gates. First, they feed AND gates 9, 10, 11, and 12, which control the paint solenoid valves. Second, they feed AND gates 13, 14, 15, and 16, which control the diverters.

AND gates 9, 10, 11, and 12 have input 1 in common with each other. Input 1 of all these paint control gates is driven by the line marked "in paint zone". This means that when the part enters the paint zone all the number 1 inputs of gates 9–12 will be driven HI. Then, depending on which classification gate is turned on, one of the four paint control gates will be enabled. This in turn will energize the proper solenoid valve. For example, if the tall/light classification gate is turned on (AND7), it will put a HI on input 2 of AND11. When the part enters the paint zone, and the "in paint zone" line goes HI, AND11 will be enabled. This will energize the red paint solenoid. The solenoid will remain energized until the "in paint zone" line goes back LO, disabling AND11.

The diverter control gates, ANDs 13, 14, 15, and 16, work the same way. Their number 1 inputs are tied in parallel and are driven by the "in diverter zone" line. When this line goes HI, one of the four diverter control gates will be enabled, which will energize the proper diverter solenoid. For example, if the tall/light classification gate (AND7) is turned on, it will apply a HI to the number 2 input of AND15. When the "in diverter zone" line goes HI, it will put a HI on input 1 of AND15. The output of AND15 will then go HI, energizing the tall/light diverter solenoid and causing the tall/light diverter in Fig. 2-4(b) to swing out over the conveyor. The diverter solenoid will remain energized until the "in diverter zone" line returns to LO, disabling AND15.

The circuits at the top of Fig. 2-10 furnish the signals which tell the location of the part as it rides the conveyor, namely the "in paint zone" and "in diverter zone" signals.

As a part enters the paint zone it actuates LS1, which applies a +15-V HI to the number 1 input of OR1. The OR1 output goes HI and seals itself in by feeding back to AND1. Thereafter, as long as input 2 of AND1 is HI, the AND gate will stay enabled, and OR1 will remain on by virtue of its number 2 input. As shown in the diagram, the OR1 output is nothing other than the "in paint zone" line.

When the part leaves the paint zone and enters the diverter zone, it actuates LS2. This applies a HI to input 1 of OR2, which causes the OR2 output to go HI. The OR2 output does several things. First, it puts a HI at the input of I1, which causes a LO at input 2 of AND1. This disables AND1 and breaks the seal on OR1. The "in paint zone" line goes back LO, and the paint solenoid valve shuts off. Second, the output of OR2 feeds into AND2. Since the number 2 input of AND2 is also HI at this time, AND2 turns on and seals up OR2. Third, the OR2 output is the "in diverter zone" signal, which goes down to the bottom of Fig. 2-10 and drives the diverter control gates, as discussed earlier.

When the part is guided off the belt and down a chute, one of the chute limit switches will be actuated, applying a HI to I2. The I2 output goes LO and applies LO signals to AND2, AND3, and AND4. The LO on AND2 breaks the seal on OR2, allowing "in diverter zone" signal to go back LO. Whichever diverter was swung out thus returns to its normal position. The LOs at AND3 and AND4 disable those gates, applying LOs to the number 2 inputs of OR3 and OR4. This breaks the seals on OR3 and OR4, if they were sealed. Therefore the height and weight circuits are reset and are prepared to test the next part on the conveyor.

2-7 INPUT DEVICES FOR SOLID-STATE LOGIC

The circuit of Fig. 2-10 shows direct-switched connections between the HI logic supply voltage and the gate inputs. For example, LS1 makes a direct connection between the +15-V dc supply line and input 1 or OR1. While this switching arrangement is theoretically acceptable, there are some practical reasons why it is a bad idea.

The primary reason is that mechanical switches never make a clean contact closure. The contact surfaces always "bounce" against each other several times before they make permanent closure. This phenomenon is called *contact bounce* and is illustrated in Fig. 2-11.

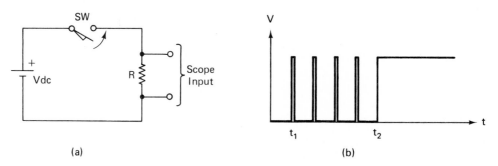

Figure 2-11. The problem of contact bounce.

In Fig. 2-11(a), when the mechanical switch closes to connect resistor R across dc supply V, the voltage waveform across R looks like Fig. 2-11(b). The elapsed time between initial contact and permanent closure ($t_2 - t_1$ in the waveform) is usually rather short, on the order of a few milliseconds or less. Although the bouncing is quite fast, logic gates respond quite fast, so it is possible for a gate to turn on and off every time a bounce takes place. The unwarranted turning on and off can cause serious malfunctions in logic circuitry.

Capacitive Switch Filters. The solution to this problem is to install some type of filter device between the switch and the logic gate. The filter device must take the bouncing input and turn it into a smooth output. One straightforward method of doing this is shown in Fig. 2-12(a).

When the limit switch first closes, capacitor C starts to charge through the Thevenin resistance of $R_1 \parallel R_2$. Because the limit switch contacts stay closed only

(a)

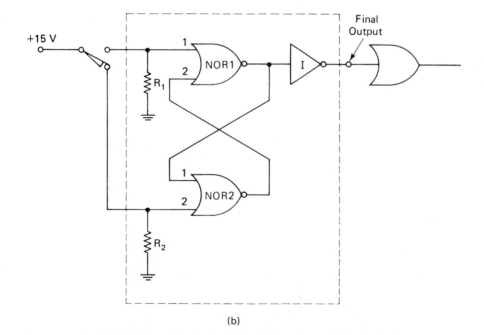

(b)

Figure 2-12. (a) *RC* switch filter to eliminate the effects of contact bounce. (b) Bounce eliminator built with solid-state gates.

a very short time on the first bounce, the charge buildup on C is not great enough to affect the gate input. The same holds true for all subsequent bounces—the switch never stays closed long enough to trip the gate because of the necessity to charge C. When permanent closure finally occurs, C can charge to the threshold voltage of the gate and turn it on. The filter of Fig. 2-12(a) also serves to reject noise signals from external sources. That is, if a high-speed noise pulse should occur on the lead coming from the switch, it will be rejected by the low-pass filter and will not appear at the gate input.

Of course, when the capacitor does charge up, it cannot charge to the full supply voltage level. It can only charge to the Thevenin voltage of the R_1-R_2 voltage divider. This is normally not a problem since solid-state gates operate reliably with an input voltage less than full supply voltage.

Bounce Eliminators. Another method of eliminating contact bounce is shown in Fig. 2-12(b). This approach differs from the one in Fig. 2-12(a) in that it triggers on the *first* contact bounce instead of waiting for the final closure. After it turns on, it ignores all subsequent bounces. A disadvantage of this circuit is that it requires a double-throw switch instead of a single N.O. contact. Here is how it works.

With the limit switch released, the N.C. contact is closed and a HI level is applied to R_2 and to input 2 of NOR2. The output of NOR2 is therefore LO, causing input 2 of NOR1 to be LO. Input 1 of NOR1 is also LO because R_1 pulls it down to ground. With both inputs of NOR1 LO, its output is HI; inverter I then makes the final output LO.

During the switching process, this is the sequence of events:

a. The N.C. contact opens first (break-before-make switch), causing the number 2 input of NOR2 to go LO. NOR2 does not change state because its number 1 input is still HI.
b. The N.O. contact closes momentarily on the first closure. This puts a momentary HI on input 1 of NOR1, causing its output to go LO. The inverter drives the final output HI. The NOR1 output feeds input 1 of NOR2, so NOR2 now has two LO inputs. Its output therefore goes HI, applying a HI to input 2 of NOR1. NOR1 has two HI inputs at this point in time.
c. The N.O. contact bounces open on the first rebound. This causes a LO at input 1 of NOR1, but input 2 maintains its HI level. Therefore NOR1 does not change states, and the final output remains HI.
d. There are several more bounces, each one changing the logic level of input 1 of NOR1. However, because the limit switch N.C. contact remains open, a HI remains on input 2 of NOR1, holding NOR1 steady.

When the limit switch is released at some later time, the bounce eliminator does the same thing in reverse, causing a jitter-free transition to the LO level at the final output. You should trace through the operation of the circuit as this happens.

Signal Converters. The capacitive filter and the bounce eliminator that we have considered both assume that the input device is switching a *logic level* voltage

(+15 V in Fig. 2-12). Since virtually all industrial logic gates use a supply voltage of 20 V or less, the input devices must operate reliably under relatively low-voltage and -current conditions in order to allow direct switching of this type. This is sometimes feasible, but there are many situations in which it is not feasible. Sometimes the information-gathering devices cannot give reliable operation under low-voltage conditions.

There are two major reasons for this unreliability. First, the input devices may be physically remote from the decision-making logic. Therefore the wire-runs between the input devices and the logic circuits are lengthy and necessarily have higher resistance than if they were shorter. Higher resistance causes a higher *IR* voltage drop along the run. If the beginning voltage is already small, large *IR* voltage drops in the wires cannot be tolerated because the logic might confuse a HI level with a LO level. It is better to start with a large voltage so the system can afford to suffer some voltage loss in the connecting wires.

Second, the contact surfaces of input devices tend to accumulate airborne dust and debris; oxides and other chemical coatings can also form on the surfaces. These things cause the contact resistance to increase, sometimes making it impossible for a small voltage to overcome the resistance. A high voltage level is needed to ensure that the increased resistance can be overcome.

Besides, the very act of switching a high voltage creates arcs between the contact halves. These arcs burn away oxides and residue and keep the surfaces clean.

Under many industrial circumstances, therefore, it is absolutely necessary to use high voltages to drive the input devices. When this is done, there must be an interface device added to convert the high-voltage input signal to a low-voltage logic signal. Such devices are called *signal converters*, *original input converters*, *logic input interfacers*, and other names. We will use the term *signal converter* in this book. A schematic symbol of a signal converter is shown in Fig. 2-13(a). A schematic diagram containing three signal converters is shown in Fig. 2-13(b).

In most industrial schematic diagrams, signal converters are drawn with two wires as shown in Fig. 2-13(b), although an actual signal converter usually has four wires attached to it. The schematic representation is simple and uncluttered, but it suggests the action of a signal converter, namely that a low-voltage logic 1 appears at the output when a high-voltage input signal is applied by the closing of the contact of the input device.

Figure 2-14 shows the internal construction of two typical signal converters for converting a 115-V ac input to a +15-V dc logic level.

Figure 2-14(a) is the familiar full-wave power supply with a center-tapped transformer. The input device switches 115 V ac to the primary winding, and the rectifier and filter circuits convert the secondary voltage to 15 V dc. Note that this type of signal converter has four wire connections even though the schematic symbol is drawn with only two wires.

This signal converter provides electrical isolation between high-voltage input circuits and low-voltage logic circuits by virtue of the magnetic coupling between the transformer windings.

Electrical isolation between the two circuits is desirable because it tends to prevent electromagnetic or electrostatic noise generated by the input circuitry

(a)

(b)

Figure 2-13. Signal converters for converting high-voltage input signals to low-voltage logic signals.

from passing to the logic circuitry. In an industrial logic system, noise pickup in the input device circuit is very often a problem. This is due to the long wire-runs between the logic panel and the input devices and the tendency to carry the wires in conduit running close to power wires. Power wires driving motors and switchgear are inherently noisy and can easily induce unwanted electrical noise in the connecting wires between input devices and logic.

The signal converter illustrated in Fig. 2-14(b) uses a reed relay. The output of the full-wave bridge energizes the relay coil, and the relay contact switches the logic supply voltage onto the signal converter's output line. The logic circuitry is isolated from the input circuitry through the relay. This type of signal converter does not produce its own logic signal voltage but must have the logic supply brought in from outside. Therefore it has five wire connections. It would still be drawn schematically as shown in Fig. 2-13(b).

Both the signal converters illustrated in Fig. 2-14 contain capacitors which serve to filter out high-frequency noise and switch bounce. Therefore they do not nor-

Figure 2-14. (a) Signal converter which uses a transformer to isolate the logic circuit from the input circuit. (b) Signal converter which uses a reed relay to isolate the logic circuit from the input circuit.

mally require any other filter circuit or bounce eliminator connected to their outputs.

A pilot light indicator can be wired into a signal converter as shown by the dashed lines in Fig. 2-14(b). This is a troubleshooting aid for maintenance personnel. The condition of the input can be seen at a glance; it is then not necessary to apply a voltmeter to tell the state of the input.

Occasionally the input devices in an industrial system are driven by a high-voltage dc supply instead of the usual 115 V ac. A large dc voltage creates more arcing across switch contacts than the same value of ac voltage. Therefore a dc voltage is even more effective at burning away deposits and residue that collects on the contact surfaces. In such cases a dc-to-dc signal converter is used. The circuit of Fig. 2-14(b) would work in such an application.

In recent years optically coupled signal converters have become very popular. Their popularity is due to their light weight, excellent reliability, and low cost. They require no transformer or relay for electrical isolation between input and logic circuits, and their isolating ability is very good. They will be discussed when photoelectric devices are covered in Chapter 11.

2-8 OUTPUT DEVICES FOR SOLID-STATE LOGIC

The solid-state logic diagram of the conveyor/classifying system (Fig. 2-10) shows the paint solenoids and diverter solenoids driven directly by AND gates. While it is possible to drive actuating devices (solenoids, motor starters, etc.) directly from logic gates, this is not the usual practice. Rather, an *output amplifier* is inserted between the logic circuit and the actuating device. The purpose of the output amplifier is to increase the low-voltage/low-current logic power to higher-voltage/higher-current output power.

The symbol for an output amplifier (sometimes called a *driver* or a *buffer*) is shown in Fig. 2-15(a). Output amplifiers as they would appear in an industrial schematic drawing are shown in Fig. 2-15(b).

The OA in the output amplifier symbol is often omitted or replaced by a D, standing for driver. As with signal converters, output amplifiers are shown schematically with only two wires, an input and an output. When the input line goes

Figure 2-15. Output amplifiers for amplifying low-power logic signals into high-power output signals.

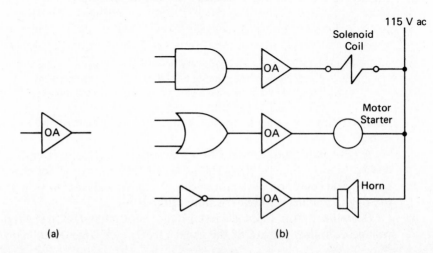

(a)   (b)

to a logic HI, the output line energizes the actuating device. In actual construction, most output amplifiers have four wires attached to them.

Most output amplifiers are designed to drive a 115-V ac load, since most industrial solenoid valves, motor starter coils, horns, etc., are designed for use with 115 V ac. This situation is represented in Fig. 2-15(b), with the common power line labeled 115 V ac.

Other output amplifiers obtain their operating power from a separate high-voltage dc supply instead of the 115-V ac line. Such amplifiers are used with actuating devices designed to operate on that particular dc voltage. Popular dc voltage levels for driving dc actuating devices are 24, 48, and 115 V dc. An example of the construction of a dc output amplifier is given in Fig. 2-16(a).

The dc output amplifier consists of a power transistor driven by a small signal transistor with an emitter resistor. The load is connected in series with the collector lead of the power transistor and is powered by the 24-V supply. The 24-V supply is referenced to the logic ground supply line by a ground connection somewhere in the control circuit cabinet. This is all shown in Fig. 2-16(a).

The logic supply voltage is brought into the output amplifier as collector supply for Q_1. When the input terminal goes HI, Q_1 turns ON, raising the R_3 voltage high enough to bias the power transistor ON. Thereafter most of the Q_1 emitter current flows into the base of the power transistor. The collector current of the power transistor energizes the actuating device. The diode in parallel with the load is placed there to suppress inductive kickback from the load when it is deenergized.

Because of the common ground, there is not complete electrical isolation between logic circuits and output circuits with the output amplifier of Fig. 2-16(a). Careful precautions must therefore be taken with wire routing to avoid injection of noise into the logic.

Figure 2-16(b) is an example of an output amplifier using a reed relay. When the amplifier's input terminal goes HI, it turns ON the transistor and picks the reed relay. The relay contact then connects the load across the 115-V ac lines. This arrangement does provide electrical isolation between logic and output circuits.

More up-to-date ac output amplifiers use solid-state devices instead of reed relays. Such output amplifiers usually have an SCR at their heart, with the SCR often triggered by a unijunction transistor (UJT). A typical design of such a solid-state ac output amplifier will be presented in Chapter 5.

2-9 SOLID-STATE LOGIC COMPARED TO RELAY LOGIC

Magnetic relays have been handling most of the logic requirements of twentieth-century industry for many years, and they will continue to be widely used. Because of improved materials of construction and improved design, relays now are capable of several million trouble-free operations under normal conditions. However, under certain conditions and in certain settings, solid-state logic is demonstrably superior to relay logic. We will discuss the conditions under which solid-state

(a)

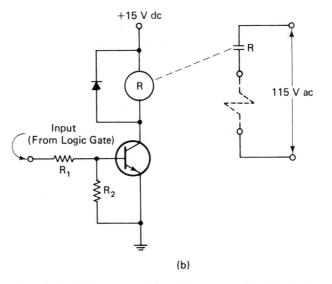

(b)

Figure 2-16. (a) Output amplifier using a power transistor to control the current through the output device. (b) Output amplifier using a relay contact to control the current through the output device.

logic is preferred and try to point out some of the important considerations used to decide between the two types of logic.

Reliability. In most industrial situations the most important consideration when selecting logic circuitry is reliable maintenance-free operation. Relays have moving mechanical linkages and contacts, which are subject to wear. Also, their coils must allow fairly large inrush currents in order to create the force necessary to move the linkages. This puts stress on the coil wire and insulation. These are the reasons the life expectancy of relays is limited to a few million operations, as mentioned above. This may seem like a remarkable life span, as indeed it is, but consider how long a relay would last if it cycled twice per minute. Two operations per minute figures out to 2880 operations per day, or over 1 million operations per year. At this rate, a relay with a life expectancy of 2 million operations would last only about 2 years. The rate of two operations per minute for 24 hrs a day is not unusual in an industrial circuit. Many relays must operate more often than that, with a corresponding reduction of trouble-free life.

Solid-state gates, on the other hand, have an unlimited life expectancy. They have no moving parts and no appreciable inrush current. Barring unexpected thermal shock or overcurrents, a solid-state device will last forever. This is an obvious advantage of solid-state logic over relay logic.

Relay components are exposed to the atmosphere. Dirt particles can therefore work their way into the mechanical apparatus and interfere with proper motion. Chemicals and dust in the atmosphere can attack the surfaces of the contacts, causing them to pit. When contact surfaces are not smooth they may weld together. Coil insulation can also be damaged by chemical action.

By contrast, solid-state gates can be and usually are sealed in packages which are impervious to the atmosphere. Chemicals and airborne particles cannot interfere with their proper operation.

Explosive Environments. The fact that relays are exposed to the atmosphere has another important consequence. Relay contacts create sparks when they operate, due to metal clash and load kickback. If there are explosive gases in the atmosphere, sparks are unacceptable. Under these conditions relays can be used only in expensive airtight enclosures.

Solid-state gates turn on and off without sparking, making them inherently safe in explosive environments.

Space Requirements. Considering physical size and weight, solid-state logic is clearly more compact. This is not usually an important factor in industrial circuits, but it occasionally can be. An example might be a situation in which a new system was being installed in the space previously occupied by an old system and space was at a premium. If the control circuit was extensive, the space conserved by using solid-state logic could be an important consideration.

Operating Speed. Concerning operating speed, it is strictly no contest between relays and logic gates. Relays operate in milliseconds, whereas most solid-state devices operate in microseconds or nanoseconds. Roughly speaking, a solid-state gate is at least 1000 times as fast as a relay. Again, this high speed is often not an important factor in industrial logic, but it might be. Operating speed becomes important if mathematical computation is required in the decision-making process.

Cost. For an extensive logic circuit containing hundreds of decision-making elements, solid-state logic is cheaper to build and operate than an equivalent relay logic circuit. This is because the low per-gate cost overrides the extra expenses associated with solid-state logic. These extra expenses include the cost of dc power supplies, signal converters and output amplifiers, and special mounting hardware for the printed circuit boards.

Solid-state logic gates consume only a small fraction of the power consumed by relays. Therefore in a large circuit the energy savings can be considerable.

Advantages of Relay Logic. On the plus side for relays are several assets not possessed by solid-state circuits. First, as implied above, relay logic is cheaper to build if the circuit is small. This is because relays require no separate power supply, require no interfacing at the information gathering (input) end or at the actuating (output) end, and mount very easily on a panel.

Second, relays are not subject to noise pickup. They cannot be fooled by an extraneous noise signal; solid-state gates can be fooled by such noise signals.

Third, relays work well at high ambient temperatures found in industrial environments. Solid-state logic must often be fan-cooled or air conditioned when used in a hot environment. This negates some of the advantages of energy conservation and reliability, since air conditioning requires energy to run, and the logic is only as reliable as the air conditioner.

Fourth, and often of critical concern, is that many maintenance personnel are thoroughly familiar with relay logic but much less familiar with solid-state logic. Given this situation, down-time may be longer for a system malfunction when solid-state logic is used. Actually, the troubleshooting process for a solid-state system is not very much different from that for a relay system, but many maintenance personnel don't realize this.

2-10 A SOLID-STATE LOGIC CIRCUIT FOR A MACHINE TOOL ROUTING CYCLE

We will now explore some more examples of circuits using logic gates. The circuit presented in this section is a simple cycling circuit using noninverting gates, ANDs and ORs. Logic circuits using noninverting gates are easier to explain and to understand than circuits using inverting gates.

The circuit presented in Sec. 2-11 is a fairly uncomplicated logic circuit using inverting gates, NANDs.

Finally, in Sec. 2-12 we will explore a more complex circuit using NAND gates. Circuits using NANDs and NORs are more confusing because of the constant necessity to invert the thinking process, but it is necessary to learn to deal with such circuits. They are popular in industrial control for the reasons given in Sec. 2-5: they are cheaper and faster and draw less current than ANDs and ORs.

Consider the machining application illustrated in Fig. 2-17. The purpose is to route two channels into the top of the workpiece, both running in the east-west direction. The first channel is toward the north edge of the piece, and the second channel is toward the south edge. This is accomplished by loading the workpiece

(a)

(b)

Figure 2-17. (a) Physical layout of a machine tool router. (b) Control circuit of the machine tool router.

onto a stationary table between two square bars that prevent it from sliding in the east-west direction but permit motion in the north-south direction. The piece is placed on the table so its north side is snug against the face block which touches the north edge of the table. The face block is loaded with powerful springs so it will not move back from the north edge of the table unless forced back by a hydraulic cylinder. Cylinder *B* must extend and push the workpiece against the face block to displace the piece a few inches to the north.

The router is mounted on a movable frame that can move east and west. When cylinder *A* extends, the router frame moves east. When cylinder *A* retracts, the router frame moves west.

The machining cycle proceeds as follows:

a. When the workpiece is properly positioned between the square bars and snug against the face block, the operator presses the START button. Cylinder A extends to the east and routes the north channel.
b. When the cylinder *A* cam hits LS2, indicating that the first channel is complete, cylinder *B* extends and moves the workpiece to the north. When cylinder *B* reaches its fully extended position, its cam actuates LS3.
c. Cylinder *A* retracts to the west and routes the south channel in the top of the piece. It stops when its cam actuates LS1.
d. Cylinder *B* retracts to the south, allowing the springs to return the workpiece to its original position. This completes the cycle.

Refer to Fig. 2-17(b) for the control schematic. Here is how the circuit works. When the workpiece is properly located between the square bars and up against the face, the "in position" contacts leading into signal converter SC4 go closed. When the START button is pressed the output of SC4 goes HI, enabling OR1. The OR1 output enables OA1, which energizes the cylinder *A* solenoid. The cylinder A hydraulic valve shifts, stroking cylinder A to the east. The first channel is routed.

OR1 seals up by putting a HI on input 1 of AND1. This causes the AND1 output to go HI since input 2 was already HI. This is so because LS3 is released at this time, causing a LO input to I1 and a corresponding HI output from the inverter.

When cylinder *A* completes its stroke and the router bit has cleared the workpiece, a cam actuates LS2, causing a HI input to OR2 from SC2. The OR2 output goes HI, energizing the cylinder *B* solenoid through OA2. The cylinder *B* solenoid valve shifts and extends cylinder *B*. Meanwhile OR2 has sealed up through AND2. This happens because OR2 supplies a HI to input 1 of AND2, and input 2 of AND2 is already HI. The HI on input 2 comes from I2, whose input is LO due to the LS1 contact being open.

When cylinder *B* has fully extended, placing the workpiece in position for the second cut, LS3 is actuated. The LS3 N.O. contact closes, applying a 115-V ac input to SC3. I1 therefore has a HI input, resulting in a LO applied to input 2 of AND1. This breaks the seal on the OR1 gate, shutting off OA1 and deenergizing the cylinder *A* solenoid. The hydraulic valve shifts back to its normal position, causing oil to flow into the rod end of cylinder *A* and causing cylinder *A* to retract to the west. As the router frame moves to the west, the router cuts the second channel.

When cylinder *A* has fully retracted it hits LS1. This applies a 115-V ac input to SC1, resulting in a LO output from I2. This LO is applied to the number 2 input of AND2, causing OR2 to lose its seal. When the OR2 output goes LO, OA2 deenergizes the cylinder *B* solenoid. Cylinder *B* retracts to the south, allowing the face block springs to shove the workpiece back into the starting position. The machining cycle is now complete, and the operator removes the piece and inserts a new one.

2-11 A LOGIC CIRCUIT FOR A FIRST FAILURE ANNUNCIATOR

Figure 2-18 shows the schematic of a *first failure annunciator*. A first failure annunciator is a circuit which tells the system operators which input device gave the warning signal which caused the system to shut down. By way of background, many industrial systems have input devices which constantly monitor conditions in the system, making sure no unsafe condition exists. If an unsafe condition should occur, these devices shut the system down to eliminate the dangerous condition and blow a horn to inform the operators. Unfortunately, by the time the operators arrive on the scene, the unsafe condition may have already corrected itself; or the act of shutting down the system may have made it impossible to tell exactly *which* unsafe condition caused the problem. In such a situation, what is needed is a circuit which can record which input device gave the initial warning and ignore any subsequent warning signals which only occurred because of the shutdown process. This is the purpose of a first failure annunciator.

As a specific system to have in mind, consider an industrial air/gas heated furnace. Three unsafe conditions which could possibly occur in such a system are (a) the natural gas supply pressure is too high, (b) the combustion air pressure is too low to adequately burn all the gas, or (c) the temperature has exceeded the maximum safe value for this particular furnace; the maximum safe temperature is commonly called the *temperature high-limit*, or just the *high-limit*.

Any one of these conditions is considered unsafe enough to warrant immediate shutdown of the furnace. By the time the human operators come over to see what's wrong, the condition may have corrected itself. The gas pressure may have temporarily surged up and then returned to normal. The air pressure may have momentarily dropped and then recovered, etc. The human operators will have no clue as to the nature of the problem, and no corrective action by them will be possible. Therefore a first failure annunciator is needed. It should be understood that Fig. 2-18 does not show the actual circuitry by which the furnace is shut down; it shows *only* the first failure annunciator circuitry.

As the circuit monitors a correctly operating furnace, the situation is this.

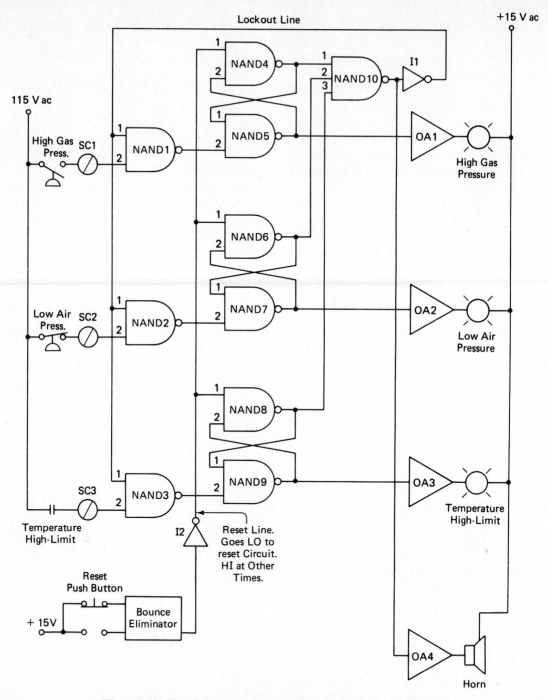

Figure 2-18. First failure annunciator which indicates the initial cause of the failure.

The high gas pressure switch is open because the gas pressure is below the set point of the pressure switch. Therefore SC1 has no 115-V ac input signal and consequently applies a logic LO to input 2 of NAND1. The N.C. low air pressure switch is open because the air pressure is above the set point of the pressure switch. Therefore SC2 has no 115-V ac input, and it applies a LO to input 2 of NAND2. Likewise, the temperature high-limit switch is open because the furnace temperature is below the maximum limit, so input 2 of NAND3 is also LO. NAND gates 1, 2, and 3 all have HI outputs because of the LOs on their number 2 inputs.

Now concentrate on NAND5. Its number 2 input is HI because NAND1 has a HI output. Its number 1 input is also HI due to the action of the RESET line at some time in the past. Here is what happened the last time the circuit was reset: When the RESET button was pushed, the output of the bounce eliminator went HI. Therefore the output of I2 (the RESET line) went LO, applying a LO to input 1 of NAND4. The LO at this input guaranteed a HI at the output of NAND4. This HI fed back around to the number 1 input of NAND5. With both inputs HI, the output of NAND5 went LO, applying a LO to input 2 of NAND4. Then when the RESET push button was released, allowing the RESET line to return to HI, the state of NAND4 did not change. Its output remained HI because of the LO at input 2.

The above paragraph shows that while the circuit is monitoring a correctly operating furnace, NAND4 has a HI output and NAND5 has a LO output. The exact same argument would apply to NAND6 and NAND7 and also to NAND8 and NAND9. Therefore NANDs 6 and 8 have HI outputs and NANDs 7 and 9 have LO outputs. Since this is true, all three output amplifiers have LO inputs from NANDs 5, 7, and 9. Therefore all three indicating lamps are turned off.

NAND10 has all inputs HI from NANDs 4, 6, and 8. Its output is therefore LO, causing a LO to OA4. The horn is deenergized. I1 has a HI output, which is applied to the number 1 inputs of NANDs 1, 2, and 3. This is the complete situation under normal furnace conditions.

Now suppose that there is a failure of the gas pressure in that it temporarily surges up too high. This causes a high-voltage input to SC1, which delivers a logic HI to input 2 of NAND1. There are now two HIs into NAND1 so its output goes LO. This causes a LO at input 2 of NAND5 which drives the output of NAND5 HI. Therefore OA1 has a HI input, and it turns on the HIGH GAS PRESSURE lamp. The NAND5 output also applies a HI to input 2 of NAND4, driving the output LO. The output goes LO because the number 1 input of NAND4 was already HI. The LO from NAND4 feeds back around to input 1 of NAND5, which seals the NAND5 output HI. That is, as long as the output of NAND4 is LO, the output of NAND5 is held HI. And as long as the output of NAND5 is HI, the output of NAND4 will be LO. The only way to break this seal is by pushing the RESET button and driving the number 1 input of NAND4 LO.

We have seen how NAND4 and NAND5 seal up and how they turn on the proper indicating light. Now look at NAND10. When its number 1 input goes

LO, its output goes HI. This HI is fed down to OA4, which causes the alarm horn to blow. Meanwhile the I1 output (the LOCKOUT line) goes LO and applies a LO to input 1 of NANDs 1, 2, and 3. The LO inputs to NAND2 and NAND3 lock these gates in their starting condition, namely, outputs HI. It does not matter what SC2 or SC3 does thereafter, because the outputs of NAND2 and NAND3 are locked HI by the LOs at their number 1 inputs. Therefore the NAND6-NAND7 combination cannot change states and the NAND8-NAND9 combination is also frozen. Thus it is impossible for any *other* indicating lights to turn on once the first one turns on.

Even if the high gas pressure switch returns to the open condition (which it would certainly do when the furnace was shut down), the high gas pressure indicating lamp will continue to glow because of the seal of the NAND4-NAND5 combination.

The foregoing explanation was based on gas pressure being the first failure, but of course the circuit action would be the same if the air pressure or temperature high-limit was the first failure. Air pressure failure would seal up the NAND6-NAND7 combination and lock NAND1 and NAND3 in starting condition with outputs HI. Temperature limit failure would seal up the NAND8-NAND9 combination and lock NANDs 1 and 2 outputs HI when the LOCKOUT line went LO.

When the RESET push button is pressed, the RESET line goes LO and breaks whatever seal has occurred. This returns the entire circuit to starting condition.

2-12 A LOGIC CIRCUIT FOR A MACHINE TOOL DRILLING CYCLE

Figure 2-19(a) shows a rough drawing of a drilling apparatus. The workpiece is brought into position and held steady by clamps. Two holes are to be drilled in the piece. One hole is vertical, and one hole is horizontal, and both must pass through the same internal point. Therefore they cannot both be drilled at the same time. The logic circuit implements the following cycle:

a. When the workpiece is clamped in position, the operator presses the START button, causing cylinder *A* to extend. The rotating drill bit descends and drills the vertical hole.
b. When cylinder *A* is fully down, LS2 is contacted, causing cylinder A to retract and the drill bit to withdraw from the piece.
c. When cylinder *A* returns to fully up, LS1 is actuated, causing cylinder *B* to extend and drill the horizontal hole.
d. When cylinder *B* is fully extended, LS3 is contacted, causing cylinder *B* to retract and withdraw the drill bit.

The control circuit is shown in Fig. 2-19(b). The description of this circuit is more complex than the descriptions of Figs. 2-17(b) or 2-18. Here is how it works.

When the piece is clamped in position, the "in position" contact closes, and the operator presses the START button. This delivers a high-voltage input to SC4 which brings input 1 of NAND1 HI. Since LS1 is contacted at this time (cylinder

(a)

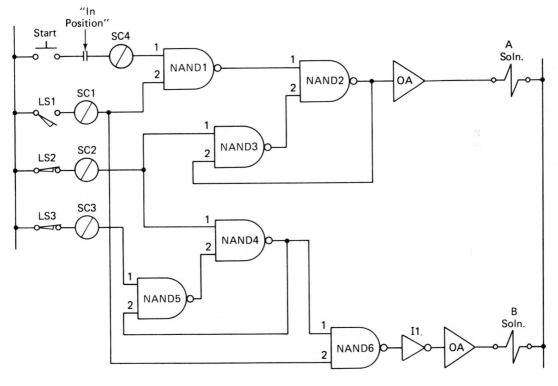

(b)

Figure 2-19. (a) Physical layout of a machine tool drilling operation. (b) Control circuitry of the machine tool drilling operation.

A is retracted at the beginning of the cycle), the number 2 input of NAND1 is also HI. Therefore the output of NAND1 goes LO and applies a LO to input 1 of NAND2. The NAND2 output goes HI, turning on OA1 and energizing the cylinder *A* solenoid. Therefore the hydraulic valve shifts, and cylinder *A* starts to descend.

Meanwhile, NAND2 has been sealed HI because of the feedback through NAND3. The NAND2 output puts a HI on input 2 of NAND3. Since LS2 is released at this time, the N.C. contact is closed, and SC2 has a HI output. Therefore input 1 of NAND3 is also HI at this time. With both inputs HI, the NAND3 output goes LO, making input 2 of NAND2 LO. Therefore it doesn't matter what happens to input 1 of NAND2 since the LO at input 2 will guarantee a HI output.

When cylinder *A* is fully extended, it contacts LS2 and opens the N.C. LS2 contact. When the 115-V ac input is removed from SC2, its output goes LO. This LO does two things:

a. It breaks the seal on NAND2. When input 1 of NAND3 goes LO, its output goes HI. The HI is applied to input 2 of NAND2. Input 1 of NAND2 is already HI because of LS1 being released (and the START button being released). With two HI inputs, NAND2 goes LO at its output and deenergizes the cylinder *A* solenoid. Cylinder *A* therefore begins its retraction stroke.
b. The LO from SC2 also arrives at input 1 of NAND4. The output of NAND4 goes HI and seals back through NAND5. This is because LS3 is released at this time, applying a HI to the number 1 input of NAND5. With two HI inputs, the NAND5 output goes LO and delivers a LO to input 2 of NAND4. This seals the NAND4 output HI, no matter what the number 1 input does. Therefore NAND4 will maintain a HI output even after LS2 is released.

The events described in the above two paragraphs all take place at the instant LS2 is contacted. Since the cylinder *A* solenoid has been deenergized, cylinder *A* immediately starts retracting, and it releases LS2. This causes HIs to reappear at input 1 of NAND3 and input 1 of NAND4, but these HIs have no effect on those gates. When cylinder *A* is all the way back it contacts LS1, and the output of SC1 goes HI. This HI appears at input 2 of NAND6. Since input 1 of NAND6 is already HI at this time (NAND4 is sealed HI), the output of NAND6 goes LO. This LO is inverted by I1, allowing OA2 to energize the cylinder *B* solenoid. Therefore cylinder *B* begins its extend stroke to drill the horizontal hole.

When cylinder *B* is fully extended it actuates LS3 and opens the LS3 N.C. contact. The SC3 output goes LO, causing the NAND5 output to go HI. This breaks seal on NAND4 since NAND4 has two HI inputs at this time. The NAND4 output returns to LO and puts a LO on input 1 of NAND6. The NAND6 output goes HI, so I1 delivers a LO to OA2. The cylinder B solenoid therefore deenergizes and cylinder *B* retracts, withdrawing the horizontal drill bit.

As the retraction starts, LS3 is again released, causing the N.C. contact to go closed. This reapplies a HI to input 1 of NAND5, but NAND5 is not affected because its number 2 input is LO at this time.

This completes the drilling cycle and the operator removes the workpiece.

2-13 FAMILIES OF SOLID-STATE LOGIC—THEIR CHARACTERISTICS AND SPECIAL FEATURES

To this point, not very much has been said about the *internal* behavior of solid-state logic gates. We have been content to view them as black boxes with a certain combination of digital inputs yielding a certain digital output. This is the most direct way to grasp the essence of solid-state logic, and this is the proper way to think of solid-state gates in most industrial situations that arise. However, it is desirable to have a *rough* understanding of the internal action of solid-state logic circuits. This will enable a logic user to recognize and/or avoid misapplications of the logic circuits, and it helps a user to understand manufacturers' specifications and recommendations for proper use.

A *logic family* represents a general method of constructing solid-state logic circuits. The substance of the notion of a logic family is that it describes *what kinds of components* are used to build the circuits and how these components tie together.

2-13-1 The RTL family

RTL (resistor-transistor logic) is the simplest logic family to understand. As the name implies, resistors and transistors are the components used to construct the circuits. To be more specific, the R signifies that the circuit input is connected to a resistor and the T signifies that the circuit output is taken from a transistor.

Basic Gate Is the NOR. We have already been exposed to RTL logic in Sec. 2-4. Figure 2-6(b) shows a three-input OR gate built on RTL principles. Another popular way of building RTL gates is to provide each input resistor with its own transistor and then to tie the transistor collectors together. This is shown in Fig. 2-20(a).

In Fig. 2-20(a), if any one of the inputs *A*, *B*, or *C* goes HI, its associated transistor will turn ON and pull the output down to ground. Therefore *any* HI input causes a LO output. This gate is a NOR. Other types of logic gates can be made in the RTL family by adding extra components.

Current-Sourcing Family. The RTL family fits into the classification of a *current-sourcing* logic family. A current-sourcing family has the property that conventional current flows *out of* the driving gate and *into* the driven gate (the load). Figure 2-20(b) shows a HI output from one gate, the driving gate, being applied to an input terminal of another gate, the driven gate. The gates have not been drawn in their entirety; only the output circuitry of the driving gate and the input circuitry of the driven gate have been drawn. As Fig. 2-20(b) shows, conventional current is flowing out of the driving gate and into the driven gate. The driving gate is a *source* of current—hence the term *current-sourcing* logic. Current-sourcing logic families are in contrast to *current-sinking* logic families, which we will encounter shortly.

Fan-out. The contrast between current sourcing and current sinking is important when it comes to the matter of *fan-out.* The fan-out of a logic gate is the number of like gates which can be connected to its output. In other words, fan-out is the number of gates that a certain gate can drive, assuming that the driven gates are all identical to the driving gate. Figure 2-20(c) shows one driving gate connected to several driven gates. The fan-out is limited, because if too many driven gates are hooked up, the current through R_C of the driving gate will become too large. That is, the driven gates' combined current draw will cause the voltage drop across R_C to become so large that the collector output voltage will no longer be distin-

Figure 2-20. (a) Schematic drawing of an RTL NOR gate. (b) RTL is a current-sourcing logic family. (c) Fan-out in the RTL family. (d) Wire-ANDing of RTL gates.

(a)

(b)

(c)

(d)

Figure 2-20 (Cont.)

59

guishable as a HI. The HI voltage will degenerate to the point that the driven gates won't recognize it as a HI. They might interpret it as a LO.

In a current-sourcing logic family, the fan-out problem (also called the loading problem) always occurs when the output goes HI, not when it goes LO. In this case it is easy to see that when the output is LO, it doesn't matter how many driven gates are connected.

The fan-out of RTL gates is usually in the range of 4 to 10.

Tying Outputs Together. Occasionally a logic user may want to tie the outputs of two gates together, as illustrated in Fig. 2-20(d). This is loosely called *wire-ANDing* or *wire-ORing*. This practice is not always permissible, namely when the gates have totem-pole outputs, as discussed in Chapter 1. RTL logic gates usually don't have totem-pole outputs, so wire-ANDing is permitted for them.

Hanging Inputs. Oftentimes in solid-state logic not every gate input is needed, and the question arises about what to do with the unused inputs. It is a good idea to tie unused inputs to some point in the circuit. They may be tied to the supply line, to the ground line, or to another input, depending on circumstances, but it is not always absolutely essential that this be done. When an unused input is left unconnected, we refer to it as a *hanging input*, or a *floating input*.

A hanging input (also called a *dangling input*) is more likely to pick up unwanted noise signals than an input which is connected to something. But if the circuit is well protected from noise to begin with, this consideration is not very important. What *is* always important is to realize how the gate itself interprets a hanging input. That is, some logic families interpret a hanging input as a LO signal, and other logic families interpret a hanging input as a HI signal. An OR gate in a logic family of the latter type could not tolerate a hanging input, for example. If the hanging input were interpreted as a HI, the OR gate output would be locked in the HI state, no matter what the other inputs did. This would render the gate useless. This is one example of a time when it is absolutely essential that an unused input be terminated somewhere in the circuit, not left hanging. In this example it would probably be grounded.

RTL gates interpret a hanging input as a LO. This is really pretty obvious, since a hanging input cannot deliver any base current into its transistor, which is the same result obtained from a LO input.

Other Features. The RTL family has fair operating speed, which is not usually of paramount importance in industrial applications. It has a typical propagation delay time of about 12 nsec, although this can vary among manufacturers.

RTL is straightforward and easy to visualize and understand. Its immunity to electrical noise is fair, and its fan-out is acceptable. At one time RTL was the most popular logic family for industrial use, but this is no longer true.

RTL logic gates appear in three physical forms. These forms are

a. Discrete components mounted on a printed circuit board
b. Discrete components encapsulated in a plastic container and hermetically sealed

c. Integrated circuits (ICs)

The physical appearance of logic circuits is shown in Fig. 2-21. A printed circuit board with discrete resistors and transistors is shown in Fig. 2-21(a).

A hermetically sealed plastic container is shown in Fig. 2-21(b). The discrete

Figure 2-21. Physical appearance of RTL logic circuits. (a) Discrete components mounted on a printed circuit board. (b) Discrete components encapsulated and sealed from the atmosphere. (*Courtesy of the Square D Company.*) (c) Dual-in-line IC. (*Courtesy of Motorola, Inc.*) (d) Flat-pack IC. (*Courtesy of Motorola, Inc.*) (e) Metal-can IC. (*Courtesy of Motorola, Inc.*)

(a)

(b)

(c)

(e)

(d)

components are mounted inside in three dimensions; they are not confined to a single plane.

Three different integrated circuit packages are shown in Figs. 2-21(c), (d), and (e). The *dual-in-line package*, abbreviated DIP, is shown in Fig. 2-21(c). It is usually made of plastic and is by far the most popular IC package in industry.

The *flat-pack package* is shown in Fig. 2-21(d). Flat packs are usually made of ceramic and can withstand extreme temperatures better than plastic DIPs.

The *metal-can package* is shown in Fig. 2-21(e). Metal cans are used only for integrated circuits having 10 output terminals or less. They are not so prevalent as they once were, because most digital logic ICs have more than 10 terminals. Fourteen and 16 terminals are most common.

2-13-2 The DTL family

DTL (diode-transistor logic) has its inputs applied to diodes, and the output is taken from a transistor. We have seen a DTL gate before in Fig. 2-6(a). A more conventional rendition of a DTL gate is shown in Fig. 2-22(a).

Basic Gate Is the NAND. The gate of Fig. 2-22(a) is a NAND. Here is how it works. If any one of the inputs A, B, or C is held LO, its associated diode will not allow the base of Q_1 to rise more than 0.6 V above ground level. The +5-V supply voltage will deliver current through the 1.75-K and 2-K resistors, and 4.4 V will drop across these resistors, with only 0.6 V appearing across the diode whose cathode is LO. With such a low voltage on the base of Q_1, it cannot turn ON, especially since there is another diode in its emitter lead which must also be broken down. With no emitter current in Q_1, there is no base current delivered to Q_2, so it turns OFF. The output rises to the supply voltage level of +5 V.

If all three inputs are brought HI, the current coming down through the 1.75-K and 2-K resistors will no longer be able to flow through the diodes to ground. Therefore it will turn and flow into the base of Q_1. An alternative view is that the voltage at the anodes will rise to 0.6 V above the voltage at the cathodes of the input diodes. This higher voltage on the base of Q_1 will turn Q_1 ON. Q_1 delivers enough current into the base of Q_2 to saturate it, bringing the output down to ground level.

Therefore the output goes LO if and only if all inputs are HI. This is a NAND function. Other types of gates can be built in DTL by adding extra components.

Current Sinking. The DTL family fits into the classification of *current-sinking* logic. A current-sinking family has the property that conventional current flows *into* the driving gate *from* the driven gate. Figure 2-22(b) shows a LO input from one gate, the driving gate, being applied to an input terminal of another gate, the driven gate. Again the gates have not been drawn in their entirety. As the figure shows, conventional current flows out of the driven gate and into the driving gate. The driving gate *sinks* the current back to ground—hence the term *current-sinking* logic.

(a)

(b)

Figure 2-22. (a) Schematic drawing of a DTL NAND gate. (b) DTL is a current-sinking logic family. (*Courtesy of Motorola, Inc.*)

Fan-out. In current-sinking logic, the fan-out problem always occurs when the output goes LO, not when it goes HI. In this case, imagine several driven gates all connected to the output of the driving gate in Fig. 2-22(b). If the output is HI, there is little or no current flow between the gates, so it doesn't matter how many driven gates are connected.

However, when the output goes LO, each one of the driven gates will dump current into the output transistor of the driving gate. If there are too many driven gates, their combined currents may force the output transistor to come out of saturation. Remember that the *greatest* amount of collector current the transistor will carry is given by

$$I_{C_{max}} = \beta I_B.$$

If the driven gates dump so much current into the collector of the transistor that the collector current gets close to $I_{C_{max}}$, the output transistor will start cutting back the current through its own 6-K collector resistor. This will cause less than 5 V to be dropped across that resistor, and the output transistor comes out of saturation. Under these circumstances the output voltage will not be a firm LO; it will no longer be distinguishable as a LO, and some of the driven gates may interpret it as a HI.

The fan-out of DTL gates is usually 8 or more.

Tying Outputs Together. DTL gates can safely be tied together, wire-ANDed, as long as they have no active pull-up transistor in the output (totem-pole output). Most DTL gates have only a collector resistor, as shown in Fig. 2-22(b), so wire-ANDing is permitted.

Hanging Inputs. DTL gates interpret a hanging input as a HI. As mentioned in Sec. 2-13-1, a DTL OR gate or NOR gate cannot tolerate a hanging input. Unused inputs should either be tied to another input which *is* being used or tied to the ground. A DTL AND gate or NAND gate can tolerate a hanging input. The choice of whether to let an unused input hang or to tie it to something is based on the noise situation and other circumstances.

DTL gates interpret a hanging input as a HI because a hanging input will not allow current to flow out of the input terminal; this is the same result obtained from a HI input.

Other Features. The DTL logic family has fairly good noise immunity, and it has higher fan-out than the RTL family. Power dissipation per gate is rather low, but this is not of prime importance in industry. A typical DTL gate might dissipate an average power of 8 mW.

DTL gates, like RTL gates, are often built of discrete components and mounted on printed circuit boards. They are also common in IC packages. DTL ICs are packaged in DIPs, flat packs, and metal cans.

2-13-3 The HTL family

HTL (high-threshold logic) is very similar in construction and operation to DTL logic. A schematic of the basic gate is shown in Fig. 2-23(a).

Basic Gate Is the NAND. The HTL gate in Fig. 2-23(a) works in exactly the same way as the DTL gate in Fig. 2-22(a) except that the input voltage required to turn ON Q_1 is much higher. The input voltage must be high enough to break down one

(a)

(b) HTL Gate

(c) DTL Gate

Figure 2-23. (a) Schematic drawing of an HTL NAND gate. (b) V_{out} vs. V_{in} transfer curve of an HTL gate. (c) Transfer curve of a DTL gate. (*Courtesy of Motorola, Inc.*)

transistor base-emitter junction plus the 7-V zener diode. (The input diode offsets one of the two transistor junctions.)

Therefore the input voltage must reach about 7.6 V in order to turn ON Q_1 and Q_2. Of course, *all* inputs must reach this level in order to turn ON the transistors and drive the output LO. The basic gate of Fig. 2-23(a) is a NAND. Other gates are possible by adding extra components.

HTL Characteristics. The important advantage that the HTL family possesses is excellent noise immunity. This excellent noise immunity is achieved because of the 7.6-V input required to turn ON the transistors. The transfer curve of an HTL gate is shown in Fig. 2-23(b). A transfer curve is a graph of output voltage versus input voltage, even though we don't expect to see continuously variable input voltages when the gate is in use.

Contrast this transfer curve with the curve for a DTL gate, which is presented in Fig. 2-23(c).

The HTL transfer curve shows that any value of input voltage between 0 and 6.5 V will be interpreted as a LO. A properly used HTL gate has a guaranteed LO output of between 0 and 1.5 V, so the input to the gate in question is guaranteed to fall in this range [$V_{in_{LO}}$ in Fig. 2-23(b)]. Therefore we have a situation in which the LO input from some other HTL gate is no greater than 1.5 V, but the gate we are investigating will interpret any voltage up to 6.5 V as a LO signal. This provides a 5.0-V safety margin (6.5 V − 1.5 V), called *LO noise margin* in Fig. 2-23(b). It means that a LO input line to an HTL gate can tolerate at least 5.0 V of unexpected noise signal without falsely triggering the gate into the opposite state.

A similar situation prevails for HI input voltages. The graph shows that an HTL gate will interpret any value of input voltage between 8.5 and 15 V as a HI. A properly used HTL gate has a guaranteed HI output between 13.5 and 15 V, so a HI input to the gate we are investigating is guaranteed to fall in this range [$V_{in_{HI}}$ in Fig. 2-23(b)]. This provides a *HI noise margin* of 5.0 V, as shown in the graph. Thus a HI input line to an HTL gate can tolerate at least 5.0 V of unwanted noise signal without falsely triggering the gate.

An inspection of the transfer curve for the DTL gate shows that it too has noise margins, but they are not so great as those of an HTL gate. A typical DTL gate has a LO noise margin of 1.2 V (1.6 V − 0.4 V) and a HI noise margin of 2.2 V (4.6 V − 2.4 V). Comparing the two families, it can be seen that an HTL gate is far less vulnerable to being falsely triggered than a DTL gate because of its much higher noise margins. This makes the HTL family attractive for use in electrically noisy industrial environments.

The higher supply voltage used with the HTL family necessitates the use of larger resistors to keep the power dissipation at an acceptable level. Power dissipation is still fairly high at about 20 mWatts per gate. The higher value resistors slow down gate response, resulting in propagation delay times of about 120 nsec. This is very slow in comparison with most other logic families, but as mentioned several times before, slow speed is no great shortcoming in industrial control. In fact, it can be a blessing, since it automatically filters out all noise signals that occur at very high frequencies.

The fan-out of an HTL gate is about 10, making it a little better than a DTL gate in this regard. HTL is a current-sinking family, like DTL. Hanging inputs are interpreted as HI.

It is permissible to wire-AND HTL gates as long as they have no active pull-up output transistor. However, pull-up transistors are frequently seen in HTL gates [the output circuit is different from Fig. 2-23(a)], so the user must be careful if he wishes to wire-AND two HTL outputs.

HTL logic gates are usually integrated circuits. HTL ICs are always packaged as DIPs.

2-13-4 The TTL family

TTL (transistor-transistor logic) has its inputs connected to transistors, and its output is taken from a transistor. The basic TTL gate is shown in Fig. 2-24.

Basic Gate Is the NAND. In Fig. 2-24, if any one of the three inputs is LO, Q_1 will turn ON. Q_1 is a multiple-emitter transistor, so if a current flow path can be established from the base through *any one* of the emitters, the transistor will turn ON. With Q_1 ON, its collector is pulled down close to ground. This removes the base drive from transistor Q_2 and causes it to turn OFF. When Q_2 turns OFF, it

Figure 2-24. Schematic drawing of a TTL NAND gate, showing the multiple-emitter input transistor.

prevents base current from reaching Q_4, so Q_4 also turns OFF. However, Q_3 can be supplied through the 1.6-K resistor, so Q_3 turns ON, pulling the output up toward 5 V. Q_3 and Q_4 are a totem-pole output pair; Q_4 is the pull-down transistor, and Q_3 is the pull-up transistor.

If all three inputs go HI, there can be no base-emitter current through any of the three emitters, so Q_1 turns OFF. The *base-collector* junction of Q_1 now becomes forward biased and passes current into the Q_2 base. Transistor Q_2 turns ON, causing Q_4 to turn ON and pull the output down LO. Meanwhile, Q_3 turns OFF because the voltage at its base drops to less than 1 V, which is not sufficient to overcome the combination of the Q_3 base-emitter junction plus the diode in the Q_3 emitter lead. The pull-up transistor therefore releases (opens) the connection between the output terminal and the $+5$-V supply.

The basic gate is thus a NAND; other gates are possible by adding extra components.

TTL Characteristics. TTL combines good noise immunity with low power dissipation and fairly high speed. TTL's noise immunity is not so good as HTL's, but it is generally better than that of DTL or RTL. Its power dissipation is low about 10 mW per gate, and it has a propagation delay of around 10 nsec. Most TTL gates have a fan-out of 10.

TTL is current-sinking logic, since conventional current flows out an emitter lead of the driven gate and into the collector of the pull-down transistor of the driving gate. In TTL, a hanging input is interpreted as a HI.

Most TTL gates have an active pull-up transistor in the output circuit, so they cannot be wire-ANDed. Special *open-collector* TTL gates are available if a user desires to tie outputs together.

The TTL family never appears in discrete component form. TTL gates are always integrated circuits. Both DIPs and flat packs are common. TTL presently has a greater selection of logic circuitry than any other logic family and is the most widely used of any logic family.

2-13-5 The CMOS family

CMOS (complementary metal oxide semiconductor) logic does not use a regular bipolar transistor for its amplifying device. Instead, it uses *field-effect transistors* (FETs), specifically metal oxide semiconductor field-effect transistors (MOSFETs). Figure 2-25 shows the schematic diagram of a CMOS NOR gate.

We do not have space to go into the details of operation of a CMOS logic gate. To do so would demand a background explanation of several nontrivial notions about FETs, such as the difference between *p*-channel and *n*-channel FETs and the difference between depletion and enhancement in an FET. For specifics on the internal behavior of CMOS logic circuits, consult any good book which deals with FETs in detail.

Suffice it to say here that the main advantage of the CMOS family is its extreme-

Figure 2-25. Schematic drawing of a CMOS NOR gate, showing the field-effect transistors.

ly low power consumption. A typical CMOS gate will consume and dissipate only about 2 nW of power when switched at 1 kHz; this is less than one-millionth the power dissipated by a TTL gate, for example. Yet in spite of its minuscule power consumption, the CMOS family is blessed with noise immunity better than that of TTL and almost as good as that of HTL. It is unusual to combine both of these assets in any electronic circuit.

Notice also that the CMOS gate schematic in Fig. 2-25 shows *no resistors or diodes.* CMOS circuits use nothing but MOSFETs as circuit components. This enables them to be manufactured as integrated circuits very cheaply. Also, because an IC MOSFET requires less area on the silicon crystal than an IC resistor or diode, the packing density can be made very high. This permits complicated and extensive circuits to be fabricated on an IC chip.

CMOS promises to become more popular as time passes and might well turn out to be the most popular logic family of all.

Other Characteristics of CMOS. CMOS cannot be classified as either current-sinking or current-sourcing logic, since virtually no current flows between the driving gate and the driven gate. This is because the dc input resistance of a MOSFET is extremely high, near 1 trillion Ω ($10^{12} \Omega$). The high input resistance of a MOSFET allows a large number of gate inputs to be connected to a single gate output. In other words, the CMOS family possesses a high fan-out. Most CMOS gates have a fan-out in excess of 100, depending on the frequency at which the gates are switched. This is another important advantage of this logic family.

It is never permissible to wire-AND CMOS gates, since they always have a pull-up FET in the output circuit.

CMOS logic gates and other CMOS digital circuits are always ICs, never discrete components. They are packaged in DIPs and flat packs.

QUESTIONS AND PROBLEMS

1. Explain the purpose of each one of the three sections of an industrial logic control circuit.

2. Name some common information-gathering (input) devices used in industrial logic systems.

3. Name some common actuating (output) devices used in industrial logic systems.

4. What two general types of information do input devices supply to an industrial logic circuit?

5. What is the most common voltage used for input circuits for industrial control in the United States?

6. Explain the difference between a normally open limit switch contact and a normally closed limit siwtch contact. Draw the schematic symbol for each.

7. Repeat Question 6 for relay contacts.

8. Explain why series-connected contacts constitute an AND circuit.

9. Explain why parallel-connected contacts constitute an OR circuit.

10. In words, explain the operation of an AND gate. Draw the truth table for a two-input AND gate. Then draw the truth table for a four-input AND gate. How many different combinations of inputs are possible when there are four inputs?

11. Repeat Question 10 for an OR gate, a NAND gate, and a NOR gate.

12. In Fig. 2-6(a), draw in all the current flow paths if lines X and Y are at $+5$ V and line Z is at 0 V.

13. Repeat Question 12 for all three inputs at $+5$ V.

14. In Fig. 2-6(b), draw in all the current flow paths if lines X and Y are at $+5$ V and line Z is at 0 V.

15. Repeat Question 14 for all three inputs at 0 V.

16. Draw the logic circuit to implement the following conditions: The solenoid is energized if LS1 and LS2 are both actuated or if LS3 is actuated. Draw the circuit using relay logic and also using solid-state logic.

17. Repeat Question 16 for these conditions: The solenoid energizes if LS1, LS2, and LS3 are all actuated or if LS1 is not actuated.

18. Repeat Question 16 for these conditions: The lamp turns on if LS1 and LS2 are both not actuated or if LS2 is actuated while LS3 is not actuated.

19. Repeat Question 16 for these conditions: The motor starter energizes if either LS1 or LS2 is actuated at the same time that either LS3 is actuated or LS4 is not actuated.

20. Explain the meaning of the term *fan-out* as applied to logic circuits in general and logic gates in particular.
21. Explain the meaning of the term *fan-in* as applied to logic circuits in general and logic gates in particular.
22. Describe the difference between a positive logic system and a negative logic system.
23. Explain the difference between a discrete circuit and an integrated circuit.
24. Why are capacitive switch filters sometimes necessary? Describe what they do, and name some of the benefits that arise from using them.
25. Repeat Question 24 for bounce eliminators.
26. What is a logic signal converter, and what does it do?
27. If a commercially built signal converter is used to interface between an input device and solid-state logic circuitry, is a bounce eliminator or switch filter also needed under normal circumstances? Why?
28. Will the signal converter of Fig. 2-14(a) work on dc as well as ac? Why?
29. Repeat Question 28 for the signal converter of Fig. 2-14(b).
30. Why is it a good idea to provide electrical isolation between input circuits and solid-state logic circuits?
31. Name some conditions under which solid-state logic is preferred to relay logic.
32. Name some conditions under which relay logic might be preferred to solid-state logic.
33. What is the purpose of an output amplifier for use with solid-state logic?
34. What is the purpose of the diode in the collector circuit of Q_2 in Fig. 2-16(a)?
35. What is the purpose of the diode in the base lead of Q_2 in Fig. 2-16(a)?
36. Does the output amplifier of Fig. 2-16(b) furnish electrical isolation between the logic circuit and the output circuit? Explain carefully.

Questions 37–39 apply to the solid-state conveyor/classifying system illustrated in Figs. 2-4 and 2-10.

37. What conditions are necessary to enable AND9?
38. What conditions are necessary to enable AND14?
39. Explain how the momentary closing of one of the chute limit switches breaks the seals on OR2, OR3, and OR4 if they were sealed.

Questions 40–42 apply to the first failure annunciator illustrated in Fig. 2-18.

40. Explain how the *first* failure causes the circuit to ignore all subsequent failures.
41. Explain how the circuit remembers which failure occurred even if the failure corrects itself.
42. Try to explain what would happen if two failures occurred at exactly the same instant. (This would be a fantastic coincidence.)

Questions 43–45 apply to the machine tool routing system illustrated in Fig. 2-17.

43. If LS3 got stuck in the closed position, what would happen when the operator pressed the START button?

44. If LS3 failed to close when cylinder *B* extended, what would happen?

45. If SC3 malfunctioned so that it could not deliver a HI output, what would happen during the machine cycle?

Questions 46–52 apply to the machine tool drilling system illustrated in Fig. 2-19.

46. The LS1 contact is normally open, but it is *drawn* in the closed position. Why is this?

47. When the START button is pressed at the beginning of the cycle, what does NAND1 do? Explain.

48. When the START button is pressed, explain how NAND2 seals up (output HI).

49. What action of the system breaks the seal of NAND2? Explain.

50. What action of the system causes NAND4 to seal up (output HI)? Explain how NAND4 seals up.

51. What two conditions are necessary to cause the output of NAND6 to go LO? Explain.

52. What action of the system breaks the seal of NAND4? Explain.

53. Explain why it is permissible to wire-AND logic gates which have a simple collector resistor in the output circuit but not permissible for gates having a pull-up transistor in the output circuit.

54. Explain why the fan-out problem occurs when the gate output goes HI for current-sourcing logic but occurs when the output goes LO for current-sinking logic.

55. What gives HTL gates their noise immunity?

56. Explain why a hanging input in a current-sourcing logic family is interpreted as a LO but a hanging input in a current-sinking family is interpreted as a HI.

3

Transistor Switches for Memory and Counting

Besides their usefulness as decision-makers, transistor switches can also be used to build logic devices that have memory and logic circuits that can count. We will see in this chapter how such devices work, and we will explore examples of their application in industrial systems.

OBJECTIVES

After completing this chapter you will be able to:

1. Explain the behavior of *RS* flip-flops and *JK* flip-flops from the black-box point of view
2. Describe the operation of flip-flops as memory devices in the control circuits presented as examples
3. Explain the action of shift registers from a black-box point of view
4. Describe how a shift register keeps track of digital data regarding a part moving on a conveyor system

5. Count to 9 in binary

6. Convert from binary to decimal for any number up to 9

7. Convert from decimal to BCD and vice versa for numbers greater than 9

8. Explain the operation of decade counters from a black-box point of view

9. Explain how cascaded decade counters work to count above 9

10. Explain the action of a 1-of-10 decoder

11. Describe in detail the operation of a carton routing and palletizing system using cascaded decade counters and 1-of-10 decoders

12. Explain the operation of a one-shot from the black-box point of view

13. Explain the uses of a digital clock and the operation of a free-running multivibrator-type clock

14. Describe in detail the operation of an automatic tank-filling system using one-shots, flip-flops, a decade counter, and a free-running clock

15. Describe the operation of a decade down-counter

16. Explain the operation of a decimal-to-BCD encoder

17. Describe the operation of time delay relays including the four different types of time delay contacts

18. Explain the operation of a solid-state timer based on a series RC charging circuit

19. Describe in detail the operation of a traveling hopper material supply system using a down-counter, an encoder, and solid-state timers

3-1 FLIP-FLOPS

A flip-flop is a digital circuit which has memory. Its output is either HI or LO, just like a logic gate, but the difference is that a flip-flop will remain in a given state even if the input signal which put it in that state is removed. A flip-flop output will change states only when specifically ordered to do so by another input signal.

The circuitry which is at the heart of every flip-flop is shown in Fig. 3-1. At all times, one transistor is turned ON and the other transistor is cut OFF. Here's how the circuit works. Suppose that transistor T_1 is saturated. If it is, the collector of T_1 will be at virtually 0 V, which will remove the base drive from transistor T_2. Therefore T_2 will be cut OFF, and its collector will tend to rise to $+V_s$. This will cause current to flow through R_{B_1} into the base of T_1, holding that transistor saturated. The collector of T_1 will stay at 0 V, while the collector of T_2 will stay at a HI level, almost V_s. The collector of T_2 is the output of the flip-flop, which is identified by the letter Q in this book. Under these conditions, Q will remain HI forever, as long as the power is kept on.

On the other hand, if we had started out by assuming that T_2 was saturated instead of T_1, the argument would have worked the other way. We would have found that the collector of T_1 would be near $+V_s$, while the collector of T_2 would

Figure 3-1. Schematic diagram of a basic flip-flop.

be at 0 V. Under this assumption we would conclude that Q will remain LO forever, as long as the power is kept on.

According to the preceding explanation then, the circuit has a memory. If it is forced by some external signal into the state where Q is HI, it will maintain that HI even if the forcing signal is removed. Likewise, if Q is forced LO, it will maintain that LO when the forcing signal is removed.

Figure 3-2(a) shows how the forcing signals could be brought in. If the input labeled S goes HI, it forces T_1 to turn ON by delivering saturating base current through the 100-K input coupling resistor. When T_1 turns ON, T_2 would turn OFF and Q would go HI. If the HI on S is then removed and the S line goes back LO, the circuit will not change states. It has "remembered" that the last input to go HI was the S input.

If the R input goes HI (after S has returned to LO), it turns ON T_2 through the other 100-K input resistor. Therefore Q goes LO. T_1 is turned OFF by the T_2 collector going to 0 V. The T_1 collector rises almost to V_s and provides another source of base current to T_2. If the HI on R is then removed, the circuit maintains the same state. It has "remembered" that the last input to go HI was R.

For the resistor sizes given in Fig. 3-2(a), the transistors must have dc betas greater than about 55 in order to saturate reliably. This is because the total resistance in the base circuit will be slightly more than 50 times the resistance in the collector circuit, plus the fact that some of the base drive voltage is sacrificed in overcoming the base-emitter junction.

The circuit of Fig. 3-2(a) is actually a simple model of a particular type of flip-flop, called an *RS flip-flop*. The letters R and S are derived from the words *reset* and *set*. The black-box schematic symbol for an RS flip-flop is shown in Fig. 3-2(b). When you see that symbol, you need not remember the internal action of the transistors and resistors. Simply recall the input-output relationships of an *RS* flip-flop which have been described. That is, Q can be either HI or LO; a HI

(a)

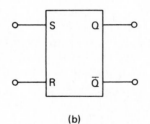

(b)

Figure 3-2. (a) Schematic diagram of an *RS* flip-flop, showing the forcing inputs. (b) Black-box symbol of the *RS* flip-flop.

on S sets Q to a HI; a HI on R resets Q to a LO. Output Q remains in a given state forever, or until it is ordered into the opposite state by a new input signal.

In almost all commercial flip-flops, there is a second output, called $\bar{Q}$ (pronounced "Q not"). This output is always the *opposite* of the Q output and is obtained by attaching a wire to the collector of transistor T_1 in Fig. 3-2(a) and bringing it out of the circuit. When Q is HI, $\bar{Q}$ is LO; when Q is LO, $\bar{Q}$ is HI. The word *complement* is often used to mean opposite in a digital circuit, so $\bar{Q}$ is sometimes referred to as the complement of Q. The $\bar{Q}$ output may or may not be actually used, depending on the particular application of the flip-flop, but it is always available for use.

From now on, we will use the *bar* symbol over a letter to mean the complement of the letter itself. Thus $\bar{A}$ would mean the complement of line A, where line A

would be a line somewhere in a digital circuit. If line A is 1 (HI), $\bar{A}$ is 0 (LO); if A is 0, $\bar{A}$ is 1.

3-2 A WELDER CONTROL CIRCUIT USING *RS* FLIP-FLOPS

Imagine a situation in which two automatic welders are powered from the same supply bus. The supply bus can deliver enough current to drive one welder, but it cannot power two welders at the same time because of the large current draw. Therefore if an automated system signals for a second welder to begin welding when the first welder is already involved in a weld, the initiation of the second weld must be postponed. When the first welder is finished, then the signal for the second welder will be honored.

To accomplish this, a circuit is needed which knows if a weld is currently being performed and can receive and *remember* input requests for a second weld. Since the circuit must remember something, it will contain flip-flops. A circuit to accomplish this action is shown in Fig. 3-3.

Figure 3-3. Welder control circuit illustrating the memory ability of *RS* flip-flops.

Here is how it works. If a weld is called for by the closing of one of the WELDER START contacts, then the appropriate flip-flop turns ON (its Q output goes HI). That is, either FF1 or FF3 will turn ON because a HI will appear on its S input. For purposes of illustration, suppose that the A welder START contact closes, causing the FF1 Q output to go HI. This will apply a HI to input 1 of AND1. If the B welder is *not* welding at this time, the number 2 input of AND1 will also be HI. This is explained in the next paragraph. In that case the output of AND1 will go HI and apply a HI to the S input of FF2. The Q output of FF2 goes HI, causing the output amplifier to energize the A welder contactor. This contactor connects the A welding transformer to the power bus and produces a weld. The preceding description applies equally well if a weld is requested on the B welder when the A welder is turned off. FF3 would turn ON, enabling AND2, which would turn ON FF4.

Therefore if a weld is called for by the system control, it takes place immediately as long as the other welder is not welding at that time. On the other hand, consider what would happen if the A welder START contact closed while welder B was welding. In that case the number 1 input of the NOR gate would be HI because it is connected to the Q output of FF4. The output of the NOR goes LO. This LO is applied to input 2 of AND1, guaranteeing a LO output from the AND1 gate and preventing FF2 from turning ON. Thus the A welder cannot start.

As soon as the B welder is finished, the B welder STOP contact will close, applying a HI to R of FF3. This turns OFF FF3 and drives the $\bar{Q}$ output HI. This HI appears at the R input of FF4, turning that flip-flop OFF. Therefore the Q output of FF4 goes LO, removing the HI from the NOR gate input at the same time it deenergizes the B contactor. The output of the NOR gate goes back HI, enabling AND1. At that time, a HI gets through to S of FF2, which turns on the A welder.

Thus a weld request is postponed if the other welder is presently operating. However, the circuit of Fig. 3-3 remembers the request and acts upon it when the other welder is free.

3-3 CLOCKED *RS* FLIP-FLOPS

Although the straightforward RS flip-flop of Sec. 3-2 has many uses in industrial electronics, a more commonly seen flip-flop is the *clocked RS flip-flop*. A clocked RS flip-flop is one which doesn't respond to the S and R inputs at the instant they arrive but responds to them only later when a *clock* signal is received at its clock input. Figure 3-4(a) shows the black-box schematic symbol for a clocked RS flip-flop and Fig. 3-4(b) shows one method of creating the clocked feature.

Figure 3-4(b) contains an *RC differentiator*. An RC differentiator is an RC circuit which has a very short time constant. When the CLK line goes HI in Fig. 3-4(b), the capacitor quickly charges to the applied voltage. During the charging

(a)   (b)

Figure 3-4. (a) Black-box symbol of a clocked *RS* flip-flop. (b) One
method of constructing a clocked *RS* flip-flop.

period a spike appears across the resistor, which is fed to the two AND gates. The
AND gates are therefore partially enabled at the instant the clock line goes from
LO to HI. If there is a HI at either *S* or *R* at that instant, the flip-flop will respond
to that HI input and turn ON or OFF as directed.

Note that the only time there is a HI on the common input to the two AND
gates is at the instant the clock line goes HI. The clock line simply resting in the HI
state will not deliver a HI to the common inputs because the capacitor will be fully
charged and there will be no voltage across the resistor. As the clock line goes
from HI to LO, a negative voltage spike appears across the resistor. A negative
spike does nothing for the AND gates. Therefore the only instant the flip-flop can
respond to its *S* and *R* inputs is at the instant the clock waveform delivers a positive-going edge, from LO to HI. For this reason such flip-flops are sometimes
called *edge-triggered* flip-flops; the flip-flop of Fig. 3-4(b) would be a *clocked
positive edge-triggered flip-flop*. In Fig. 3-4, the small triangle drawn inside the box
at the CLK terminal indicates that the flip flop is edge-triggered.

Many clocked flip-flops are *negative* edge triggered. They behave exactly as
discussed above except that they respond to *S* and *R* inputs only at the instant the
clock line goes from HI to LO. In an industrial setting, you can expect to see both
positive and negative edge-triggered flip-flops. However, to avoid confusion, we
will assume throughout this book that all clocked flip-flops are negative edge
triggered. Schematically, negative edge-triggered flip flops are distinguished from
the positive edge-triggered variety by small circles drawn outside the box by the
CLK terminal. Look ahead to Fig. 3-5 to see examples of this notation.

Flip-flops are often described by truth tables. Flip-flop truth tables are interpreted a little differently than logic gate truth tables, but they have the same purpose. They tell the output of the flip-flop for every combination of inputs. Table
3-1 is the truth table of a clocked *RS* flip-flop.

This is how to interpret the truth table. The binary digit in the *S* column tells
the logic level of the *S* input at the instant the clock edge arrives. As usual, 1
means HI and 0 means LO. The digit in the *R* column tells the level at the *R* input

TABLE 3-1. The truth table of a clocked *RS* flip-flop, showing the input-output relationship.

S	R	Q_{n+1}
0	0	Q_n
0	1	0
1	0	1
1	1	UND

when the clock edge arrives. The Q_{n+1} heading in the last column stands for the state of the *Q* output just *after* the edge has arrived. The $n + 1$ subscript suggests that the flip-flop has been triggered "one more time."

Starting with the top row in the table, we see that if *S* and *R* both equal 0 when the edge arrives, the state of the *Q* output is Q_n. This means that *Q after* the clock edge (Q_{n+1}) is the same as *Q before* the clock edge (Q_n), so the flip-flop has not changed states at all. If *Q* was HI, it remains HI; if *Q* was LO, it remains LO. Some truth tables have the words "no change" entered instead of Q_n. This is just what we would expect the flip-flop to do under these conditions, since we know that it will maintain its present state unless specifically ordered to change.

The second row indicates that if *S* is 0 and *R* is 1 when the edge arrives, the *Q* output will go to 0. The flip-flop is said to be OFF.

The third row indicates that if *S* is 1 and *R* is 0 when the edge arrives, the *Q* output will be 1 after the edge has occurred. The flip-flop is said to be ON.

The fourth row indicates that if *S* and *R* are both HI when the clock edge arrives, the output state is undecided. There is no telling what state the flip-flop will go to. This input combination is sometimes called an "illegal" combination. It would be the designer's job to make sure that this input combination never occurs in the actual circuit.

3-4 OSCILLATING MACHINING TABLE USING CLOCKED *RS* FLIP-FLOPS

Imagine a machining operation in which a table is moved back and forth by a reversing motor. This might arise in a planing operation in which the planing tool remained stationary and the workpiece was mounted on an oscillating table. Figure 3-5(a) shows such a layout. When the motor spins in one direction, the rack and pinion assembly moves the table to the right; when the motor spins in the other direction, the rack and pinion move the table to the left. When the table has moved to the extreme right, it contacts RIGHT LS, which signals the control circuit to

Figure 3-5. (a) Physical appearance of the oscillating planer. (b)
Control circuit of the oscillating planing system, illustrating the
application of clocked *RS* flip-flops.

reverse the motor and run to the left; when it moves to the extreme left it contacts
LEFT LS, which signals the circuit to run the motor to the right. This action continues as long as necessary to complete the planing operation.

When the operator is satisfied that the planing is complete, he throws a selector switch into the STOP PLANING position. The table then continues in motion until it has returned to its extreme left position.

Here is how the circuit works. Assume that the table is running to the right and the two-position selector switch is in the CONTINUE PLANING position. The fact that the switch is drawn closed in Fig. 3-5(b) means that the contact is closed when the operator selects that position. Conversely, the contact is open when the operator selects the STOP PLANING position.

If the table is running to the right, it is because the RUN TO RIGHT motor starter is energized, which implies that FF1 is ON. Also, since the RUN TO LEFT motor starter coil is necessarily deenergized, it follows that FF2 is OFF. Therefore the situation is this: The HI on Q_1 puts a HI on S of FF2 and a LO on R of FF2 through I2; the LO on Q_2 puts a LO on input 1 of the AND gate, which applies a LO to S FF1 and a HI to R of FF1.

When the RIGHT LS is contacted, it turns on its signal converter, which applies a HI on input 2 of the NOR gate. The NOR gate output goes LO and delivers a negative edge to both CLK inputs. Since the FF1 inputs are ordering FF1 to turn OFF, it does just that; the RUN TO RIGHT starter coil deenergizes. The FF2 inputs at the instant the edge arrives are $S = 1$, $R = 0$, so FF2 turns ON. When Q_2 goes HI it enables its output amplifier and energizes the RUN TO LEFT starter coil. The table therefore reverses direction and moves to the left.

When the table reaches the extreme left, the LEFT LS is actuated. At this time the opposite conditions hold true. Q_2 is HI and Q_1 is LO, so FF1 has $S = 1$, $R = 0$, and FF2 has $S = 0$, $R = 1$. This is so as long as the SS (selector switch) is in the CONTINUE PLANING position and is applying a HI to the number 2 input of the AND gate. When LEFT LS closes and the clock input terminals receive their negative edges from the NOR, FF1 turns ON and FF2 turns OFF. The motor reverses again and begins moving the table to the right.

Now suppose that the operator decides to terminate the planing operation. At some point in the travel, he throws the SS to the STOP PLANING position. This removes the HI from input 2 of the AND gate, forcing the output to go LO. Therefore FF1 has a LO on S and a HI on R, no matter what the state of the number 1 input of the AND. The next time the table contacts LEFT LS, both FF1 and FF2 will be turned OFF because both flip-flops will have $S = 0$ and $R = 1$. This will be true of FF1 because of the AND gate output being LO; it will be true of FF2 because Q_1 will be LO while the table is moving to the left. With both flip-flops OFF, Q_1 and Q_2 are both LO, and the two motor starters are deenergized. The motor stops, leaving the table in the left position.

If the table had been running to the right at the time the operator changed the SS, the reversing of the motor would take place as usual when the table hit RIGHT LS, because FF2 is free to turn ON regardless of the condition of the AND gate. The table will always stop in the left position.

You may wonder how the cycle is ever started once a new workpiece is placed on the table. This problem has been left as an exercise at the end of the chapter.

When trying to understand the action of clocked RS flip-flops in a circuit, it is important to focus on the conditions at the *S* and *R* inputs *at the exact instant the clock edge arrives.* In many instances, the very act of triggering a flip-flop causes an almost instantaneous change in the state of the inputs. This is the case in Fig. 3-5(b). Do not pay any attention to the fact that the inputs change state immediately after delivery of the clock edge. The only concern of a flip-flop is the state of the inputs at the exact instant the edge appears.*

To keep this idea clear, it is convenient to think of the clock negative edge as being infinitely fast. That is, it goes from HI to LO in absolutely zero time. If this were true, then any change at the inputs due to triggering the flip-flop would occur too late, since the negative edge is already over and done with by the time the change happens.

Of course, no real-life clock edge can have a fall time of absolutely zero, but this notion helps us to explain and understand the behavior of clocked flip-flops. It prevents confusion in those situations where the inputs change when the flip-flop is triggered.

3-5 *JK* FLIP-FLOPS

The most widely used flip-flop is the *JK flip-flop*. It has two inputs, just like the *RS* flip-flop, but the inputs are referred to as *J* and *K*. The action of a *JK* flip-flop is quite similar to a clocked *RS* flip-flop, the only difference being that the *JK* flip-flop has what is called a *toggling mode*.

Table 3-2 shows the truth table for a *JK* flip-flop. Here is how to interpret it. The first row tells us that if the *J* and *K* inputs are both LO when the clock negative edge arrives, the state of *Q* *after* the edge (Q_{n+1}) is the same as the state of *Q* *before* the edge (Q_n). The flip-flop stays in whatever state it happens to be.

TABLE 3-2. The truth table of a *JK* flip-flop.

J	*K*	Q_{n+1}
0	0	Q_n
0	1	0
1	0	1
1	1	$\bar{Q}_n$

The second and third rows tell us the same thing we have already seen for a clocked *RS* flip-flop. If $J = 0$ and $K = 1$, the *Q* output goes to 0. If $J = 1$ and $K = 0$, the *Q* output goes to 1.

*This idea is valid only for flip-flops built on the so-called *master-slave* principle. In this book we will assume that all our flip-flops are built this way.

The last row shows the new possibility. Both inputs being HI is now a legal combination. If both J and K equal 1, the flip-flop will change to the opposite state, or *toggle*. That is, the Q output after the edge is the complement of the Q output before the edge ($Q_{n+1} = \bar{Q}_n$). To say it another way, if J and K are both HI when the clock edge arrives, the flip-flop will turn ON if it was OFF and will turn OFF if it was ON.

Figure 3-6(a) shows the black-box schematic symbol for a *JK* flip-flop, and Fig. 3-6(b) shows one way of constructing a *JK* flip-flop from an *RS* flip-flop.

(a)   (b)

Figure 3-6. (a) Black-box symbol of a *JK* flip-flop. (b) One method of constructing a *JK* flip-flop.

It can be seen from Fig. 3-6(b) why the *JK* flip-flop toggles if J and K are both HI. If the flip-flop is OFF, $\bar{Q}$ is HI and Q is LO. AND1 will therefore be enabled when a positive spike arrives from the *RC* differentiator. This takes place when the clock line delivers a negative edge, due to the inverter. When AND1 goes HI, the S input goes HI and turns ON the flip-flop.

On the other hand, if the flip-flop is already ON, Q is HI and $\bar{Q}$ is LO. Since K is also HI, AND2 will be enabled when the positive voltage spike arrives from the differentiator. This applies a HI to R and turns OFF the flip-flop.

The toggling ability of a *JK* flip-flop makes it extremely useful in numerous applications, especially those which require counting. We shall see in Sec. 3-7 how *JK* flip-flops can be used to count.

Preset and Clear. One more feature of flip-flops is important to know. Many of them have what are called *preset* and *clear* inputs. These inputs are usually shown coming into the top and bottom of a flip-flop, as in Fig. 3-7(a).

Preset and clear are inputs which are independent of the J and K inputs; they do not require a clock edge in order to drive the flip-flop. The preset input forces the flip-flop into the ON state regardless of what is happening at the J, K, and CLK terminals. The clear input forces the flip-flop into the OFF state, again regardless of J, K, and CLK.

In most flip-flops, the preset input must go LO to drive the flip-flop ON and the clear input must go LO to drive the flip-flop OFF. However, this is not univer-

(a)

(b)

Figure 3-7. (a) *JK* flip-flop with preset and clear inputs. (b) Setting
the initial condition of a string of flip-flops.

sal, and some flip-flops respond to their preset and clear inputs when they go HI.
To avoid confusion, we will assume in this book that the preset and clear inputs
must go LO to affect the flip-flop. We will use a small circle drawn outside the
box at the preset or clear terminal to remind ourselves that the flip-flop responds
to a LO logic level.

One obvious reason for needing preset and clear inputs on flip-flops is that
flip-flops go into random states when power is first turned on. Thus the circuit
designer is never sure of his starting conditions when the power is applied after
having been off. Preset and clear inputs are needed to put the various flip-flops
into the proper starting states before the circuit operation can begin. In such cases,
the preset and clear inputs are usually controlled by manual switches.

For example, if a circuit contained six flip-flops and the starting conditions
called for flip-flops 1, 3, and 5 to be ON and flip-flops 2, 4, and 6 to be OFF before

the circuit could begin functioning, the situation would be handled as shown in Fig. 3-7(b). When the INITIALIZE push-button switch is pressed, the output of the bounce eliminator goes LO. The LO is applied to the six flip-flops as indicated, presetting 1, 3, and 5 (turning them ON) and clearing 2, 4, and 6 (turning them OFF).

Usually, preset and clear commands will override input signals from the *J*, *K*, and CLK inputs. That is, if a *JK* flip-flop should get a signal to turn ON from its *J*, *K*, and CLK inputs while the CL input is held LO, it will ignore the clocked signal and obey the clear signal.

Initialization when power comes on is not the only use for preset and clear inputs. These inputs are often useful during the normal functioning of the flip-flop circuit. We will see examples of this as we proceed.

3-6 SHIFT REGISTERS

A shift register is a string of flip-flops which transfer their contents from one to another. The best way to understand the operation of a shift register is to look at its schematic diagram and observe how it works.

3-6-1 Shift registers constructed of *JK* flip-flops

Figure 3-8 shows four *JK* flip-flops connected together so that the outputs of one flip-flop drive the inputs of the next one. That is, Q_1 and $\bar{Q}_1$ are connected to *J* and *K* of FF2, Q_2 and $\bar{Q}_2$ are connected to *J* and *K* of FF3, and so on. This circuit is a four-bit shift register; it is called *four-bit* because it has four memory elements (flip-flops) and can therefore store four pieces of binary information, or *bits*.

When a negative edge appears on the CLK line, it is applied to the CLK terminals of all four flip-flops simultaneously. At this instant, all the flip-flops respond to the input levels at their *J* and *K* inputs. However, since the *J* and *K* inputs of one flip-flop are just the *Q* and *Q̄* outputs of the neighboring flip-flop, the result is that all information is transferred, or shifted, one place to the right. Therefore if FF1 is

Figure 3-8. Shift register built with *JK* flip-flops.

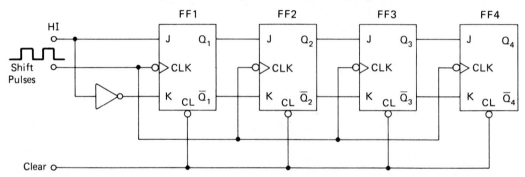

ON at the instant the negative edge hits the CLK terminals, FF2 will be turned ON. If FF2 is OFF at the instant the negative edge hits, FF3 will be turned OFF. The only flip-flop which doesn't respond this way is FF1, which must have signals applied to its J and K inputs from some external circuit.

As a specific example, suppose that the CLEAR line in Fig. 3-8 goes LO to initialize all flip-flops at the OFF state. Assume also that J of FF1 is wired to a 1 and K is wired to a 0, as shown in that drawing. Now let us see what happens as the pulses start arriving on the SHIFT line.

As the very first negative edge hits the register, FF4 is told to turn OFF because it has $J = 0$, $K = 1$. This is so because FF3 is already OFF, making $Q_3 = 0$ and $\bar{Q}_3 = 1$. Since FF4 is already OFF, the signal to turn OFF does not affect it; it just stays OFF.

FF3 is signaled to turn OFF via Q_2 and $\bar{Q}_2$, and it also stands pat. The same is true for FF2, which is signaled by Q_1 and $\bar{Q}_1$. It remains OFF also. FF1, however, turns ON because of the 1 on J and 0 on K. Therefore at the completion of the first shift pulse, the state of the shift register, reading from left to right, is

1000.

Now consider what happens when the second negative edge hits the CLKs. FF4 is told to turn OFF by FF3 because FF3 is OFF at this instant. FF3 is likewise told to turn OFF by FF2. FF2, though, is told to turn ON because its J input is held HI by Q_1 and its K input is held LO by $\bar{Q}_1$. FF2 turns ON at this instant. FF1 still has $J = 1$ and $K = 0$ from outside, so it turns ON again, or, in other words, it maintains its ON state. The state of the register is now

1100.

What is happening here is that all the stored information in the flip-flops is shifted one place to the right whenever the shift command occurs. Meanwhile an external circuit keeps feeding 1s into the leading flip-flop.

After the third negative edge, the condition would be

1110,

and after the fourth shift command the state would be

1111.

Any subsequent shift commands will have no effect on the contents of the shift register since they will merely cause one 1 to be lost on the far right (FF4) while another 1 comes in from the far left (into FF1).

3-6-2 A conveyor/inspection system using a shift register

The shift register finds widespread use in industrial applications involving conveyor systems, where each flip-flop in the shift register represents one zone on the conveyor system. The state of a particular flip-flop, ON or OFF, stands for some characteristic of the piece that is in that particular zone. The characteristic must be a digital characteristic, one that can be represented by a binary 1 or 0. The most obvious example is pass/fail; either the part passes inspection and is routed to the next production location or it fails inspection and is rejected.

Think of a conveyor which is broken up, mentally at least, into four physical zones. Each time a part moves from one zone to the next, it causes a shift command to be delivered to the shift register. Thus, the binary characteristic of the piece moves to the next flip-flop as the piece itself moves to the next zone.

When the piece leaves the fourth conveyor zone, the binary bit leaves the fourth flip-flop of the shift register. When a new piece enters the first conveyor zone, a new binary bit is fed into the first flip-flop of the shift register. The shift register thereby keeps track of information about the pieces on the conveyor.

In most situations, as information is shifted from one flip-flop to another, it will reach a certain flip-flop where it is acted upon by a detecting circuit. The detecting circuit reads the binary bit at a certain flip-flop and causes some action to be performed at that zone in the industrial system.

Here is a specific example. Suppose we have a production setup in which evenly spaced parts come down a conveyor and are inspected by a person. We shall call the location where the inspection takes place zone 1. Further work is to be done on the parts by other workers in zones 2 and 3 farther down the conveyor. However, if the parts do not meet inspection standards in zone 1, it is useless to waste effort by performing more work in zones 2 and 3. This is why they are inspected in zone 1; if they fail inspection in zone 1, they are not worked on as they pass through zones 2 and 3.

But, because of certain physical constraints, parts that fail the inspection cannot be removed from the conveyor and placed in the reject bin directly from zone 1. Instead, they continue down the conveyor just like good parts, until they reach zone 4. In zone 4, a diverter swings out and diverts bad parts into the reject bin. Good parts leave zone 4 in the normal manner and continue on their way.

The inspector decides if a part passes or fails inspection. If a part fails, he presses a REJECT button while the part is still in his zone; he also marks the part for the benefit of the other workers in zones 2 and 3. This might be done by dabbing some paint on it with a brush, or by tipping it over, or whatever. The mark signals the workers in zones 2 and 3 not to do any work on the part because it is to be rejected.

When a part leaves zone 1, the shift register keeps track of whether it passed inspection or failed. As the part proceeds through the zones, the pass/fail information keeps pace in the shift register. When the part enters zone 4, the shift register signals the diverter whether or not to swing out to dump it into the reject bin.

Such a layout is shown in Fig. 3-9(a), and the control circuit is shown in Fig. 3-9(b).

The circuit of Fig. 3-9(b) is really quite simple. If the REJECT button is pressed while the part is in zone 1, the output of I1 goes LO and pulls the preset input of FF1 down LO. This presets FF1 into the ON state (Q_1 goes HI). Remember, we are assuming that flip-flops respond to a LO preset signal. This may be confusing

(a)

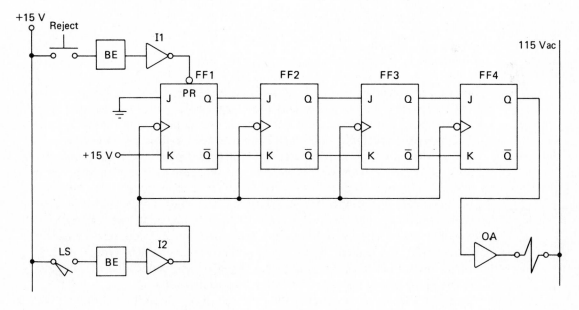

(b)

Figure 3-9. (a) Physical layout of a conveyor/inspection system. (b) Control circuit of the conveyor/inspection system, illustrating the use of a shift register to keep track of the progress of parts through the system.

since the more natural expectation is that they would respond to a HI. However, response to a LO is the more usual case.

As the bad part leaves zone 1 and enters zone 2, the limit switch is momentarily contacted. This causes the output of I2 to go LO, delivering a negative edge to all the CLK terminals. FF2 turns ON at this instant because Q_1 is applying a 1 to J and $\bar{Q}_1$ is applying a 0 to K. Therefore as the bad part enters zone 2, the information about its badness enters flip-flop 2. A bad part in a zone is indicated by that flip-flop being ON.

FF1 turns back OFF when the negative edge hits the CLK terminals because of the hard-wired LO at J and HI at K.

Since the parts are evenly spaced, every part on the conveyor moves into a new zone at the time the limit switch contact closes in response to a part passing from zone 1 to zone 2. Thus as the bad part enters zone 3, LS closes again because the following part is entering zone 2. This causes another clock edge to appear, which turns ON FF3. As the bad part enters zone 4, LS causes another clock edge, which turns ON FF4. When Q_4 goes HI, it energizes the diverter solenoid and swings the diverter out. As the conveyor continues to move, the bad part is guided into the reject bin by the diverter.

As the following part enters zone 4, FF4 turns back OFF if the part is good. The diverter immediately swings back into normal position before the part can run into it.

3-6-3 Prepackaged shift registers

So far, our illustrations of shift registers have shown several flip-flops serially tied together. To be sure, this is exactly how prepackaged shift registers are internally built, but they are not always illustrated this way. A prepackaged shift register is usually shown as a box having a clock input (CLK), a clear input (CL), preset inputs for each bit (PR_n), shift inputs for the first bit (J and K), and outputs for each bit (Q_n and $\bar{Q}_n$). This symbol is shown in Fig. 3-10(a) for a four-bit shift register.

The most common lengths for prepackaged shift registers are four, five and eight bits. If a longer shift register is needed, two or more small ones can be cascaded as shown in Fig. 3-10(b). In that figure, two 4-bit shift registers are cascaded to make an 8-bit shift register. As the drawing shows, this is done by tying the CLK inputs together, tying the clear (CL) inputs together, and connecting the outputs of the last bit to the inputs of the first bit of the next register.

There are many different types of shift registers. They all exhibit the same basic behavior, that of shifting binary bits from one location to the next. Their secondary features differ from one another and from what we have discussed. For example, some shift registers can shift either to the right or the left. Naturally such shift registers have more input terminals than are shown in Fig. 3-10(a) because they must be instructed as to the direction of the shift. Some shift registers have a special

Figure 3-10. (a) Black-box symbol of a four-bit shift register with a common clear input and individual preset inputs. (b) Tying two shift registers together (cascading).

LOAD input terminal for signaling when bits are to be preset or "loaded" into the register. To avoid confusion, we will stay with only one type, the type illustrated in Fig. 3-10(a).

3-7 COUNTERS

A *digital counter* is a circuit that counts and remembers the number of input pulses that have occurred. Every time another input pulse is delivered to the CLK terminal of a counter, the number stored in the circuit advances by one.

This sounds simple, but there is a catch. Since digital counters are built out of flip-flops and logic gates, the only two digits they can use are 0 and 1. This is so because the transistors in flip-flops and logic gates are always either all the way

ON or all the way OFF, as we know. There is no middle ground. Therefore it is not possible for them to store the digits 2, 3, 4, 5, 6, 7, 8, or 9, since storing other digits would demand other possible states of the transistors.

This is not an insurmountable problem. It only means that instead of working in the *decimal* number system that we human beings find so comfortable, electronic digital counters must work in the *binary* number system.

3-7-1 The binary system

We will not have time to go into any great detail about the structure of number systems in this book. If you already understand the philosophical basis on which number systems are constructed, so much the better. If you are not familiar with the construction of number systems in general and the binary system in particular, you may want to do some reading on the subject. There are many good books available which explain the binary system, how to count in it, how to convert back and forth from decimal to binary, etc. However, to understand digital counters in industrial control, all you really need to know about binary numbers are two things:

a. How to count to 9 in binary
b. How to interpret binary-coded-decimal (BCD) numbers

3-7-2 Counting in binary

Counting in binary is really quite easy if you recognize that moving one column (place) to the left just increases the power of 2. We said that we only needed to know how to count to 9 though, so here is the code in Table 3-3.

In Table 3-3(a) the equivalent binary number is written alongside each of the decimal digits 0 through 9. This number sequence must be thoroughly memorized. You should be able to mentally convert from a decimal digit to the binary number very quickly. You should also be able to convert in the other direction, from binary to decimal.

In Table 3-3(b), each binary number has been shown having four places, even though the number doesn't require all four places. In other words, the unneeded zeros have not been eliminated from the left columns but have been carried along as fillers. The reason for doing this will become clear as we go on. Be prepared to recognize binary numbers written in either format.

3-7-3 Binary-coded decimal (BCD)

Decimal numbers larger than 9 can be written in strict binary, and this is what is done for computing systems. In most industrial systems, though, larger decimal numbers are handled by binary-coded-decimal representation (BCD). In BCD, every decimal digit is represented by its own four-bit binary number. Thus, decimal

TABLE 3-3

Decimal Number	Binary Number
(a) Binary Code from 0 to 9 with Leading Zeros Eliminated	
0	0
1	1
2	10
3	11
4	100
5	101
6	110
7	111
8	1000
9	1001
(b) Binary Code with Leading Zeros Included	
0	0000
1	0001
2	0010
3	0011
4	0100
5	0101
6	0110
7	0111
8	1000
9	1001

47 would be

$$0100 \quad 0111$$

in BCD. Decimal 92 would be

$$1001 \quad 0010$$

in BCD.

This explains the reason for carrying along the leading zeros in binary numbers. If every decimal digit is to be represented by its own binary equivalent, we must reserve four spaces in order to be able to handle the digits 8 and 9. If four spaces have been set aside but the decimal number doesn't require four bits, something has to fill in the unused spaces. That something is zeros.

3-7-4 Flip-flops used to count

Figure 3-11 shows how four *JK* flip-flops can be connected together to form a counter. The best way to understand this counter is to apply one count pulse at a time to the CLK input of FF*A* and watch carefully what happens. Before we start, note some slight differences in the way the flip-flops are shown. The *J* and *K* inputs are now drawn on the right-hand side of the box and the outputs are on the left

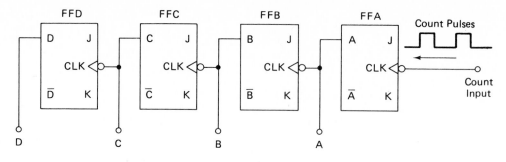

Figure 3-11. Counting circuit built with *JK* flip-flops.

side. Also, instead of calling the outputs Q and $\bar{Q}$, they are now called by the identifying letter of the flip-flop. That is, the outputs of FFA are called A and $\bar{A}$, the outputs of FFB are called B and $\bar{B}$, and so on.

None of the flip-flop J and K inputs are connected to anything. They are hanging. We will assume that this is a logic family which treats a hanging input as a HI. Therefore all Js are 1 and all Ks are 1. Under this condition, each flip-flop will change to the opposite state, or toggle, whenever it sees a negative-going edge at its CLK terminal.

The clock terminals are not all tied in parallel as they were for the shift register. In the counter of Fig. 3-11, the clock terminal of one flip-flop is driven by the Q output of the neighboring flip-flop to its right. Under this condition, a given flip-flop will toggle if and only if the flip-flop to its right goes from ON to OFF. That is, FFB will toggle when A goes from HI to LO. FFC will toggle when B goes from HI to LO. FFD will toggle when C goes from HI to LO.

Assuming that all four flip-flops are cleared to begin the counting, let one count pulse arrive at the CLK terminal of FFA, hereafter called the COUNT INPUT. When the negative edge occurs, FFA toggles from OFF to ON. This applies a positive edge to the CLK terminal of FFB because A goes HI. Positive edges have no effect, though, so FFB stays as it was, namely OFF. Since B did nothing, FFC and FFD do nothing. If the starting condition was

$$DCBA = 0000,$$

the condition of the counter after the first count pulse is

$$DCBA = 0001.$$

Now let a second pulse arrive at the COUNT INPUT. The negative edge causes FFA to toggle, so A goes to 0. This causes a negative edge to appear at CLK of FFB. Flip-flop B toggles, so B goes HI. C and D again do nothing because the positive edge delivered to C is useless. The state of the counter is now

$$DCBA = 0010.$$

Let a third pulse occur at COUNT INPUT. FFA toggles, but B does nothing because it receives a positive edge. C and D also do nothing. The state of the counter is now

$$DCBA = 0011.$$

Let a fourth pulse arrive at COUNT INPUT. The negative edge causes FFA to toggle. Therefore output A goes from HI to LO. This delivers a negative edge to FFB, so it toggles too. The B output goes from HI to LO. This in turn delivers a negative edge to CLK of FFC, so FFC also toggles. C then goes from LO to HI. FFD does nothing on the positive edge. The condition of the counter is now

$$DCBA = 0100.$$

If we make a table of the state of the counter after each pulse, it looks like Table 3-4. Table 3-4 exactly matches the first half of Table 3-3(b). Therefore the circuit of Fig. 3-11 is counting properly. The binary number stored in the counter's flip-flops equals the number of count pulses that have occurred.

TABLE 3-4. The state of the counter after each input pulse.

After this pulse	The state of the counter is			
	D	C	B	A
0	0	0	0	0
1	0	0	0	1
2	0	0	1	0
3	0	0	1	1
4	0	1	0	0

Continue delivering count pulses one at a time and trace the operation of the counter through nine pulses. You should be able to verify that the circuit counts correctly all the way to 9.

3-7-5 Decade counters

The counter shown in Fig. 3-11 is fine, but it has a serious flaw from our point of view. It won't stop counting at 9. Instead, it would continue to count up to the binary equivalent of 15 and would return to 0 on the sixteenth pulse. As mentioned earlier, in a strict computing situation this may be desirable, but in an industrial control situation it is usually undesirable. Industrial circuits are usually confined to binary-coded-decimal numbers. In BCD, the largest binary number we ever want to see is 1001, or 9. Therefore we need a counter that can count to 9 and return

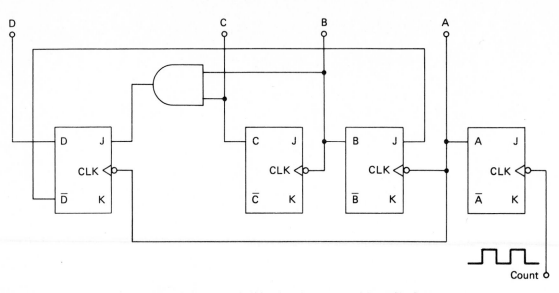

Figure 3-12. One way to build a decade counter with *JK* flip-flops and an AND gate.

to 0 on the tenth count pulse. A counter that does this is called a *decade counter*. One way to build a decade counter is shown in Fig. 3-12.

The operation of the counter in Fig. 3-12 is somewhat more difficult to understand than the one in Fig. 3-11 because some of the flip-flops are not being used in a straight toggling mode. Specifically, FFB and FFD no longer have their *J* and *K* inputs both HI (hanging). *J* of FFB is tied to the $\bar{D}$ output, and *J* of FFD is driven by the AND gate. The *K* inputs of all four flip-flops are hanging and are effectively HI. To understand the operation of this counter, let us make some observations about each flip-flop. We can then refer to these observations as we watch the counting process take place.

First, FFA. This flip-flop receives a negative edge on every count pulse. Since *J* and *K* are both HI, FFA toggles on every negative edge.

Second, FFB. This flip-flop receives a negative edge only when FFA turns OFF. *J* is connected to $\bar{D}$, so *J* is HI if FFD is OFF. Therefore FFB will toggle if FFD is OFF at the time the negative edge arrives. FFB will turn OFF if FFD is ON when the edge arrives, because then *J* will be LO and *K* will be HI.

Third, FFC. This flip-flop receives a negative edge when *B* goes from HI to LO, that is, when FFB turns OFF. Since both *J* and *K* are HI, FFC toggles every time it receives a negative edge.

Last, FFD. This flip-flop receives a negative edge when FFA turns OFF. If both *B* and *C* are HI, the AND gate will bring *J* HI. Therefore FFD will toggle if FFB and FFC are both ON when the edge arrives. If either *B* or *C* is LO, *J* will be LO. *K* is always HI, so FFD will turn OFF if either FFB or FFC is OFF when the edge arrives.

Armed with all the information in the preceding four paragraphs, we can now begin delivering count pulses one at a time and watch the progression of the circuit. Assuming that all flip-flops are cleared to start, the initial state of the counter is

$$DCBA = 0000.$$

The first negative edge appears as the first count pulse ends. FFA toggles into the ON state. No other flip-flop receives a negative edge, so no other flip-flop does anything. The state of the counter after one pulse is

$$DCBA = 0001.$$

When the second edge arrives, FFA toggles to OFF. FFB receives a negative edge as A goes LO. Since FFD is OFF at this time, FFB toggles into the ON state. FFD also receives a negative edge as A goes LO. FFC is OFF at this time, so FFD receives a signal to turn OFF. Since it is already OFF, it just stays there. The state of the counter after two pulses is

$$DCBA = 0010.$$

When the third edge arrives, FFA toggles to ON. Flip-flops B, C, and D get no negative edges, so they all stay where they are. The state of the counter after three pulses is

$$DCBA = 0011.$$

When the fourth pulse arrives, FFA toggles to OFF. FFB receives a negative edge as A goes LO, and it toggles to OFF because FFD is OFF at this time. FFC receives a negative edge as B goes LO, and it toggles to ON. FFD also receives a negative edge as FFA turns OFF. FFD gets the signal to turn OFF since C was LO *at the instant the negative edge arrived.* As mentioned in Sec. 3-4, do not allow yourself to be confused by changing conditions right *after* the edge has arrived. All that matters is that C was LO and was keeping the AND gate LO at the instant the negative edge hit the CLK terminal of FFD. Since FFD gets a signal to turn OFF, it just stays OFF. The state of the counter after four pulses is

$$DCBA = 0100.$$

When the fifth pulse arrives, FFA toggles to ON. No other flip-flops receive a negative edge, so they all stay put. The state of the counter after five pulses is

$$DCBA = 0101.$$

When the sixth pulse arrives, FFA toggles to OFF. This applies a negative edge to FFB and FFD. FFB toggles to ON because FFD is OFF at this time. FFD gets the signal to turn OFF because B was LO at the instant the edge arrived. FFC receives no negative edge and does nothing. The state of the counter after six pulses is

$$DCBA = 0110.$$

When the seventh pulse arrives, FFA toggles to ON. No other flip-flops receive a negative edge, so they stay where they are. The state of the counter after seven pulses is

$$DCBA = 0111.$$

When the eighth pulse arrives, FFA toggles to OFF. As A goes LO, it delivers a negative edge to FFB and FFD. FFB toggles to OFF because FFD is OFF at the time the edge arrives. FFD will toggle to ON because both B and C are HI at the instant the edge arrives. As B goes LO, it delivers a negative edge to FFC, so FFC toggles to OFF.

If this action is confusing, look at it this way. Inspect the state of the counter *before* the negative edge arrives. From that state, identify what all the J and K inputs are. Then simply use the truth table of a JK flip-flop to tell what any flip-flop will do when the negative edge strikes.

Before the eighth pulse occurs, the state of the counter is $A = 1$, $B = 1$, $C = 1$, $D = 0$. With B and C both equaling 1, the AND gate is enabled and is putting a 1 on J of FFD. Thus both J and K are 1s on FFD. Therefore when the next edge strikes, FFD will toggle. Again noting that the state of the counter is $A = 1$, $B = 1$, $C = 1$, $D = 0$ before the eighth pulse occurs, the J input of FFB is HI, since $\bar{D} = 1$. With J and K both HI, FFB will toggle when the next edge strikes. No matter which way we look at it, the state of the counter after the eighth pulse is

$$DCBA = 1000.$$

When the ninth pulse arrives, FFA toggles to ON. All other flip-flops maintain their same states. The state of the counter after the ninth pulse is

$$DCBA = 1001.$$

When the tenth pulse arrives, FFA toggles to OFF. This delivers a negative edge to FFB and FFD. FFB gets the signal to turn OFF because FFD is ON at the instant the edge arrives. Since FFB is already OFF, it just stays OFF. FFD gets the signal to turn OFF since B (and C) is LO at this time. Therefore FFD turns OFF. The state of the counter after the tenth pulse is

$$DCBA = 0000.$$

Table 3-5 summarizes the output states of the decade counter. The crucial feature that makes a decade counter so useful is the fact that it returns to zero or "resets" on the tenth pulse.

3-7-6 Cascading decade counters

The black-box symbol for a decade counter is shown in Fig. 3-13(a). The counter resets to $DCBA = 0000$ when the clear line (CL) goes LO. If the clear line is HI, the count advances by one every time a negative edge appears at the clock terminal. Often the CLK terminal is labeled COUNT or COUNT INPUT.

To count higher than decimal 9, decade counters can be cascaded as shown in Fig. 3-13(b). The first decade counter then contains the binary representation of the *units* number, the second decade counter contains the binary code for the *tens*

TABLE 3-5. State of the decade counter after each input pulse, showing that the counter resets to zero on the tenth pulse.

Number of count pulses	Output state of counter			
	D	C	B	A
0	0	0	0	0
1	0	0	0	1
2	0	0	1	0
3	0	0	1	1
4	0	1	0	0
5	0	1	0	1
6	0	1	1	0
7	0	1	1	1
8	1	0	0	0
9	1	0	0	1
10	0	0	0	0

number, and the third decade counter contains the binary code for the *hundreds* number.

The D output of the units counter drives the CLK input of the tens counter as shown. The D output of the units counter goes from HI to LO as the counter goes from 9 (1001) to 0 (0000) on the tenth input pulse. As it does this, it delivers a negative edge to the CLK input of the tens counter, causing its count to advance by one. Therefore the tens counter advances by one every time the units counter goes through its complete range and returns to zero.

The relationship between the hundreds counter and the tens counter is exactly the same.

To illustrate with numbers, after 9 clock pulses have been delivered to COUNT INPUT in Fig. 3-13(b) the output state of the entire three-stage counter is

$$0000 \quad 0000 \quad 1001.$$

When the tenth count pulse is delivered, the output state goes to

$$0000 \quad 0001 \quad 0000.$$

(a)

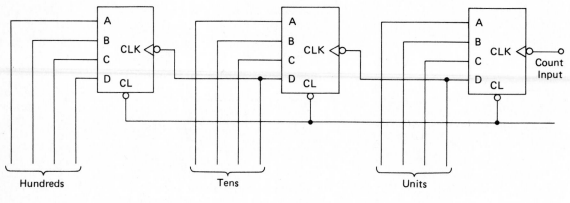

(b)

Figure 3-13. (a) Black-box symbol of a decade counter. (b) Cascading decade counters together to count higher than 9.

After 99 count pulses have been delivered, the output state is

0000   1001   1001,

and on the one-hundredth count pulse, the output state becomes

0001   0000   0000.

The three-stage counter shown in Fig. 3-13(b) can count as high as 999. If a fourth stage were added, the highest possible count would be 9999.

If 257 count pulses were delivered to COUNT INPUT in Fig. 3-13(b), the outputs of the counters would be

0010   0101   0111.

If 890 count pulses were delivered, the output state would be

1000   1001   0000.

You should be able to write the output state of the three-stage counter for any number of count pulses between 0 and 999.

3-8 DECODING

In many industrial applications using decade counters, the system operators set a 10-position selector switch to "watch" the counter and take some sort of action when the state of the counter matches the setting on the switch. This idea is illustrated in Fig. 3-14.

Figure 3-14. The combination of a decade counter, a 1-of-10 decoder, and a 10-position selector switch. This combination is often seen in industrial cycle control circuits.

The decade counter has four output lines, D, C, B, and A, which contain the binary code for the number in the counter. The box in between the counter and the selector switch in Fig. 3-14 is called a *decoder* because it takes binary-coded information and converts it to decimal information that human beings can understand. That is, if the binary information represents the decimal digit 2 ($DCBA = 0010$), the decoder drives the 2 output line HI. If the binary information represents the decimal digit 3 ($DCBA = 0011$), the decoder drives the 3 output line HI, and so on. Since it converts coded numbers to uncoded decimal numbers, it is called a decoder.

101

In Fig. 3-14, if the output of the decoder is the same as the setting of the selector switch, the common terminal of the switch will go HI. Overall, the OUTPUT of the circuit goes HI when the counter reaches the setting of the 10-position selector switch. The HI on the output could then be used to accomplish some action in the system. This is how a manually operated selector switch can "watch" a counter and take action when it reaches a certain count.

The most straightforward way of constructing a decoder is shown in Fig. 3-15. Look first at Fig. 3-15(a). It shows 10 four-input AND gates, each gate with a different combination of inputs. Each combination of inputs represents one of the possible states of the decade counter. Therefore, for any one of the 10 possible output states, one of the AND gates will be enabled. For example, if the state of the counter is $DCBA = 0101$ (decimal 5), then all four inputs to AND gate number 5 will go HI. Figure 3-15(a) shows that the inputs to that gate are $\bar{D}$, C, $\bar{B}$, and A. If the counter state is $DCBA = 0101$, then $\bar{D}C\bar{B}A = 1111$; when all four inputs are HI, the output goes HI. Thus if the counter has counted to 5, the number 5 output of the decoder goes HI. You should verify that the decoder works properly for the other counter output states.

Figure 3-15(b) shows the same circuit as Fig. 3-15(a). The only difference is that all the interconnections are shown. This schematic drawing drives home the point that there are really only four inputs to the decoder and that these four are decoded to a decimal number from 0 to 9.

The decoder of Fig. 3-15 is called a *BCD-to-decimal decoder* or a *BCD-to-1-of-10 decoder*. There are other types of packaged decoders available (BCD-to-seven-segment, Gray-code-to-decimal, excess-three-code-to-decimal, etc.), but for our purposes the term *decoder* will refer to a BCD-to-decimal decoder unless stated otherwise.

3-9 A PALLETIZING SYSTEM USING DECADE COUNTERS AND DECODERS

Consider a situation in which cartons are sealed by a carton sealing machine and then taken away by conveyor to one of two palletizers. Palletizers are machines which stack cartons in an orderly manner on pallets. When one pallet is completely loaded, a diverter swings over and begins routing cartons to the other palletizer. While the second pallet is being loaded, the first loaded pallet is removed and an empty pallet brought into its place.

Because the system handles cartons of different sizes, the number of cartons per pallet will vary. Therefore the operators must be able to easily change the per-pallet carton count. The general layout is depicted in Fig. 3-16(a), and the control circuitry is shown in Fig. 3-16(b).

As can be seen from the drawing, LSA is actuated just prior to a carton being loaded by palletizer A, and LSB is actuated just prior to a carton being loaded by palletizer B. When the preset number of cartons have been loaded onto either

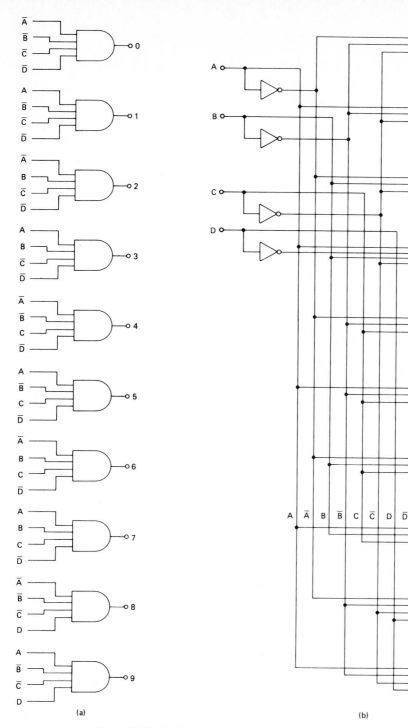

Figure 3-15. (a) Construction of a 1-of-10 decoder, showing the inputs of each decoding gate. (b) 1-of-10 decoder showing the actual wire connections.

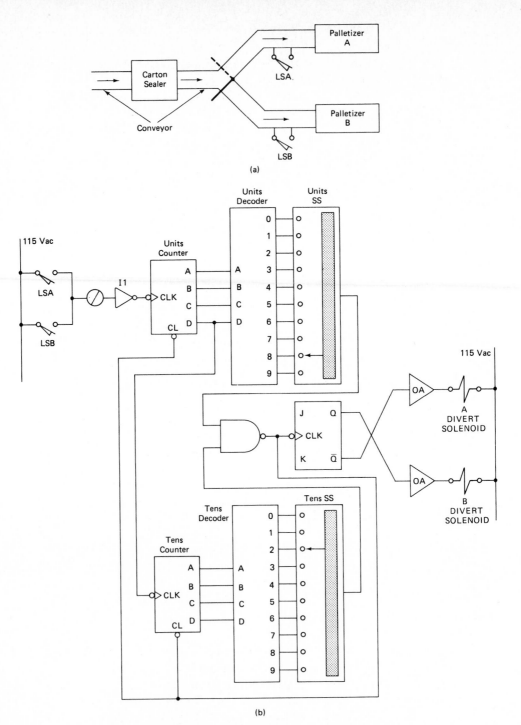

Figure 3-16. (a) Top view of the palletizing system. (b) Control circuit of the palletizing system, showing the operation of decade counters, 1-of-10 decoders, and 10-position selector switches.

palletizer, the diverter swings into the opposite position. The cartons that follow are then routed to the opposite palletizer.

The control circuit works as follows. Assume that the diverter is routing cartons to palletizer *A*. This implies that *A* DIVERT SOLENOID is energized, which means that the *JK* flip-flop is OFF. As the cartons pass by LSA, they momentarily close the N.O. contact, causing the I1 input to go HI. As the I1 output goes LO, it delivers a negative edge to the units decade counter, which advances its count by one. The selector switches as drawn are set to 8 on the units switch and to 2 on the tens switch. Therefore the pallet will be loaded with 28 cartons. As the twenty-eighth carton passes LSA, the decade counters go to the states

Tens	Units
0010	1000

At this instant the tens decoder is holding the 2 output HI, and the units decoder is holding the 8 output HI. Therefore both SS common terminals go HI, which causes both NAND inputs to go HI. As the NAND output goes LO, it delivers a negative edge to the flip-flop. With *J* and *K* both HI the flip-flop toggles to ON. The $\bar{Q}$ output goes LO, deenergizing the *A* DIVERT SOLENOID, and the *Q* output goes HI, energizing the *B* DIVERT SOLENOID. This swings the diverter into the dotted position of Fig. 3-16(a), so succeeding cartons are routed to palletizer *B*.

Meanwhile, the NAND output has pulled the CL terminals down LO on both the units and the tens counter. This immediately resets both counters to 0000, in preparation for beginning the count of cartons to palletizer *B*. As the counters are cleared, the NAND inputs go back LO. The output of the NAND goes HI, which removes the LO clear signal, putting the counters right back into counting condition.

When 28 cartons have been loaded onto pallet *B*, the flip-flop toggles back to OFF, and the counters are cleared again. The system then begins all over again, loading pallet *A*.

Whenever a different-sized carton is to be run, the operators just set the selector switches to a different number. Any number of cartons from 0 to 99 can be selected.

3-10 ONE-SHOTS

The *one-shot* (formal name: monostable multivibrator) is a very useful circuit in digital industrial controls. Its output temporarily goes HI when the circuit is triggered; then it returns to LO after a certain fixed time. A one-shot is used whenever the situation calls for a certain line (the output) to go HI *for a short while* if another line (the input) changes state.

Figure 3-17 illustrates the action of a one-shot. We will assume that one-shots are triggered by a negative edge at the *T* (trigger) terminal. Actually, some one-

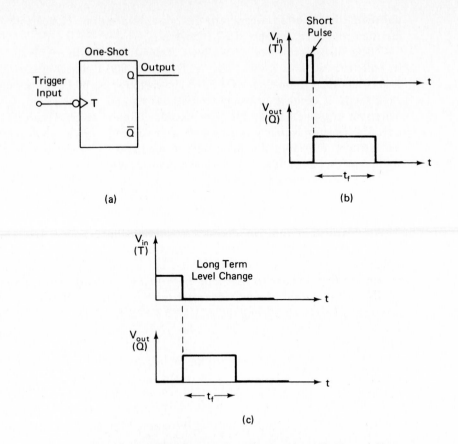

Figure 3-17. (a) Black-box symbol of a one-shot. (b) V_{in} and V_{out} waveforms when the one-shot is triggered by a short pulse. (c) Input-output waveforms when the one-shot is triggered by a long-term level change of a logic signal.

shots are triggered by a positive edge at T, but for simplicity we will consider only negative edge-triggered one-shots.

The black-box schematic symbol for a one-shot is shown in Fig. 3-17(a). Notice that it has two outputs, Q and $\bar{Q}$. The $\bar{Q}$ output is the complement of the Q output, just like the outputs of a flip-flop. When the one-shot is triggered, the Q output goes HI while the $\bar{Q}$ output goes LO. After a period of time has elapsed (called the *firing time, t_f*), the Q output returns to LO and the $\bar{Q}$ output returns to HI.

The waveforms in Fig. 3-17(b) show how a one-shot behaves when it is triggered by a short pulse. When the negative-going edge of the short pulse occurs, the one-shot triggers, or fires. The Q output quickly goes HI and remains HI for a length of time equal to t_f. The firing time t_f is usually adjustable by adjusting a resistor or capacitor in the circuit.

When driven by a short pulse as in Fig. 3-17(b), a one-shot may be acting as a pulse stretcher; that is, a short-duration input pulse is converted into a longer-duration output pulse. Or it may be acting as a delay device; that is, when a negative edge occurs at T, another negative edge appears at Q, but delayed by a time t_f. Or it may just be used to clean up a ragged input pulse; that is, the output pulse of a one-shot is well shaped in that it has steep positive-going and negative-going edges, regardless of the condition of the input pulse.

When a one-shot is driven by a long-term level change as shown in Fig. 3-17(c), it is acting rather like a pulse shrinker. One-shots are frequently used in this mode to clear a counter (or a flip-flop) when a certain line changes levels. For example, it is often necessary to begin counting in a counter, shortly after a certain level change takes place, but it is necessary for the count to begin at zero. If the level change persists after the next counting sequence is supposed to begin, then we cannot use the level change *itself* to clear the counter. This is because the changed level would hold the counter in the reset state. What is needed is a circuit which can *temporarily* apply a reset signal to the counter when the level change takes place. The reset signal must then go away in time for the next round of counting to begin. A one-shot performs this action perfectly.

We have assumed in this book that counters (and flip-flops) are reset by a LO level applied to the clear terminal. The fact that the waveforms in Fig. 3-17 show a HI-level output pulse may therefore be a matter of concern. However, we have seen that one-shots also have a $\bar{Q}$ output, which delivers a LO-level pulse during the time the Q output is delivering a HI level. The $\bar{Q}$ output would be used to reset a counter in a situation such as that described in the preceding paragraph.

There are many ways to build one-shots, but Fig. 3-18 shows one popular way.

It works as follows. When the circuit is at rest, T_2 is turned ON and saturated. Its base current is supplied through R_{B_2}. The collector of T_2 is virtually at ground potential, so the Q output is LO. The base of T_2 is only 0.6 V above ground potential because of the forward-biased base-emitter junction.

T_1 is cut OFF because it has no base drive. Its base resistor, R_{B_1}, is connected to the T_2 collector, which is at 0 V. Therefore R_{C_1} is completely disconnected from the grounded emitter of T_1 and is free to carry current to charge capacitor C. Since C is connected to the base of T_2, which is close to ground potential, it will charge up almost to the supply voltage V_s. The polarity of the charge on C is plus ($+$) on the left and minus ($-$) on the right, as shown.

Now, let a negative edge appear at T. The inverter causes a HI to be applied to the RC differentiator, which applies a positive spike to the base of T_1. This turns T_1 ON and pulls the collector of T_1 down to ground. Since the charge on C cannot disappear instantly, the voltage across the capacitor plates is maintained. With the $+$ side of the capacitor pulled down to 0 V by T_1, the $-$ side goes to a voltage far below ground potential. This applies a negative voltage to the base of T_2, shutting it OFF. The collector of T_2 rises toward V_s and is now capable of supplying base current to T_1. Therefore T_1 remains turned ON even after the positive spike from the differentiator goes away. Q is now HI, and $\bar{Q}$ is LO.

As time goes by, charging current flows onto the plates of C. The flow path is

Figure 3-18. Schematic diagram of a one-shot.

down through R_{B_2}, through C, and through collector to emitter of T_1 into ground. As can be seen, this path seeks to charge C to the opposite polarity; what happens is that the voltage across C gets smaller. When the capacitor voltage crosses through zero and reaches 0.6 V in the opposite polarity, it bleeds a small amount of current into the base of T_2. This small base current causes collector current to flow in T_2, lowering the collector voltage. The reduced collector voltage causes a reduction in base current to T_1. This in turn causes a reduction in T_1 collector current. The T_1 collector voltage rises slightly, thereby raising the base of T_2 higher yet. This action is regenerative; once it begins, it avalanches. In the end, T_2 is saturated once again, and T_1 is cut OFF. Q is LO, and $\bar{Q}$ is HI, and the circuit has returned to its original state.

One-shots are often prepackaged circuits, having the schematic symbol given in Fig. 3-17(a). They usually have provision for the user to connect an external resistor and/or capacitor to set the firing time. The manufacturers of packaged one-shots furnish graphs which show the relationship between t_f and the size of external resistor and capacitor.

One-shots are classified as either *retriggerable* or *nonretriggerable*. Retriggerable means that if a *second* negative edge occurs during the firing time of the one-shot, the output pulse resulting from the first negative edge will be extended beyond its normal duration. We will assume that our one-shots are nonretriggerable; they ignore triggering edges occurring during an output pulse. Several examples of one-shots in industrial controls will be presented in Secs. 3-12 and 3-14.

3-11 CLOCKS

Often in industrial digital circuits it is necessary to keep various digital devices synchronized with each other. In other situations, a continuous train of pulses is needed to supply count pulses to a counter if the system does not generate count pulses naturally as it performs its functions. In either case, what is required is a circuit which supplies a continuous stream of square-edged pulses. Such circuits are referred to as *clocks*.

A black-box symbol of a clock is shown in Fig. 3-19(a). The output waveform shown is a square wave. It can also be thought of as a pulse train with a 50% duty cycle. Many clocks have just such an output; some clocks have duty cycles other than 50%.

The output frequency (pulse repetition rate) of a clock is set by resistor, capacitor, or inductor sizes internal to the circuit. In the case of a crystal-controlled clock,

Figure 3-19. (a) Black-box symbol of a clock. (b) A clock combined with a frequency divider to obtain a signal at a different frequency.

(a)

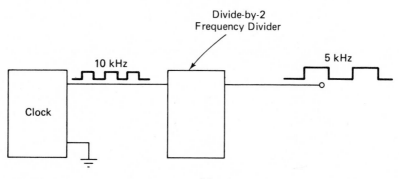

(b)

the frequency is determined by the cut of the crystal; crystal-controlled clocks are very frequency-stable. Some clocks have frequency dividers connected to their outputs. A frequency divider takes the clock pulse frequency, divides it by some integer number, and produces an output pulse train at the lower frequency. Some systems need two or more clock signals of different frequencies in order to synchronize events properly.

The simplest way to build a clock when frequency stability is not important is illustrated in Fig. 3-20. We will not go into the details of the operation of this

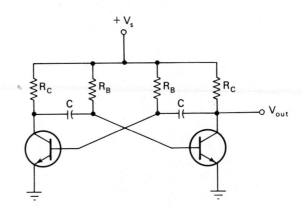

Figure 3-20. Schematic diagram showing one way of building a clock.

circuit. Suffice it to say that if both sides of the circuit have identical components, the output is a square wave at a frequency given approximately by

$$f = \frac{0.7}{R_B C}. \tag{3-1}$$

The operation of this clock is thoroughly explained in many books, if it is of interest to you. If not referred to in the index by the name *clock*, the circuit may be indexed by its formal name, *astable multivibrator*, or by the name *free-running multivibrator*.

3-12 AN AUTOMATIC TANK FILLER USING A CLOCK AND ONE-SHOTS

Consider the system illustrated in Fig. 3-21. The four tanks are refilled from a main tank when their liquid levels drop below a certain setting. That is, if the level in tank 2 drops below its low setting, valve 2 will automatically open and refill tank 2 until the liquid level reaches its high setting. Due to certain system restrictions, it is important that only one tank be refilling at one time.

The circuit to control this system utilizes a clock and several one-shots and is shown in Fig. 3-21(b). The abbreviation OS is used for one-shot.

Here is how it works. Each tank has two limit switches, one that closes on low liquid level and one that closes on high liquid level.

If all tank levels are satisfactory, FF5 is OFF. Its $\bar{Q}$ output is HI, so input 1 of AND5 is HI. The clock is delivering square-wave pulses to the AND gate, so the AND gate output is also a square wave at the same frequency as the clock. Therefore the decade counter is counting merrily along. As the counter proceeds through its various count states, the decoder steps through *its* output states. That is, first the 1 output goes HI, then the 2 output goes HI while 1 returns to LO, then 3 goes HI while 2 returns to LO, and so on. However, when the counter gets to 5, the 5 output of the decoder causes a negative edge to be delivered to the T terminal of OS5 via the inverter. The one-shot fires for a few microseconds, applying a LO to the CL terminal of the counter. The counter is immediately reset to zero when this occurs. The very next count pulse from AND5 causes the counter to count from 0 to 1, since the clear signal has long since departed by the time the count pulse arrives. This is an illustration of a one-shot resetting a counter and then removing the clear signal in time for the next count; this application was suggested in Sec. 3-10.

Figure 3-21. (a) Physical layout of an automatic tank-filling system. (b) Control circuit of the automatic tank filler, showing the use of one-shots and a free-running clock.

(a)

(b)

Figure 3-21. (Cont.)

Therefore the counter is continually counting through states 0–4; when it reaches 5 it stays in that state only long enough for the clear signal to reset it to 0.

The 1 output of the decoder partially enables AND1. The 2 output of the decoder partially enables AND2, and so on. AND gates 1–4 are partially enabled in succession as the decoder goes through its output states.

If a low liquid level limit switch goes closed, the AND gate it controls will be fully enabled. For example, suppose the low-level limit switch in tank 3 closes. Then, as soon as decoder output 3 goes HI, AND3 will go HI. This applies a HI to the *S* input of FF3, which turns FF3 ON. The *Q* output of FF3 signals OA3, which opens valve 3 to refill the tank.

Meanwhile the Q output of FF3 has applied a HI to input 3 of OR gate 1. This causes the OR gate to apply a HI to S of FF5, turning ON that flip-flop. When $\bar{Q}$ of FF5 goes LO, AND5 is disabled and the counter no longer receives count pulses. The counter therefore freezes in its present state.

As the liquid level in tank 3 rises, the low limit switch opens, disabling AND3 and removing the HI from S of FF3. The flip-flop remains ON because of its memory ability. Tank 3 continues refilling until the high liquid level limit switch closes. This applies a HI to the R input of FF3, causing it to turn OFF. When the Q output goes LO, it shuts off valve 3 and disables OR1. The HI is therefore removed from S of FF5. Also, as Q of FF3 goes LO, it delivers a negative edge to the trigger input of OS3, causing that one-shot to fire. The Q output of OS3 goes HI for a few microseconds, enabling OR2. The output of OR2 goes HI temporarily and applies a HI to R of FF5. The flip-flop turns OFF, and its $\bar{Q}$ output returns HI. When this happens, the clock pulses are once again gated into the counter, and counting picks up where it left off.

If you consider the problem of resetting FF5 when the filling operation is finished, you will see why one-shots 1–4 are necessary. OR2, which resets FF5, cannot be driven directly by the $\bar{Q}$ outputs of flip-flops 1–4. With this design, even if one of the flip-flops turned ON, the other three $\bar{Q}$ outputs would hold R of FF5 HI. This would prevent FF5 from ever turning ON, so it would not work. Instead, it is necessary to *temporarily* apply a HI to the R input of FF5 when any one of flip-flops 1–4 turns OFF. One-shots are the best means of doing this.

3-13 DOWN-COUNTERS AND ENCODERS

3-13-1 Decade down-counters

The counters discussed in the preceding sections have all counted in the *up* direction. That is, whenever a count pulse was delivered, the count increased by one. In many cases in industrial control, it is very useful to have a counter which counts in the *down* direction. That is, every time a count pulse is delivered, the number stored in the counter *decreases* by one. This type of counting is especially desirable when it is necessary to produce an output signal after a certain presettable number of counts and also to produce another output signal a fixed number of counts *earlier*. In Sec. 3-15 we will see an example of a down-counter used in just such a manner.

A decade down-counter is shown schematically in Fig. 3-22(a). It is similar in operation to a decade up-counter except that it counts in the down direction. When its contents are zero, the next count input pulse places it in the 9 state ($DCBA = 1001$).

The down-counter has A, B, C, and D *inputs* as well as outputs in order to preset a number into the counter. When the LOAD input terminal goes LO, the BCD number which appears at the A, B, C, and D inputs is preset or loaded into the counter. During loading, any count pulses appearing at CLK are ignored.

Input Count Pulses	Counter Output			
	D	C	B	A
Preset	1	0	0	1
1	1	0	0	0
2	0	1	1	1
3	0	1	1	0
4	0	1	0	1
5	0	1	0	0
6	0	0	1	1
7	0	0	1	0
8	0	0	0	1
9	0	0	0	0
10	1	0	0	1

(a) (b)

Figure 3-22. (a) Black-box symbol of a decade down-counter. (b) The state of the down-counter after each input pulse.

When the LOAD terminal returns to HI, the *A*, *B*, *C*, and *D* inputs become disabled, and the count pulses at CLK begin stepping the counter.

Figure 3-22(b) shows the output states of the down-counter for 10 successive input count pulses, assuming the counter was preset to 9. If the counter were preset to a lower number, naturally it would reach 0 in a smaller number of counts. When it reaches 0, the next count pulse sends it back to 9.

Some counters can be made to count either up or down. They are called *up/down-counters* and have a special control input to tell them in which direction they are supposed to count.

Down-counters can be cascaded just like up-counters. A cascadable down-counter normally has a special output, which tells the neighboring counter when it is going from 0000 to 1001. This way the tens counter, for instance, can count down by one every time the units counter goes through a complete range of values and returns to 9.

3-13-2 Decimal-to-BCD encoders

A down-counter often has an *encoder* associated with it, so encoders will be discussed now. An encoder is a device which takes in a decimal number and puts out a binary number. It is the reverse of a decoder.

There are various types of encoders available, but we will concentrate on the type that converts a 1-of-10 decimal input to BCD output. Such an encoder is shown schematically in Fig. 3-23(a), and its truth table is given in Fig. 3-23(b).

As can be seen from the truth table, the output is the binary equivalent of the

(a)

Truth
Table

Inputs										Outputs			
0	1	2	3	4	5	6	7	8	9	D	C	B	A
1	0	0	0	0	0	0	0	0	0	0	0	0	0
0	1	0	0	0	0	0	0	0	0	0	0	0	1
0	0	1	0	0	0	0	0	0	0	0	0	1	0
0	0	0	1	0	0	0	0	0	0	0	0	1	1
0	0	0	0	1	0	0	0	0	0	0	1	0	0
0	0	0	0	0	1	0	0	0	0	0	1	0	1
0	0	0	0	0	0	1	0	0	0	0	1	1	0
0	0	0	0	0	0	0	1	0	0	0	1	1	1
0	0	0	0	0	0	0	0	1	0	1	0	0	0
0	0	0	0	0	0	0	0	0	1	1	0	0	1

(b)

Figure 3-23. (a) Black-box symbol of a 1-of-10 encoder. (b) The truth table of the encoder, showing the output state for every legal combination of inputs.

decimal input. The truth table as given implies that there are never two inputs HI at the same time. It is the responsibility of the control circuit designer to make sure this is so.

There is always the possibility that two or more inputs might go HI at the same time, through some malfunction in the input circuitry to the encoder. If it is important to know what the encoder will do in such a case, the manufacturer's specification sheet will explain. Most packaged encoders obey the higher-number input if that problem ever occurs.

Encoders very often receive their input from a 10-position selector switch. The SS is manually set by the system operator, and the number selected appears at the output of the encoder in BCD form. The output of the encoder can then be wired to the input of a down-counter for presetting the down-counter. When the LOAD terminal of the down-counter goes LO, the SS setting is preset into the counter. This arrangement is illustrated in Fig. 3-24.

Figure 3-24. The combination of a 10-position selector switch, a decimal-to-BCD encoder, and a decade down-counter. This combination is often seen in industrial system control.

Care must be taken with the input lines to an encoder. Some logic families interpret a hanging input to be HI, as we know. If a particular encoder belongs to such a logic family, the simple input method of Fig. 3-24 will not work because all the disconnected inputs will be considered as HI. In such logic families, the manufacturer often gets around this problem by building the encoder to respond to a LO input level instead of a HI input level. That is, whichever one of the 10 input lines that is pulled LO is considered to be the desired input number. To simplify our discussion from now on, we will assume that all our packaged encoders respond to a LO input level, and we will use the small circles drawn outside the box to remind ourselves of that fact. Therefore the truth table in Fig. 3-23(b) should be visualized with all the input 0s being 1s, and all the input 1s being 0s. The logic level applied to the common terminal of the selector switch then becomes a LO (a ground connection), rather than the HI level (+V) that is indicated in Fig. 3-24.

3-14 TIMERS

In industrial control, it is often necessary to inject a time delay between the occurrence of two events. For example, consider a situation in which two large motors are to be started at about the same time. If both motors are powered from the same supply bus, it is bad practice to switch them both across the lines at the same instant, because large motors draw quite large inrush currents at the instant of starting and continue to draw current far in excess of their normal rated current for several seconds after starting. The motor current drops to its normal rated value only when the motor armature has accelerated up to normal running speed. During the time that the motor is drawing this excessive current, the current capability of the supply bus may be strained. Such a time is no time to require the supply bus to start *another* large motor. Fuses or circuit breakers in the supply lines may open, disconnecting the entire bus. Even if that does not happen, the combination of two starting currents may very well cause excessive voltage drop along the supply lines, resulting in a lower terminal voltage applied to the motors. This prolongs the acceleration period and may cause overheating of the motor windings themselves.

As can be seen from the preceding argument, when two large motors are powered by the same bus, there should be a *time delay* between their starting instants. This can be accomplished with *time delay relays*, as shown in Fig. 3-25.

3-14-1 Time delay in relay circuits

In Fig. 3-25(b), two large three-phase ac induction motors are driven by a common 460-V supply bus. The contacts that switch the motor windings across the lines are controlled by motor starter *A* (MS*A*) and motor starter *B* (MS*B*). The control situation calls for motor *A* and motor *B* to start at approximately the same time, but it is not necessary that they start at *exactly* the same time.

When the initiating contact in Fig. 3-25(a) closes, it energizes the coil of MS*A* and also energizes the coil of relay 1 (R1). The MS*A* contacts in the high-voltage supply circuit start MOTOR *A*. MOTOR *A* proceeds to draw a large inrush current, perhaps as much as 1000% of rated full load current. The contact controlled by relay R1 in Fig. 3-25(a) does not close immediately. It delays closing until a certain amount of time has elapsed. By the time it does close to energize MS*B*, MOTOR *A* has reached full speed and has relaxed its current demand.

The delayed closing of the relay contact can be accomplished by several methods. The most popular method has been the use of a pneumatic dashpot attached to the moving member of the relay. When the relay coil energizes, a spring exerts a force on the moving member, attempting to close the contact, but a pneumatic (air-filled) dashpot prevents the movement from taking place. As the trapped air bleeds past a needle valve out of the dashpot, the necessary movement occurs, and the contacts close. Thus the normally open contacts will not close

Figure 3-25. (a) A simple relay circuit with a time delay contact. (b) The motor power circuit associated with the relay control circuit in part (a).

instantly when the relay is picked. They close after a certain time delay, which is adjustable by adjusting the needle valve. The abbreviation N.O.T.C. in Fig. 3-25(a) stands for "normally open timed closing." The unusual symbol in that figure is the accepted Joint Industry Conference symbol for an N.O.T.C. contact.

Other types of timed contacts are also commonly used. Table 3-6 gives the names, symbols, and brief explanations of each type of contact. The top two types are sometimes called *on delay* contacts, and relays that have such contacts are called *on delay* relays, because the delayed action takes place as the relay energizes. The bottom two contacts and the relays that contain them are sometimes described as *off delay* because the delay action takes place as the relay deenergizes.

Note that a time delay contact always delays in one direction only. In the other direction, it acts virtually instantaneously, just like a normal relay contact.

An example of the use of an N.C.T.C. contact is illustrated in Fig. 3-26. In Fig. 3-26(a), a wagon is to be filled with powder from an overhead hopper. The wagon is moved underneath the hopper outlet pipe; then the solenoid is energized to open a valve. When the wagon is sufficiently full, the solenoid is closed, and the wagon is moved away. However, there will be some powder remaining in the fill pipe for a few seconds after the solenoid valve closes. To give this powder a chance

TABLE 3-6. The four types of time delay relay contacts.

	Name	Abbreviation	Symbol	Description
Delay Upon Energization (On Delay)	Normally open timed closing	N.O.T.C.		When the relay energizes, the N.O. contact delays before it closes. When the relay deenergizes, the contact opens instantly.
	Normally closed timed opening	N.C.T.O.		When the relay energizes, the N.C. contact delays before it opens up. When the relay deenergizes, the contact closes instantly.
Delay Upon Deenergiaztion (Off Delay)	Normally open timed opening	N.O.T.O.		When the relay energizes, the N.O. contact closes instantly. When the relay deenergizes, the contact delays before it returns to the open condition.
	Normally closed timed closing	N.C.T.C.		When the relay energizes, the N.C. contact opens instantly. When the relay deenergizes the contact delays before it returns to the closed condition.

to drain out into the wagon, the movement of the wagon is delayed for a few seconds after the valve goes closed. A relay circuit to accomplish this is given in Fig. 3-26(b). When the solenoid deenergizes, RA drops. A little while later, the N.C. contact of RA returns to its closed position; this energizes MSW, which starts a motor to move the wagon away.

3-14-2 Series resistor-capacitor circuits: Time constants

The previous examples have shown time delay injected into a control circuit by the action of the contacts of a relay. It is also possible to delay the energization or deenergization of the relay itself. This is usually done by taking advantage of the fact that a certain time must elapse to charge a capacitor through a resistor.

Recall that when a capacitor is charged by a dc source through a series resistor, the charging action is described by the universal time constant curve. Briefly, the rate of charge buildup (voltage buildup) is rapid when the charge on the capacitor is small, but the charging rate decreases as the charge (voltage) on the capacitor gets larger. The behavior of any resistor-capacitor series circuit can be conveniently

(a)

(b)

Figure 3-26. (a) Physical layout of a wagon being filled from a hopper. (b) A simple relay control circuit, illustrating the use of a time delay contact to allow the powder to drain out of the supply tube into the wagon before the wagon is moved.

described in terms of how many *time constants* have elapsed. A time constant for a series *RC* circuit is defined by the formula

$$\tau = RC \tag{3-2}$$

where τ stands for the time constant, measured in seconds; R stands for the resistance in ohms; and C stands for the capacitance, measured in farads, the basic capacitance unit.

Once the idea of a time constant is accepted, the behavior of *all* series *RC* circuits can be described by the universal time constant curve and by certain well-known rules. The most widely used rules are

a. A time equal to five time constants is necessary to charge a capacitor to 99.5% of full supply voltage (99.5% is generally agreed to represent a full charge).

b. In one time constant, a capacitor will charge to 63% of full supply voltage.

The meaning of these rules is graphically illustrated by the universal time constant charging curve in Fig. 3-27.

In our discussion of solid-state logic timers, references will be made to the rules given above for series *RC* circuits. These rules will also prove useful when we discuss other time-related circuit action in later chapters.

3-14-3 Solid-state timers

In a solid-state control system, the action of time delay relays is duplicated by solid-state timers. The black-box symbol for a solid-state timer, with its input-output waveform, is shown in Fig. 3-28(a). Also shown in Figs. 2-27(b), (c), and

Figure 3-27. The universal time constant curve. This curve illustrates in detail how a capacitor is charged by a dc source. It also represents many other natural phenomena.

(d) are methods of altering the waveforms to duplicate the actions of the various types of time delay relay contacts.

One method of building a solid-state timer is shown in Fig. 3-29. Here is how it works. When the input is LO, there is no current into the base of T_1, so T_1 is turned OFF. Its collector is near V_s, causing it to turn ON T_2 and T_4 through R_3 and R_{10}. With T_4 ON, its collector is LO, so the output of the overall circuit is virtually 0 V. T_2 comprises a transistor switch which is shorted to ground at this time. It discharges any charge on C_t through D_1. Therefore the voltage at the top of C_t is virtually 0 V, ensuring that zener diode D_2 is an open circuit. No current can flow into the base of T_3 through R_7 because of the zener diode. No current flows into the base of T_3 through R_6 either, because R_6 is connected to 0 V. Therefore T_3 is OFF, and its collector voltage is near V_s. The T_3 collector delivers current to the base of T_4 through R_4, comprising a second source of base current to hold T_4 ON.

When the input goes HI, it drives the collector of T_1 down to ground. This turns OFF T_2 and also removes one of the sources of base current to T_4. T_4 remains ON because it continues to receive base current through R_9. When T_2 turns OFF, it opens the transistor switch which was preventing timing capacitor C_t from charging up. Therefore C_t begins to charge with a time constant equal to $(R_f + R_t)C_t$. The subscript f on R_f is chosen because it is a fixed resistor. The subscript t on R_t is chosen because it is a timing adjustment resistor.

As C_t continues to charge, it will eventually reach a voltage which can break down the zener diode. If the reverse breakdown voltage of the zener diode is symbolized by V_z, the voltage necessary to force current through the zener diode

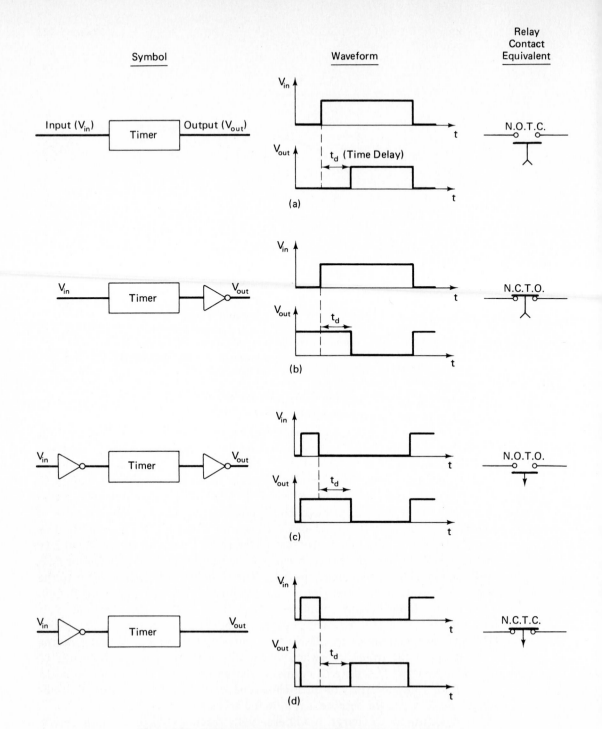

Figure 3-28. Solid-state timers and their input-output waveforms. This figure shows the equivalence between the four timer configurations and the four types of time delay relay contacts.

Figure 3-29. Schematic diagram showing one way of building a solid-state timer based on an *RC* charging circuit.

is 0.6 V greater than V_z, because any current through zener diode D_2 must go to ground through the base-emitter junction of T_3.

When the top of C_t reaches the necessary voltage, it begins to bleed a little bit of current into T_3 through D_2 and R_7. This causes T_3 to begin carrying a bit of collector current, causing the collector voltage to fall a little bit. This reduces the base current through R_9, causing the collector of T_4 to rise a little bit. The rise in T_4 collector voltage reinforces the original base current into T_3, causing it to turn on harder. Therefore the action is self-reinforcing, and it avalanches. Regenerative switching action drives the output level HI in very short time, so the positive-going edge of the output waveform is steep. Thus the output goes HI a certain time after the input goes HI. The elapsed time depends on how long it takes C_t to reach the zener's breakdown point. This time depends on the charging time constant, which is adjusted by potentiometer R_t.

When the input returns to LO, T_1 turns OFF, causing its collector voltage to rise. This immediately turns ON T_4 through R_{10}, so the output level immediately goes LO. The T_1 collector also turns ON T_2, closing the switch across the C_t-D_1 combination. When the T_2 switch closes, C_t immediately dumps its positive charge through D_1, through T_2 to ground. This removes the current source for R_7. The current source for R_6 is already removed because the output has gone LO. Therefore T_3 turns OFF, and everything is back to starting condition.

Example 3-1

In Fig. 3-29, $V_s = 20$ V, zener voltage $V_z = 12$ V, $C_t = 50 \ \mu\text{F}$, $R_f = 10$ K, and R_t is a 100-K pot. What range of time delays is possible?

Solution. To bleed current through D_2 into the base of T_3, the capacitor voltage must reach 12.6 V. This is given by

$$V_c = V_z + 0.6\ \text{V} = 12.0 + 0.6 = 12.6\ \text{V}.$$

12.6 volts is exactly 63% of the full capacitor voltage of 20 V. According to rule b in Sec. 3-14-2, it takes one time constant for a capacitor to charge to 63% of full voltage. Therefore it will take one time constant to turn ON T_3 after the input goes HI. The time delay is equal to one time constant. τ is given by

$$\tau = (R_f + R_t)C_t.$$

The minimum time constant occurs when R_t is dialed completely out. In that case,

$$\tau_{\min} = (10\ \text{K} + 0)(50\ \mu\text{F}) = 0.5\ \text{sec}.$$

The maximum time constant occurs when R_t is dialed completely in. In that case,

$$\tau_{\max} = (10\ \text{K} + 100\ \text{K})(50\ \mu\text{F}) = 5.5\ \text{sec}.$$

The range of possible delay times is therefore from **0.5 to 5.5 sec**.

Example 3-2

In the timer of Fig. 3-29, suppose that a different type of zener diode has been substituted, this one having $V_z = 16$ V. To what value should R_t be adjusted to give a time delay of 8 sec?

Solution. In this case, the C_t voltage must reach 16.6 V to turn ON T_3. On a percent basis this is

$$\frac{16.6\ \text{V}}{20.0\ \text{V}} = 83\%$$

of full supply voltage. From the universal time constant curve in Fig. 3-27, it can be seen that about 1.8 time constants are required to charge a capacitor to 83% of full voltage. Therefore

$$(1.8)(\tau) = 8\ \text{sec},$$

$$\tau = \frac{8\ \text{sec}}{1.8} = 4.44\ \text{sec}.$$

Since it will require a 4.44-sec time constant in order to produce a delay time of 8 sec, R_t can be found by

$$\tau = (R_f + R_t)(C_t)$$

$$R_t = \frac{\tau}{C_t} - R_f = \frac{4.44}{50 \times 10^{-6}} - 10 \times 10^3$$

$$R_t = 88.8 \times 10^3 - 10 \times 10^3 = 78 \times 10^3$$

$$R_t = \textbf{78 K}$$

3-15 A BIN-FILLING SYSTEM USING A DOWN-COUNTER, AN ENCODER, AND TIMERS

The system illustrated in Fig. 3-30 is an efficient method of keeping many material bins full. In this example there are nine material bins which are refilled by a traveling hopper which moves on overhead rails. The traveling hopper is itself loaded from the supply tube in the home position. The operator then sends it to whichever material bin is in need of replenishing. When it has dumped its material into that bin, the traveling hopper automatically returns home for another load.

The operation must be performed quickly in order to keep the system efficient. Therefore there is a two-speed motor driving the wheels of the traveling hopper. When the hopper leaves the home position with a full load of material, it starts off at the slow speed. After its inertia has been overcome, it changes to high speed. It travels at high speed until it is one position away from its destination. At that time, it switches back to slow speed for the final approach. When it reaches its destination, it stops and opens its dumping doors to dump the material into the bin.

The dumping doors remain open for a certain preset time and then close up. The lightened hopper then returns home at high speed.

The control circuit for this cycle must generate two electrical outputs. One output must occur when the hopper reaches its destination, to cause the motor to stop running. The other output must occur a *fixed distance earlier*, to cause the motor to slow down. This is the type of situation in which a down-counter is so useful, as mentioned in Sec. 3-13. In Fig. 3-30(b) the down-counter keeps track of the location of the traveling hopper by counting pulses generated as the hopper passes through the nine filling positions. There is an actuating cam mounted on the hopper. As the hopper moves to the right, it actuates a limit switch every time it passes through a new position

Here is how the control circuit works. The operator receives a signal that a certain bin needs material; the signaling method is not shown. He then causes the

Figure 3-30. (a) Physical layout of the bin-filling system with a traveling hopper. (b) Control circuit for the bin-filling system, illustrating the use of a 10-position selector switch, a decimal-to-BCD encoder, a down-counter, and timers.

(a)

Figure 3-30. (Cont.)

(b)

hopper to be filled with the proper material from the supply tubes. This mechanism is not shown either. When the hopper is loaded, the operator dials in the destination on the 10-position selector switch. For example, if bin 7 is the one that needs replenishing, he sets the SS to 7. After that, he presses the DELIVER push button, and the control circuit takes over.

The output of the DELIVER signal converter goes HI, causing FF1 to turn ON. Q_1 goes HI, energizing the forward motor starter coil, labeled FOR MS. This applies power to the motor, which drives the traveling hopper in the forward direction, to the right in Fig. 3-30(a). When the motor is running in the forward direction, its speed depends on which of the two contactors, SLOW or FAST, is energized. When the OR output is HI, the SLOW contactor is energized, and the motor runs slow. When the OR output is LO, the I2 inverter causes the FAST contactor to be energized, and the motor runs fast.

When the Q_1 output initially goes HI, it applies a HI to input 1 of the NAND gate. The output of TIMER1 remains LO for a certain time setting, so input 2 of the NAND remains LO for a while. The NAND output delivers a HI to the OR gate, causing the OR output to be HI. The motor therefore starts in the slow speed. After a few seconds, whatever time is set on the timer, the output of TIMER1 goes HI, causing the NAND output to go LO. This removes the HI from input 1 of the OR gate. Input 2 of the OR is probably LO at this time also; we will consider this more carefully later. With both OR inputs LO, the output goes LO, and the motor switches to high speed.

Meanwhile, back at I1, its output goes LO when the operator presses the DELIVER button. This causes a negative edge to appear at T of OS1. When OS1 fires, its $\bar{Q}$ output goes LO, applying a LO to the LOAD input of the down-counter. The BCD number which appears at the output of the encoder is thereby loaded into the down-counter. When the output pulse from OS1 goes away, the LOAD input goes back HI, and the down-counter is ready to begin counting when pulses arrive at its CLK input. All this takes place in a fraction of a millisecond, so there is absolutely no chance that the down-counter will miss any count pulses generated as the traveling hopper actuates the various COUNT limit switches, LS1–LS9.

As the motor accelerates the traveling hopper, it approaches LS1. When it hits LS1, the COUNT signal converter delivers a positive pulse. I3 converts this to a negative edge, and the down-counter counts once. Let us assume that the preset number was 7 (0111). After the traveling hopper contacts LS1, the contents of the down-counter are 0110, or 6.

As the hopper continues to move to the right at high speed, it delivers another negative edge to the down-counter every time it contacts another limit switch. The counter is therefore counting backwards toward zero. As the hopper passes through position 5 and actuates LS5, the fifth count pulse is delivered to the down-counter. This causes its contents to become 0010 (2) since it started at 7 and has received five count pulses. The hopper continues moving to the right at high speed until it hits LS6. The sixth count pulse causes the counter to step into the state $DCBA = 0001$. The decoder immediately recognizes this as the binary code for 1 and accordingly sends its 1 output HI. This HI appears at input 2 of the OR gate and drives the OR output HI. The motor therefore drops into slow speed.

As the hopper proceeds at slow speed, it arrives at its destination above bin 7. It contacts LS7 and delivers the seventh pulse to the down-counter. The counter steps into the state $DCBA = 0000$. The decoder recognizes this as 0, so it sends its 0 output HI. I4 inverts this HI and fires OS2. The Q output of OS2 goes HI and appears at R of FF1. The flip-flop turns OFF, thereby deenergizing the forward motor starter and stopping the motor. The heavily loaded hopper has low momentum since it was traveling slow, so it does not coast very far. It comes to a halt in position above bin 7.

The Q output from OS2 appears at input S of FF2, turning it ON. Q_2 goes HI, energizing the dump solenoid and starting TIMER2. The dumping doors of the hopper open and allow the material to fall into bin 7. After a sufficient time has elapsed to get rid of all the material in the hopper, TIMER2 times out, and its output goes HI. This HI appears at R of FF2 and S of FF3. FF2 turns OFF, closing the dumping doors, and FF3 turns ON. Q_3 goes HI and energizes the reverse motor starter, REV MS. This causes the motor to run in the reverse direction at high speed. Therefore the traveling hopper turns around and heads back toward home. When it contacts the HOME LS, the HOME signal converter applies a HI to R of FF3. The flip-flop turns OFF and drops out REV MS, so the hopper comes to a stop in the home position.

We said we would return to carefully consider the status of OR input 2 as the hopper is getting under way. We assumed earlier that it would be LO at that time. This assumption is correct as long as the destination is bins 2–9. If the destination is any one of those bins, the number loaded into the down-counter will *not* be 1 (0001). Therefore as the hopper is starting out, the decoder is not receiving an input of 1, so decoder output 1 will not be HI; it will be LO. Therefore OR input 2 is LO.

However, if bin 1 *is* the destination, then the down-counter was preset to $DCBA = 0001$, and the decoder output 1 will be HI as the hopper starts out. Under this condition, the motor never does change into fast speed; it makes the entire journey to bin 1 at slow speed. Trace out the circuit behavior and verify this for yourself.

QUESTIONS AND PROBLEMS

1. Explain why a flip-flop will maintain its present state forever unless ordered to change states by an outside signal.
2. In Fig. 3-3, explain why the B welder contactor cannot be energized if the A welder contactor is already energized.
3. Carefully explain the difference between an RS flip-flop and a *clocked RS* flip-flop.
4. Explain the difference between a positive edge-triggered flip-flop and a negative edge-triggered flip-flop.

5. What input combination is not legal for a clocked *RS* flip-flop? Why is it illegal?

6. In Fig. 3-5(b), is it ever possible for both *S* and *R* of FF1 to be HI at the same time? Why?

7. Explain why the oscillating table of Fig. 3-5 always stops in the extreme left position, never in the extreme right position.

8. Make the necessary additions to Fig. 3-5 to allow the operator to get the oscillations started after he has installed a workpiece.

9. What is the chief difference between a *JK* flip-flop and a clocked *RS* flip-flop?

10. For a flip-flop, does a clear signal override a signal to turn ON from the clocked inputs?

11. If the shift register of Fig. 3-8 starts in the state 0000 and two shift pulses are applied with the FF1 input terminal HI, what is the new state of the shift register?

12. Show how the shift register of Fig. 3-8 could be made to *circulate* its information; that is, the information in FF4 is not lost when a shift pulse arrives, but is recycled.

13. In Fig. 3-9, why does FF1 always contain a 0 unless the inspector presets a 1 into it?

14. If it was desired to build a 10-bit shift register, how many packaged 4-bit shift registers are needed? Draw a schematic showing all the interconnections between the packages.

15. In simple terms, why is it difficult to build electronic circuits which operate in the *decimal* number system?

16. Why is the binary number system so popular for electronic digital circuits?

17. Express the number 45 in BCD. Repeat for the numbers 761, 834, and 902.

18. Convert the following BCD numbers to decimal:
    a. 0100 0111
    b. 1000 0101
    c. 0100 0001 0011

19. For the counter shown in Fig. 3-11, what is the state of the counter after three count pulses? After eight count pulses?

20. By analyzing the circuit behavior, tell the state of the Fig. 3-11 counter after 12 count pulses, after 15 count pulses, and after 16 count pulses.

Questions 21–25 refer to the decade counter of Fig. 3-12.

21. Explain why FF*B* does not turn ON when the tenth count pulse arrives.

22. What causes FF*D* to turn OFF on the tenth count pulse?

23. Suppose the AND gate failed so that its output was always LO. Write a table showing the count sequence of the counter under this condition.

24. Suppose that when power comes on, the flip-flops land in random states so that $DCBA = 1110$. If the counter is not cleared before the count pulses start arriving, what will the counter do. Make a table showing the count sequence.

25. Repeat Question 24 for the flip-flops coming up randomly in the state $DCBA = 1010$.

26. Explain in words the action of a BCD-to-decimal decoder.

27. Referring to Fig. 3-15(a), what decoder output line will go HI if the input situation is $\bar{D}C\bar{B}A = 1111$. Repeat for $D\bar{C}\bar{B}A = 1111$.

Questions 28–30 refer to the palletizing system of Fig. 3-16.

28. What assures that the decade counters start counting from zero when a new pallet is begun?

29. Suppose the cartons were being loaded six layers high, eight cartons to a layer. What would be the settings of the selector switches? What would be the BCD output of the two counters which would cause the diverter to change positions?

30. Repeat Question 29 for cartons loaded 12 to a layer, seven layers high.

31. In words, explain what a one-shot does.

32. One-shots are sometimes called *delay elements*. Why do you think they are called this?

33. In Fig. 3-18, if the size of C were increased, what effect would this have on firing time? Explain why.

34. What means are utilized to adjust the firing time of a prepackaged one-shot?

35. Suppose we have two one-shots, one of them retriggerable and the other nonretriggerable, both with a firing time of 10 msec. A fast pulse is applied to both trigger inputs at the same instant. Seven milliseconds later, another fast pulse is applied to both trigger inputs. Make a sketch showing the output waveforms of both one-shots.

36. Consider the clock of Fig. 3-20. If $R_B = 50$ K and $C = 0.05$ μF, what is the approximate output frequency?

37. If the clock of Question 36 had a divide-by-4 frequency divider attached to its output, what is the frequency of the resulting waveform?

38. In very general terms, what is the purpose of an industrial timer?

39. In words, explain the behavior of each of the four types of time delay relay contacts, N.O.T.C., N.O.T.O., N.C.T.C., and N.C.T.O.

40. What are the standard symbols for each of the four contacts in Question 39?

41. Consider a series RC circuit with $R = 1.5$ K and $C = 0.1$ μF. How long would it take to fully charge the capacitor with a dc source?

42. For a series RC circuit with $R = 6.8$ K and $C = 100$ μF, if a 30-V dc source is applied to it, how long will it take to charge to 18.9 V (63% of 30 V)?

43. Referring to Fig. 3-29, suppose $(R_f + R_t)C_t = \tau = 0.2$ sec. If $V_s = 30$ V and $V_z = 15$ V, what is the time delay of the timer? Use the universal time constant curve of Fig. 3-27.

44. Repeat Question 43 for $V_z = 24$ V.

45. Explain each of the time delay circuits shown in Fig. 3-28. That is, explain why each circuit has the V_{in}-V_{out} waveform shown.

Questions 46–51 refer to the bin-filling system of Fig. 3-30.

46. The common terminal of the 10-position selector switch is tied to ground. Explain why this is correct (as opposed to tying it to the logic dc supply).

47. Why is the LOAD terminal of the down-counter connected to the $\bar{Q}$ output of OS1 instead of the Q output?

48. Describe the entire process of presetting the down-counter to the proper number.

49. What causes FOR MS to deenergize? Explain the process by which the circuit deenergizes this motor starter.

50. Why is it necessary to slow down the traveling hopper before it arrives at its destination?

51. Explain how the circuit slows down the hopper before it arrives at the destination.

52. Why is it necessary to have D_1 in Fig. 3-29?

53. What idealistic assumption is always made about the positive-going or negative-going edges of clock pulses which trigger flip-flops?

4
SCRs

There are numerous industrial operations which require the delivery of a variable and controlled amount of electrical power. Lighting, motor speed control, electric welding, and electric heating are the four most common of these operations. It is always possible to control the amount of electrical power delivered to a load by using a variable transformer to create a variable secondary output voltage. However, in high power ratings, variable transformers are physically large and expensive and need frequent maintenance. So much for variable transformers.

Another method of controlling electrical power to a load is to insert a rheostat in series with the load to limit and control the current. Again, for high power ratings, rheostats are large, expensive, need maintenance, and they waste energy to boot. Rheostats are not a desirable alternative to variable transformers in industrial power control.

Since 1960, an electronic device has been available which has none of the faults mentioned above. The SCR is small and relatively inexpensive, needs no maintenance, and wastes very little power. Some modern SCRs can control currents of several hundred amperes in circuits operating at voltages higher than 1000 V.

For these reasons, SCRs are very important in the field of modern industrial control. We will investigate SCRs in this chapter.

OBJECTIVES

After studying this chapter and performing the suggested laboratory projects, you will be able to:

1. Explain the operation of an SCR power control circuit for controlling a resistive load
2. Define firing delay angle and conduction angle and show how they affect the average load current
3. Define some of the important electrical parameters associated with SCRs, such as gate trigger current, holding current, forward ON-state voltage, etc., and give the approximate range of values expected for these parameters
4. Calculate approximate resistor and capacitor sizes for an SCR gate trigger circuit
5. Explain the operation and advantages of breakover trigger devices used with SCRs
6. Construct an SCR circuit for use with a 115-V ac supply and measure the gate current and gate voltage necessary to fire the SCR
7. Construct a zero-point switching circuit and explain the advantages of zero-point switching over conventional switching

4-1 THEORY AND OPERATION OF SCRs

A *silicon controlled rectifier* (SCR) is a three-terminal device used to control rather large currents to a load. The schematic symbol for an SCR is shown in Fig. 4-1, along with the names and letter abbreviations of its terminals.

Figure 4-1. Schematic symbol and terminal names of an SCR.

An SCR acts very much like a switch. When it is turned ON, there is a low-resistance current flow path from anode to cathode; then it acts like a closed switch. When it is turned OFF, no current can flow from anode to cathode; then it acts like an open switch. Because it is a solid-state device, the switching action of an SCR is very fast.

The average current flow to a load can be controlled by placing an SCR in series with the load. This arrangement is shown in Fig. 4-2. The supply voltage in Fig. 4-2 is normally a 60-Hz ac supply, but it may be dc in special circuits.

Figure 4-2. The circuit relationship among the voltage supply, an SCR, and the load.

If the supply voltage is ac, the SCR spends a certain portion of the ac cycle time in the ON state and the remainder of the time in the OFF state. For a 60-Hz ac supply, the cycle time is 16.67 msec. It is this 16.67 msec which is divided between the time spent ON and the time spent OFF. The amount of time spent in each state is controlled by the gate. How the gate does this is described later.

If a small portion of the time is spent in the ON state, the average current passed to the load is small. This is because current can flow from the supply through the SCR to the load only for a relatively short portion of the time. If the gate signal is changed to cause the SCR to be ON for a larger portion of the time, then the average load current will be larger. This is because current now can flow from the supply through the SCR to the load for a relatively longer time. In this way the current to the load can be varied by adjusting the portion of the cycle time the SCR is switched ON.

As its name suggests, the SCR is a rectifier, so it passes current only during positive half cycles of the ac supply. The positive half cycle is the half cycle in which the anode of the SCR is more positive than the cathode. This means that the SCR in Fig. 4-2 cannot be turned ON more than half the time. During the other half of the cycle time the supply polarity is negative, and this negative polarity causes the SCR to be reverse biased, preventing it from carrying any current to the load.

4-2 SCR WAVEFORMS

The popular terms used to describe how an SCR is operating are *conduction angle* and *firing delay angle*. Conduction angle is the number of degrees of an ac cycle during which the SCR is turned ON. The firing delay angle is the number of degrees of an ac cycle that elapses *before* the SCR is turned ON. Of course, these terms are based on the notion of total cycle time equaling 360 degrees (360°).

Figure 4-3 shows waveforms for an SCR control circuit for two different firing delay angles.

Let us interpret Fig. 4-3(a) now. At the time the ac cycle starts its positive

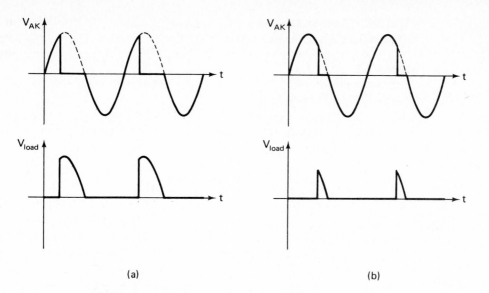

Figure 4-3. Ideal waveforms of SCR main terminal voltage (V_{AK}) and load voltage. (a) For a firing delay angle of about 60°, conduction angle of 120°. (b) For a firing delay angle of about 135°, conduction angle of 45°.

alternation, the SCR is turned OFF. Therefore it has an instantaneous voltage across its anode to cathode terminals equal to the supply voltage. This is just what would be seen if an open switch were put in the circuit in place of the SCR. Since the SCR is dropping the entire supply voltage, the voltage across the load (V_{load}) is zero during this time. The extreme left of the waveforms of Fig. 4-3(a) illustrates these facts. Farther to the right on the horizontal axes, Fig. 4-3(a) shows the anode to cathode voltage (V_{AK}) dropping to zero after about one third of the positive half cycle; this is the 60° point. When V_{AK} drops to zero, the SCR has "fired" or turned ON. Therefore in this case the firing delay angle is 60°. During the next 120° the SCR acts like a closed switch with no voltage across its terminals. The conduction angle is 120°. Firing delay angle and conduction angle always total 180°.

The load voltage waveform in Fig. 4-3(a) shows that when the SCR fires, the supply voltage is applied to the load. The load voltage then follows the supply voltage through the rest of the positive half cycle, until the SCR again turns OFF. Turning OFF occurs as the supply voltage passes through zero.

Overall, these waveforms show that before the SCR fires the entire supply voltage is dropped across the SCR terminals, and the load sees zero voltage. After the SCR fires, the entire supply voltage is dropped across the load, and the SCR drops zero voltage. The SCR behaves just like a fast-acting switch.

Figure 4-3(b) shows the same waveforms for a different firing delay angle. In these waveforms, the firing delay angle is about 135° and the conduction angle about 45°. The load sees the supply voltage for a much shorter time as compared to Fig. 4-3(a). The average current is smaller as a result.

Example 4-1

Which condition would cause the larger average load current in Fig. 4-2, a firing delay angle of 30° or a firing delay angle of 45°?

Solution. The firing delay angle of 30°, because the SCR would then spend a greater portion of the cycle time in the ON state. The more time spent in the ON state, the greater the average load current.

Example 4-2

If the conduction angle of an SCR is 90° and it is desired to double the average load current, what new conduction angle is necessary? The supply is an ac sine wave.

Solution. 180°. In this case, doubling the conduction angle doubles the average load current, because the first 90° of a sine wave is the image of the second 90°. However, in general, it is *not* true that doubling the conduction angle will double the average current.

4-3 SCR GATE CHARACTERISTICS

An SCR is fired by a short burst of current into the gate. This gate current (i_G) flows through the junction between the gate and cathode and exits from the SCR on the cathode lead. The amount of gate current needed to fire a particular SCR is symbolized I_{GT}. Most SCRs require a gate current of between 0.1 and 20 mA to fire ($I_{GT} = 0.1 - 20$ mA). Since there is a standard *pn* junction between gate and cathode, the voltage between these terminals (V_{GK}) must be slightly greater than 0.6 V. Figure 4-4 shows the conditions which must exist at the gate for an SCR to fire.

Figure 4-4. Gate to cathode voltage (v_{GK}) and gate current (i_G) needed to fire an SCR.

Once an SCR has fired, it is not necessary to continue the flow of gate current. As long as current continues to flow through the main terminals, from anode to cathode, the SCR will remain ON. When the anode to cathode current (i_{AK}) drops below some minimum value, called *holding current*, symbolized I_{HO}, the SCR will shut OFF. This normally occurs as the ac supply voltage passes through zero into its negative region. For most medium-sized SCRs, I_{HO} is around 10 mA.

Example 4-3

For the circuit of Fig. 4-5, what voltage is required at point X to fire the SCR? The gate current needed to fire a 2N3669 is 20 mA under normal conditions.

Figure 4-5. SCR with a 150-Ω resistor in the gate lead and its cathode terminal connected to circuit ground.

Solution. The voltage between point X and the cathode must be sufficient to forward-bias the junction between points G and K and also to cause 20 mA to flow through 150 Ω. The forward bias voltage is about 0.6 V. From Ohm's law, $V_{XG} = (20 \text{ mA})(150 \, \Omega) = 3.0$ V. Therefore total voltage = 3.0 + 0.6 = 3.6 V.

4-4 TYPICAL GATE CONTROL CIRCUITS

The simplest type of gate control circuit, sometimes called a triggering circuit, is shown in Fig. 4-6. This is an example of using the same voltage supply to power both the gate control circuit and the load. Such sharing is common in SCR circuits. The positions of the SCR and load are reversed from those of Fig. 4-2, but this makes no difference in operation.

Figure 4-6. A very simple triggering circuit for an SCR.

In Fig. 4-6, if the supply is ac, operation is as follows. When the switch is open, it is impossible to have current flow into the gate. The SCR can never turn ON, so it is essentially an open circuit in series with the load. The load is therefore deenergized.

When SW is closed, there will be current into the gate when the supply voltage goes positive. The firing delay angle is determined by the setting of R_2, the variable resistance. If R_2 is low, the gate current will be sufficiently large to fire the SCR when the supply voltage is low. Therefore the firing delay angle will be small, and average load current will be large. If R_2 is high, the supply voltage must climb higher to deliver enough gate current to fire the SCR. This increases the firing delay angle and reduces average load current.

The purpose of R_1 is to maintain some fixed resistance in the gate lead even when R_2 is set to zero. This is necessary to protect the gate from overcurrents. R_1 also determines the minimum firing delay angle. In some cases a diode is inserted in series with the gate to protect the gate-cathode junction against high reverse voltages.

One disadvantage of this simple triggering circuit is that the firing delay angle is adjustable only from about 0° to 90°. This fact can be understood by referring to Fig. 4-7, which shows that the gate current tends to be a sine wave in phase with the voltage across the SCR.

In Fig. 4-7(a), i_G barely reaches I_{GT}, the gate current needed to trigger the SCR. Under this circumstance the SCR fires at 90° into the cycle. It can be seen that if

Figure 4-7. Ideal waveforms of SCR main terminal voltage and gate current. The dashed line represents the gate current necessary to fire the SCR (I_{GT}). (a) Gate current is low, resulting in a firing delay angle of about 90°. (b) Gate current is greater, resulting in a firing delay angle of nearly 0°.

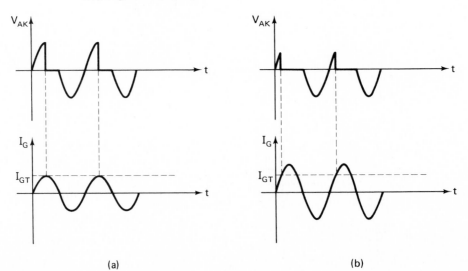

i_G were any smaller, the SCR would not fire at all. Therefore, firing delays past 90° are not possible with such a gate control circuit.

In Fig. 4-7(b), i_G is quite a bit larger. In this case, i_G reaches I_{GT} relatively early in the cycle, causing the SCR to fire early.

It should be understood that the i_G waveforms of Fig. 4-7 are idealized. As soon as the SCR of Fig. 4-6 fires, the voltage from anode to cathode drops almost to zero (actually 1 to 2 V for most SCRs). Since the gate voltage is derived from the anode to cathode voltage, it also drops virtually to zero, shutting off the gate current. Furthermore, since the gate is reverse biased when the ac supply is negative, there is really no negative gate current as shown in Fig. 4-7. In reality then, the i_G curve is a sine wave in phase with the supply voltage *only* in the region between 0° and the triggering point. At other times i_G is nearly zero.

One more point bears mentioning. Prior to triggering, the v_{AK} waveform is virtually identical to the ac supply waveform, because the voltage dropped across the load in Fig. 4-6 is negligible prior to triggering. The load voltage is so small because the load resistance in such circuits is much lower than the resistance in the gate control circuit. Load resistance is almost always less than 100 Ω and often less than 10 Ω. The fixed resistance in the gate control circuit is typically several thousand ohms. When these two resistances are tied together in series, as they are prior to triggering, the voltage across the small load resistance is naturally very low. This causes almost the entire supply voltage to appear across the SCR terminals.

Example 4-4

For Fig. 4-6, assume the supply is 115 V rms, $I_{GT} = 15$ mA, and $R_1 = 3$ K. The firing delay is desired to be 90°. To what value should R_2 be adjusted?

Solution. At 90°, the instantaneous supply voltage is

$$(115 \text{ V})(1.41) = 162 \text{ V}.$$

Neglecting the load voltage drop and the 0.6-V drop across the gate-cathode junction (they are both negligible compared to 162 V), the total resistance in the gate lead is given by

$$\frac{162 \text{ V}}{15 \text{ mA}} = 10.8 \text{ K}.$$

Therefore,

$$R_2 = 10.8 \text{ K} - 3 \text{ K} = \textbf{7.8 K}.$$

Example 4-5

In Fig. 4-6, if the resistance of the load is 40 Ω and the supply is 115 V rms (103.5 V_{avg})*, how much average power is burned in the SCR when the firing delay angle is 0°? Assume that the forward voltage across the SCR is constant at 1.5 V when it is turned ON and that reverse leakage current through the SCR is so small as to be negligible. (Reverse leakage current is less than 1 mA for most SCRs.)

Solution. Since the power burned in the SCR is zero during the negative half cycle (reverse leakage current is negligible), the overall average power is half the average power of the

*Recall that $V_{avg} = (.90) V_{rms}$

positive half cycle. The average power burned during the positive half cycle equals the forward voltage, V_T, multiplied by the average forward current during the positive half cycle ($I_{T_{avg}}$):

$$P_{(+\text{ half})} = (V_T)(I_{T_{avg}})$$

$$I_{T_{avg}} = \frac{103.5 \text{ V} - 1.5 \text{ V}}{40 \text{ }\Omega} = 2.55 \text{ A}$$

$$P_{(+\text{ half})} = (1.5 \text{ V})(2.55 \text{ A}) = 3.83 \text{ W}$$

$$P_{avg} = \tfrac{1}{2}(3.83 \text{ W}) = 1.91 \text{ W}$$

It can be seen from the above example that SCRs are very efficient devices. In Example 4-5 the SCR controlled a load current of several amps while wasting only about 2 W of power. This is much better than a series rheostat, for comparison.

The reason for the remarkable efficiency of SCRs is that when they are OFF, their current is very nearly zero, and when they are ON, their voltage is very low. In either case, the product of current and voltage is very small, resulting in low power dissipation.

It is this low power dissipation which enables the SCR to fit into a physically small package, making it economical. Economy and small size are the two most attractive features of SCRs.

Dc Supply Operation. Refer to Fig. 4-6 again; if the supply voltage is dc, the circuit operates as follows. When SW is closed, the SCR fires. The resistance in the gate lead would be designed so that this will occur. Once fired, the SCR will remain ON and the load will remain energized until the supply voltage is removed. The SCR stays ON even if SW is reopened, because it is not necessary to continue the flow of gate current to keep an SCR turned ON.

Although simple, this circuit is very useful in alarm applications. In an industrial alarm application, the SW contact could be closed when some malfunction occurs in an industrial process. As a burglar alarm, SW could be closed by the opening of a door window or by interruption of a light beam.

4-5 OTHER GATE CONTROL CIRCUITS

4-5-1 Capacitors used to delay firing

The simplest method of improving gate control is to add a capacitor at the bottom of the gate lead resistance, as shown in Fig. 4-8. The advantage of this circuit is that the firing delay angle can be adjusted past 90°. This can be understood by focusing on the voltage across capacitor C. When the ac supply is negative, the reverse voltage across the SCR is applied to the RC triggering circuit, charging the capacitor negative on the top plate and positive on the bottom plate. When the supply enters its positive half cycle, the forward voltage across the SCR tends to charge C in the opposite direction. However, voltage buildup in the new direction is delayed until the negative charge is removed from the capacitor plates. This delay in applying positive voltage at the gate can be extended past the 90° point. The larger the potentiometer resistance, the longer it takes to charge *C* positive on the top plate and the later the SCR fires.

Figure 4-8. An SCR gate control circuit which is an improvement on the circuit of Fig. 4-6. The capacitor provides a greater range of adjustment of the firing delay angle.

This idea can be extended by using either of the triggering circuits of Fig. 4-9. In Fig. 4-9(a), a resistor has been inserted into the gate lead, requiring the capacitor to charge higher than 0.6 V to trigger the SCR. With the resistor in place, capacitor voltage must reach a value large enough to force sufficient current (I_{GT}) through the resistor and into the gate terminal. Since *C* must now charge to a higher voltage, triggering is further delayed.

Figure 4-9. Improved SCR gate control circuits. Either one of these circuits provides a greater range of adjustment of the firing delay angle than the circuit of Fig. 4-8.

(a)   (b)

Figure 4-9(b) shows a double RC network for gate control. In this scheme, the delayed voltage across C_1 is used to charge C_2, resulting in even further delay in buildup of gate voltage. The capacitors in Fig. 4-9 usually fall in the range from 0.01 to 1 μF.

For given capacitor sizes, the minimum firing delay angle (maximum load current) is set by fixed resistors R_1 and R_3, and the maximum firing delay angle (minimum load current) is set mostly by the size of variable resistance R_2.

The manufacturers of SCRs provide detailed curves to help in sizing the resistors and capacitors for the gate control circuits of Fig. 4-9. In general terms, when these gate control circuits are used with a 60-Hz ac supply, the time constant of the RC circuit should fall in the range of 1–30 msec. That is, for the single RC circuit of Fig. 4-9(a), the product $(R_1 + R_2)C_1$ should fall in the range of 1×10^{-3} to 30×10^{-3}. For the double RC gate circuit of Fig. 4-9(b), $(R_1 + R_2)C_1$ should fall somewhere in that range, and R_3C_2 should also fall in that range.

This approximation method will always cause the firing behavior to be in the right ball park. The exact desired firing behavior can then be experimentally tuned in by varying these approximate component sizes.

Example 4-6

Suppose that it has been decided to use $C_1 = 0.068$ μF and $C_2 = 0.033$ μF in the gate control circuit of Fig. 4-9(b).

a. Approximate the sizes of R_1, R_2, and R_3 to give a wide range of firing adjustment.

b. If you then built the circuit and discovered that you could not adjust the firing delay angle less than 40°, what resistor would you experimentally change to allow adjustment below 40°?

Solution

a. The time constant $(R_1 + R_2)C_1$ should fall in the range of about 1×10^{-3} to 30×10^{-3}. To provide a wide range of adjustment, the time constant should be adjustable over a large part of that range. As an estimate, we might try for an adjustment range of 2×10^{-3} to 25×10^{-3}.

The minimum time constant occurs when R_2 is dialed out, so

$$(R_1 + 0)(0.068 \times 10^{-6}) = 2 \times 10^{-3},$$

$$R_1 = 29.4 \text{ K}.$$

The nearest standard size is **27 K**.

The maximum time constant (and maximum firing delay) occurs when R_2 is completely dialed in, so

$$(R_2 + 27 \times 10^3)(0.068 \times 10^{-6}) = 25 \times 10^{-3},$$

$$R_2 = 340 \text{ K}.$$

The nearest available pot size is **300 K**.

Experience has shown that the second time constant, R_3C_2, should fall somewhere toward the lower end of the suggested range. Let us assume 5 msec. Therefore

$$(R_3)(0.033 \times 10^{-6}) = 5 \times 10^{-3},$$

$$R_3 = \text{about } \mathbf{150 \text{ K}}.$$

b. Either R_1 or R_3 should be made smaller to allow lower firing delay angles, because the capacitors will charge faster with smaller resistors (smaller time constants). You would probably try R_3 first.

4-5-2 Using a breakover device in the gate lead

The circuits of Figs. 4-6, 4-8, and 4-9 all share two disadvantages:

a. Temperature dependence
b. Inconsistent firing behavior between SCRs of the same type

Regarding disadvantage **a**, an SCR tends to fire at a lower gate current when its temperature is higher (I_{GT} is lowered). Therefore, with any of the triggering circuits discussed so far, a change in temperature causes a change in firing angle and a consequent change in load current. In many industrial situations, this is unacceptable.

The second problem is that SCRs, like transistors, exhibit a wide spread in electrical characteristics within a batch. That is, two SCRs of a given type may show great differences in characteristics. The variation in I_{GT} is the most serious of these differences.

Figure 4-10 shows how these difficulties can be eliminated.

Figure 4-10. SCR gate control circuit using a four-layer diode (Shockley diode). The four-layer diode provides consistency of triggering behavior and reduces the temperature dependence of the circuit.

The *four-layer diode* in Fig. 4-10 has a certain breakover voltage. If the voltage across the capacitor is below that breakover point, the four-layer diode acts like an open switch. When the capacitor voltage rises to the breakover point, the four-layer diode fires and acts like a closed switch. This causes a burst of current into the gate, which provides sure triggering of the SCR.

The advantages of the four-layer diode are that it is relatively independent of temperature and that the breakover voltage can be held consistent from one unit to another. Therefore the imperfections of the SCR are of no importance, since it is the four-layer diode which determines the trigger point.

There are other devices which can be inserted in the gate lead to accomplish the same effect. They all have operating characteristics similar to those of the four-layer diode, and they are all temperature independent and have small spreads in breakover voltage. Some of the common triggering devices are the SUS (silicon unilateral switch), the SBS (silicon bilateral switch), the diac, and the UJT (unijunction transistor). All of these devices will be discussed in detail in Chapters 5 and 6.

4-6 ALTERNATIVE METHODS OF CONNECTING SCRS TO LOADS

4-6-1 Unidirectional full-wave control

Figure 4-11(a) shows how two SCRs can be combined with a center-tapped transformer to accomplish full-wave power control. This circuit behaves much like a full-wave rectifier for a dc power supply. When the secondary winding is in its positive half cycle, positive on top and negative on bottom, SCR_1 can fire. This connects the load across the top half of the transformer secondary. When the secondary winding is in its negative half cycle, SCR_2 can fire, connecting the load across the bottom half of the secondary winding. The current through the load always flows in the same direction, just as in a full-wave dc power supply. Figure 4-11(b) shows waveforms of load voltage and ac line voltage for a firing delay angle of about 45°.

Figure 4-11(a) indicates two separate trigger circuits for the two SCRs. Often these two circuits can be combined into a single circuit designed around one of the triggering devices mentioned in Sec. 4-5. Such a design ensures that the firing delay angle is identical for both half cycles.

4-6-2 Bidirectional full-wave control

Another common SCR configuration is shown in Fig. 4-12(a). In this circuit, SCR_1 can fire during the positive half cycle and SCR_2 during the negative half cycle. The current through the load is not unidirectional. Figure 4-12(c) shows a waveform of load voltage for a firing delay angle of about 120°. Figure 4-12(b) shows the same circuit redrawn in a more popular manner.

4-6-3 Bridge circuits containing an SCR

A single SCR can control both alternations of an ac supply when connected as shown in Fig. 4-13(a). When the ac line is in its positive half cycle, diodes A and C are forward biased. When the SCR fires, the line voltage is applied to the load. When the ac line is in its negative half cycle, diodes B and D are forward biased.

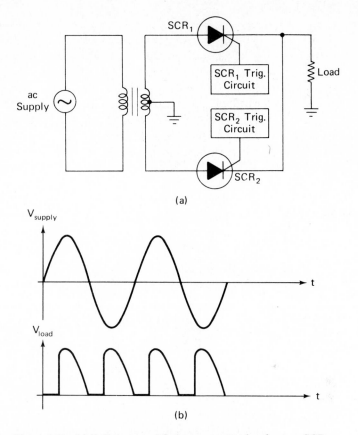

(a)

(b)

Figure 4-11. (a) Full-wave rectified power control, using two SCRs and a center-tapped winding. (b) Supply voltage and load voltage waveforms. Both ac half cycles are being used to deliver power, but the load voltage has only one polarity (it is rectified).

Again the ac line voltage is applied to the load when the SCR fires. The load waveform would look like the waveform shown in Fig. 4-12(c).

Figure 4-13(b) shows a bridge rectifier controlled by a single SCR, this time with the load wired in series with the SCR itself. The load current is unidirectional, with a waveform like that illustrated in Fig. 4-11(b).

4-7 SCRs IN DC CIRCUITS

When an SCR is used in a dc circuit, the automatic turn-OFF does not occur, because, of course, the supply voltage doesn't pass through zero. In this situation, some other means must be used to stop the SCR main terminal current (reduce it below I_{HO}). One obvious method is to simply disconnect the dc supply. In most cases this is impractical.

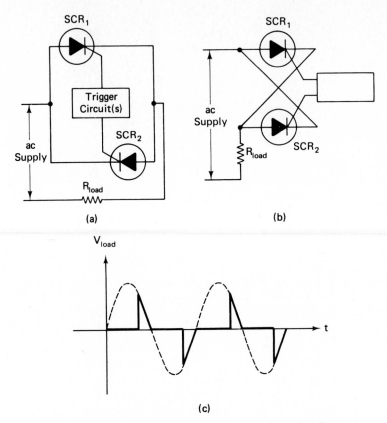

Figure 4-12. (a) Full-wave unrectified power control, using two SCRs. (b) The same circuit drawn another way. (c) Load voltage waveform. Both ac half cycles are being used to deliver power, and the load voltage is unrectified.

Often, main terminal current is stopped by connecting a temporary short circuit from the anode to the cathode. This is illustrated in Fig. 4-14(a), in which a transistor switch is connected across the SCR. When the SCR is to be turned OFF, the trigger circuit pulses the transistor, driving it into saturation. The load current is temporarily shunted into the transistor, causing the SCR main terminal current to drop below I_{HO}. The transistor is held ON just long enough to turn OFF the SCR. This normally takes a few milliseconds for a medium-sized SCR. The trigger circuit then removes the base current, shutting the transistor OFF before it can be destroyed by the large load current.

In this arrangement the trigger circuit is responsible for both turning ON and turning OFF the SCR. Such trigger circuits are naturally more complex than those discussed in Sec. 4-5, which were responsible only for turn-ON.

More reliable turn-OFF can be accomplished by actually reverse-biasing the SCR. This is shown in Fig. 4-14(b). In this circuit the capacitor charges with the

(a)

(b)

Figure 4-13. A full-wave bridge combined with an SCR to control both halves of the ac line. (a) With the load inserted in one of the ac lines leading to the bridge, the load voltage is unrectified, as in Fig. 4-12(b). (b) With the load inserted in series with the SCR itself, the load voltage is rectified, as in Fig. 4-11(b).

polarity indicated when the SCR is turned ON. To turn OFF, the trigger circuit again saturates the transistor, which effectively places the capacitor in parallel with the SCR. Since the capacitor voltage cannot change instantly, it applies a temporary reverse voltage across the SCR, shutting it OFF.

Example 4-7

In Fig. 4-14(a), imagine that the dc supply is 48 V and the trigger circuit behaves as follows:

1) It delivers a turn-ON pulse to the gate of the SCR;

2) 6.0 msec later it delivers a pulse to the base of the transistor;

3) It repeats this cycle at a frequency of 125 Hz.

a. Describe the load waveform. Neglect V_T.

b. If the load resistance is 12 Ω, how much average power is delivered to the load?

(a)   (b)

Figure 4-14. SCR commutation circuits. (a) The transistor switch shorts out the SCR, thereby turning it OFF. (b) The transistor switch puts a charged capacitor in parallel with the SCR for reverse-bias turn-OFF.

Solution

a. For a cycle frequency of 125 Hz, the period is

$$T = \frac{1}{f} = \frac{1}{125\ \text{Hz}} = 8\ \text{msec},$$

so the load waveform would be a rectangular wave, 48 volts tall, spending 6 msec up (at 48 V) and 2 msec down (at 0 V).

b.
$$P_{\text{ON state}} = \frac{V_{\text{load}}^2}{R_{\text{load}}} = \frac{48^2}{12} = 192\ \text{W}$$

$P_{\text{avg}} = (0.75)(P_{\text{ON state}})$, because the SCR is ON for 75% of the total cycle time. Therefore,

$$P_{\text{avg}} = (0.75)(192\ \text{W}) = \textbf{144 W}.$$

QUESTIONS AND PROBLEMS

1. The letters SCR stand for silicon controlled rectifier. Explain the use of the word *rectifier* in the name.
2. What two things must happen to cause an SCR to fire?
3. In words, explain what each of the following symbols stands for:
   a. I_{GT}
   b. $I_{T_{rms}}$
   c. I_{HO}
   d. V_T
4. What two important benefits arise from using breakover-type devices to trigger SCRs?
5. Name some of the common breakover-type devices.
6. Roughly speaking, how much gate current is needed to trigger a medium-power SCR?
7. Roughly speaking, how much voltage appears across the anode-cathode terminals of a medium-power SCR after it has fired?

8. Describe the methods used to turn OFF SCRs in dc circuits.

9. In Project 4-1, the instructions say that the oscilloscope chassis must be isolated from the earth if the ac supply is not isolated from the earth. Using drawings, explain carefully why this is necessary.

10. Explain why an SCR is superior to a series rheostat for controlling and limiting current through a load.

11. What effect does an increase in anode current have on V_T? Specifically, if anode-cathode current is doubled, does V_T also double?

12. After an SCR has fired, what effect does the gate signal have on the SCR?

13. In Fig. 4-6, the supply is 115 V rms, 60 Hz. The SCR has an I_{GT} of 35 mA; $R_1 = 1$ K; what value of R_2 will cause a firing delay of 90°?

14. If R_2 is set to 2.5 K in Question 13, what will be the firing delay angle? What is the conduction angle?

15. The gate control circuit of Fig. 4-9(a) is used with a switched dc voltage supply of 60 V. The low-resistance load is connected as shown in Fig. 4-8. $R_1 = 1$ K, $R_2 = 2.5$ K, $R_3 = 1$ K, and $C = 0.5$ μF. The I_{GT} of the SCR is 10 mA. If the dc supply is suddenly switched ON, how much time will elapse before the SCR fires? *Hint:* Use the universal time constant curve in Chapter 3, and use Thevenin's Theorem.

16. For the circuit of Question 15, what value of C will cause a time delay of 70 msec between closing the switch and firing the SCR?

17. For the circuit of Fig. 4-8, the supply is 220 V rms, 60 Hz. The load resistance is 16 Ω. Neglect the V_T of the SCR.
    a. How much power is delivered to the load if the firing delay angle = 0°?
    b. How much if the firing delay angle = 90°?
    c. If the firing delay angle = 135°, will the load power be less than one half or more than one half of the amount delivered for a 90° delay angle? Explain.

18. For Fig. 4-9(a), $C = 0.47$ μF. Find approximate sizes of R_1 and R_2 to give a wide range of adjustment of firing delay angle.

19. In Fig. 4-9(a), if $R_1 = 4.7$ K and $R_2 = 100$ K, choose an approximate size of C which will permit the firing delay angle to be adjusted very late.

SUGGESTED LABORATORY PROJECTS

PROJECT 4-1: AN SCR POWER CONTROL CIRCUIT

Purpose

a. To observe the operation and waveforms of an SCR driving a resistive load
b. To determine the electrical characteristics of a particular SCR
c. To observe temperature and batch stabilization by using a breakover device (a four-layer diode)

Procedure

Construct the gate control circuit of Fig. 4-9(a), with $R_1 = 47$ K, $R_2 = 500$-K pot, $R_3 = 1$ K, and $C = 0.1$ μF. The load resistor and ac supply are wired as shown in Fig. 4-8. The ac supply should be 115 V ac, *isolated from earth ground.*

If it is impossible to obtain 115 V ac isolated from earth ground, then the oscilloscope can be isolated from earth ground by a two-prong adapter. Special care is then necessary, since the oscilloscope chassis will be "hot" during parts of this laboratory project. That is, the chassis will have a large potential difference relative to earth ground and must not be touched. Many people consider this to be an unacceptable practice because it is too dangerous. The problem can also be solved by other methods: 1) Use a step down transformer, about 48 V secondary voltage, to isolate the circuit from earth ground, and operate the entire circuit at reduced voltage. 2) Check the polarization of the ac line cord to make sure the SCR cathode is connected to earth ground. Then, using a differential scope, connect the scope ground permanently to the SCR cathode and use the differential input to measure the load voltage and the gate resistor voltage.

Use a sensitive-gate medium SCR, T1C106B or equivalent. The load should be a 100 Ω 100-W power resistor or a 100-W light bulb. Insert a 0–1-A ammeter in series with the load.

a. Place the oscilloscope across the load resistor.
   1. Measure and record the minimum firing delay angle and the maximum firing delay angle.
   2. Record the average load current under both conditions. Does this agree with your understanding of the relationship between firing delay angle and load current?
   3. In which direction must you turn the 500-K pot to increase the firing delay angle? Explain why this is so.
   4. Draw the load waveform for some intermediate firing delay angle.

b. Without disturbing the potentiometer setting of part 4 above, connect the oscilloscope across the anode to cathode of the SCR.
   1. Draw the SCR voltage waveform for the same intermediate firing delay angle as before.
   2. Compare the SCR voltage waveform to the load voltage waveform. Does the comparison make sense?
   3. Measure the voltage which exists across the SCR after firing (V_T). Is it fairly constant? Is it about as large as you expected?

c. Place the oscilloscope across the 1-K gate resistor. The current flowing into the gate terminal can now be found by using Ohm's law for the 1-K resistor.
   1. Measure the gate current necessary to fire the SCR (I_{GT}). How much does it change as the firing delay angle is changed?

d. Place the oscilloscope across the load and adjust to some intermediate firing delay angle. Heat the SCR and observe the reaction of the firing delay angle. A soldering iron or lighted match held against the case of a plastic SCR for a few seconds will heat it sufficiently. Do not touch the case of a metal SCR with a grounded soldering iron. What effect does increased temperature have on an SCR circuit?

e. Install a four-layer diode (IN5793 or equivalent) in series with the 1-K gate resistor. Repeat steps a and d. What important difference do you notice?

f. If several four-layer diodes of the same type are available, substitute different ones and repeat step e. What can you conclude about the batch spread between four-layer diodes?

PROJECT 4-2: SCR CONTROL WITH
A DOUBLE RC GATE
TRIGGER CIRCUIT

Purpose

a. To observe larger firing delay angles possible with a double RC gate control circuit

b. To observe the nonsinuisoidal waveforms that occur when an SCR drives a motor or other inductive load

Procedure

Construct the gate control circuit of Fig. 4-9(b). The load and ac supply are wired as shown in Fig. 4-8. Again, the 115-V ac source should be isolated from the earth, but if that is not possible, follow the suggestions given in Project 4-1.

Use the following component sizes: $R_1 = 4.7$ K, $R_2 = 100$-K pot, $R_3 = 10$ K, $C_1 = 0.5$ μF, $C_2 = 0.05$ μF. Place a rectifier diode in the gate lead, along with a 1-K resistor to protect the gate and limit the gate current. The SCR should be a 200-V, medium current SCR, such as a C106B. As a load, use a small universal dc motor, such as a $\frac{1}{4}$-hp drill motor.

Observe the load voltage waveform by connecting the oscilloscope across the motor terminals. Try to explain why the SCR does not turn OFF exactly when the ac line passes through zero going into its negative region.

PROJECT 4-3: AN SCR ZERO-POINT
(SOFT-START) SWITCH

Zero-point switching is the technique of always switching an SCR ON at the instant when the ac supply voltage is zero. It is desirable for two reasons. (a) It prevents the large inrush of current which occurs when a rather high voltage is suddenly applied to a very low-resistance load. It therefore prevents thermal shock to the load. (b) It eliminates electromagnetic interference which results from the sudden surges of inrush current.

Figure 4-15 shows a zero-point switching circuit. The average load current is controlled by the duty cycle (pulse width) of the rectangular wave from the pulse generator.

Observe the V_{load} waveform on an oscilloscope. If a dual-trace scope is available, display both V_{load} and the output wave of the pulse generator on the screen at the same time.

Note that the load voltage always appears in *complete* half cycles but that the number of half cycles spent ON versus the number spent OFF can be varied. This is the essence of zero-point power control.

Bring an inexpensive AM radio up close to the zero-point control circuit. Do you hear any electromagnetic interference on the radio? Repeat this test for any of the circuits built in Project 4-1 or 4-2. Comment on the difference.

Can you explain how this circuit works? *Hint:* The 0.22-μF capacitor charges during the negative half cycle. The capacitor is then the energy source for firing SCR_2 as the ac line passes through zero going positive.

151

Figure 4-15. A zero-point switching power control circuit. Load power is controlled by the variable pulse width.

5
UJTs

The unijunction transistor (UJT) is a breakover-type switching device. Its characteristics make it very useful in many industrial circuits, including timers, oscillators, waveform generators, and, most importantly, gate control circuits for SCRs and triacs. In this chapter we will present the operating characteristics and theory of UJTs and examples of how they are used in such circuits. A more extensive description of UJTs used as gate trigger devices for triacs is given in Chapter 6.

OBJECTIVES

After completing this chapter, you will be able to:

1. Interpret the voltage-current characteristic curve of a UJT and identify the peak voltage, peak current, valley voltage, and valley current
2. Relate the UJT variables of peak voltage (V_P), intrinsic standoff ratio (η), and interbase voltage (V_{B1B2}), and calculate any one of these, given the other two

3. Explain the operation of UJT relaxation oscillators and UJT timers, and properly size the timing resistors and capacitors in these circuits

4. Explain the problem of UJT latch-up, why it occurs, and how to avoid it

5. Explain the operation of a line-synchronized UJT trigger circuit for an SCR, and properly size the timing and stabilizing components

6. Explain in detail the operation of a sequential load switching circuit using UJTs

7. Explain the operation of a solid-state logic output amplifier built with an SCR triggered by a UJT

5-1 THEORY AND OPERATION OF UJTS

5-1-1 Firing a UJT

The UJT is a three-terminal device, the three terminals being labeled emitter, base 1, and base 2. The schematic symbol and terminal locations are shown in Fig. 5-1(a). It is not a good idea to try to mentally relate the terminal names of the UJT to the terminal names of the common bipolar transistor. From a circuit operation viewpoint, there is no resemblance between the emitter of a UJT and the emitter of a bipolar transistor. The same is true for the relationship between the UJT base terminals and the bipolar transistor base terminal. To be sure, these terminal names make sense from an internal viewpoint which considers the action of the charge carriers, but internal charge carrier action is not an important concern to us.

In simplest terms, the UJT operates as follows. Refer to Fig. 5-1(b).

Figure 5-1. (a) Schematic symbol and terminal names of a UJT. (b) A UJT connected into a simple circuit. This drawing shows the emitter current (I_E), the emitter to base 1 voltage (V_{EB1}), and the base 2 to base 1 voltage (V_{B2B1}).

a. When the voltage between emitter and base 1, V_{EB1}, is less than a certain value called the *peak voltage*, V_P, the UJT is turned OFF, and no current can flow from E to $B1$ ($I_E = 0$).

b. When V_{EB1} exceeds V_P by a small amount, the UJT fires or turns ON. When this happens, the E to $B1$ circuit becomes almost a short circuit, and current can surge from one terminal to another. In virtually all UJT circuits, the burst of current from E to $B1$ is short-lived, and the UJT quickly reverts back to the OFF condition.

As Fig. 5-1(b) shows, an external dc voltage is applied between $B2$ and $B1$, with $B2$ the more positive terminal. The voltage between the two base terminals is symbolized V_{B1B2}, as indicated. For a given type of UJT, the peak voltage V_P is a certain fixed percent of V_{B2B1}, plus 0.6 V. That fixed percent is called the *intrinsic standoff ratio*, or just the *standoff ratio*, of the UJT, and is symbolized η.

Therefore the peak voltage of a UJT can be written as

$$V_P = \eta V_{B2B1} + 0.6\,\text{V}, \tag{5-1}$$

where 0.6 V is the forward voltage drop across the silicon *pn* junction which exists between the emitter and base 1.

Example 5-1

If the UJT in Fig. 5-1(b) has a standoff ratio of $\eta = 0.55$ and an externally applied V_{B2B1} of 20 V, what is the peak voltage?

Solution. From Eq. (5-1),

$$V_P = 0.55(20\,\text{V}) + 0.6\,\text{V} = \textbf{11.6 V.}$$

In this case V_{EB1} would have to exceed 11.6 V in order to fire the UJT.

Refer again to the circuit of Fig. 5-1(b). The capacitor would begin charging through resistor R_E at the instant the switch was closed. Since the capacitor is connected directly between E and $B1$, when the capacitor voltage reaches 11.6 V the UJT will fire (assuming $\eta = 0.55$ as in Example 5-1). This will allow the charge built up on the plates of C_E to discharge very quickly through the UJT. In most UJT applications, this burst of current from E to $B1$ represents the output of the circuit. The burst of current can be used to trigger a thyristor,* or turn ON a transistor, or simply to develop a voltage across a resistor inserted in the base 1 lead.

**5-1-2 Voltage-current characteristic
          curves of a UJT**

There is a certain internal resistance existing between the two base terminals $B2$ and $B1$. This resistance is about 5–10 K ohms for most UJTs and is shown as r_{BB} in Fig. 5-2(a). In the physical structure of a UJT, the emitter lead contacts the main body of the UJT somewhere between the $B2$ terminal and the $B1$ terminal.

---

*A generic term which includes *both* SCRs and triacs.

Thus a natural voltage divider is created, because r_{BB} is divided into two parts, r_{B2} and r_{B1}. This construction is suggested by the equivalent circuit in Fig. 5-2(a). The diode in this figure indicates the fact that the emitter is *p*-type material, while the main body of a UJT is *n*-type material. Therefore a *pn* junction is formed between the emitter lead and the body of the UJT.

Figure 5-2. (a) Equivalent circuit of a UJT. The total resistance between *B*2 and *B*1 is called r_{BB}. It is divided into two parts, r_{B2} and r_{B1}. The emitter is connected through a diode to the junction of r_{B2} and r_{B1}. (b) Voltage vs. current characteristic curve of a UJT (V_{EB1} vs. I_E). The four important points on this curve are called peak voltage (V_P), peak current (I_P), valley voltage (V_V), and valley current (I_V).

(a)

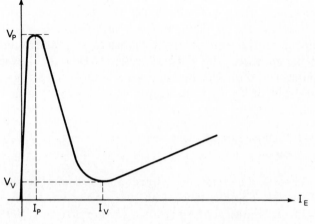

(b)

The total applied voltage, V_{B2B1}, is divided between the two internal resistances r_{B2} and r_{B1}. The portion of the voltage which appears across r_{B1} is given by

$$V_{r_{B1}} = \frac{r_{B1}}{r_{B1} + r_{B2}} V_{B2B1},$$

which is simply the equation for a series voltage divider, applied to the circuit of Fig. 5-2(a).

To fire the UJT, the E to $B1$ voltage must become large enough to forward-bias the diode in Fig. 5-2(a) and bleed a small amount of current into the emitter lead. The V_{EB1} required to accomplish this must equal the sum of the diode forward voltage plus the voltage drop across r_{B1}, or

$$V_{EB1} = V_D + \frac{r_{B1}}{r_{B1} + r_{B2}} V_{B2B1},$$

in order to fire the UJT. Comparing this to Eq. (5-1) shows that the standoff ratio is just the ratio of r_{B1} to the total internal resistance, or

$$\eta = \frac{r_{B1}}{r_{B1} + r_{B2}} = \frac{r_{B1}}{r_{BB}}. \tag{5-2}$$

The total internal resistance r_{BB} is called the *interbase resistance*.

Example 5-2
a. If the UJT in Fig. 5-1(b) has an r_{B1} of 6.2 K and an r_{B2} of 2.2 K, what is its standoff ratio?
b. How large is the peak voltage?

Solution
a. From Eq. (5-2),

$$\eta = \frac{r_{B1}}{r_{B1} + r_{B2}} = \frac{6.2 \text{ K}}{6.2 \text{ K} + 2.2 \text{ K}} = \textbf{0.74}.$$

b. From Eq. (5-1),

$$V_P = (0.74)(20 \text{ V}) + 0.6 \text{ V} = \textbf{15.4 V}.$$

The mechanism by which the UJT fires is suggested in Fig. 5-2(b). When the emitter to base 1 voltage rises to the peak voltage V_P and a small emitter current starts to flow, the UJT "breaks back" to a smaller voltage between the emitter and base 1 terminals. This smaller voltage is called the *valley voltage* and is symbolized V_V in Fig. 5-2(b). This breaking back occurs because of a drastic increase in the number of charge carriers available in the $B1$ region when emitter current starts to trickle into the main body of the device. From the external viewpoint, it appears as though r_{B1} drops to almost zero ohms in a very short time.

It is convenient to think of r_{B1} as a resistance whose value varies drastically, from its original OFF-state value all the way to nearly zero ohms. The resistance of r_{B2}, on the other hand, is fixed at its original OFF-state value. When r_{B1} drops to nearly zero ohms, the emitter to base 1 circuit allows an external capacitor to dump its charge through the device. Because r_{B2} maintains its original high resistance at this time, there is no unmanageable surge of current out of the dc power supply from $B2$ to $B1$.

The external capacitor quickly discharges to the point where it can no longer deliver the minimum current required to keep the UJT turned ON. This minimum required current is called the *valley current* and is symbolized I_V, as shown in Fig. 5-2(b). When the current flow from emitter to base 1 declines to slightly less than the valley current, the UJT reverts to the OFF state. Once it has returned to the OFF state, no current flows from E to $B1$, and V_{EB1} must again climb to V_P in order to fire the device a second time.

5-2 UJT RELAXATION OSCILLATORS

The relaxation oscillator is the heart of most UJT timer and oscillator circuits. It is essentially the same circuit as shown in Fig. 5-1(b), except that resistors are added to the $B1$ and $B2$ leads in order to develop output signals. These external resistors are rather small compared to the internal resistance of the UJT, r_{BB}. The external resistors are usually symbolized R_2 and R_1. Typical component sizes for a relaxation oscillator are given in Fig. 5-3(a).

The oscillator works on the principles discussed in Sec. 5-1. When power is applied, C_E charges through R_E until the capacitor voltage reaches V_P. At this point, the UJT will fire, as long as R_E is not too large. The limitation on R_E occurs because a certain minimum amount of current must be delivered from the dc power supply into the emitter to successfully fire the UJT, even given that V_P is reached. Since this current must arrive at the emitter terminal by way of R_E, the resistance of R_E must be small enough to permit the necessary current to flow. This minimum current is called *peak point current* or *peak current*, symbolized I_P, and is just a few microamps for most UJTs. I_P is shown graphically on the characteristic curve of Fig. 5-2(b).

The equation which given the maximum allowable value of R_E is easily obtained by applying Ohm's law to the emitter circuit.

$$R_{E_{max}} = \frac{V_s - V_P}{I_P}. \tag{5-3}$$

In Eq. (5-3), V_s represents the dc source voltage. The quantity $V_s - V_P$ is the voltage available across R_E at the instant of firing.

When the UJT fires, the internal resistance r_{B1} drops to nearly zero, allowing a pulse of current to flow from the top plate of C_E into R_1. This causes a voltage spike to appear at the $B1$ terminal, as shown in Fig. 5-3(b). At the same time the positive spike appears at $B1$, a negative-going spike occurs at $B2$. This happens because the sudden drop in r_{B1} causes a sudden reduction in total resistance between V_s and ground and a consequent increase in current through R_2. This increase in current causes an increased voltage drop across R_2, creating a negative-going spike at the $B2$ terminal, as illustrated in Fig. 5-3(c).

Figure 5-3. (a) Schematic drawing of a relaxation oscillator. For a given UJT (given η), the oscillation frequency depends on R_E and C_E. (b) Base 1 to ground voltage (V_{B1}) waveform for the relaxation oscillator. (c) Base 2 to ground voltage (V_{B2}) waveform. (d) Emitter to ground voltage (V_E) waveform.

At the emitter terminal, a sawtooth-shaped wave occurs, shown in Fig. 5-3(d). The sawtooth is not linear on its run-up, because the capacitor does not charge at a constant rate. Also, the bottom of the waveform is not exactly zero volts. There are two reasons for this:

a. The emitter to base 1 voltage never reaches 0 V, only V_V, as Fig. 5-2(b) indicates.
b. There is always some voltage drop across R_1 due to the current flow through the main body of the UJT. That is, there is always a complete circuit by which current can flow out of the dc supply terminal, through R_2, through the body of the UJT, through R_1 and into ground.

We saw earlier that in a relaxation oscillator R_E must not be too large, or the UJT will not be able to fire. Likewise, there is a limit as to how *small* R_E may be, to be assured of turning the UJT back OFF after firing. Recall that the reason a UJT turns back OFF is that the capacitor C_E discharges to the point where it cannot deliver an emitter current equal to I_V, the valley current [see Fig. 5-2(b)]. The implication here is that the UJT must not be able to draw enough emitter current through R_E either. Therefore R_E must be large enough to prevent the passage of current equal to I_V. The equation which expresses the minimum value for R_E is

$$R_{E_{min}} = \frac{V_s - V_V}{I_V}, \tag{5-4}$$

which is just Ohm's law applied to the emitter resistor. The quantity $V_s - V_V$ is the approximate voltage across R_E after firing. This is true because upon firing, the emitter to ground voltage drops to about V_V (neglecting the small voltage across R_1).

The frequency of oscillation of a relaxation oscillator of the type shown in Fig. 5-3(a) is given very approximately by

$$f = \frac{1}{T} = \frac{1}{R_E C_E}. \tag{5-5}$$

Equation (5-5) is fairly accurate as long as the UJT has an η in the neighborhood of 0.63, which is usually the case. As the η strays above or below 0.63, Eq. (5-5) becomes less accurate.

An intuitive feel for Eq. (5-5) can be gained by remembering that an RC circuit charges through 63% of its total voltage change in one time constant. If $\eta = 0.63$, C_E must charge to about 63% of V_s in order to fire the UJT. This requires a charging time of one time constant, or, in other words,

$$t_{charge} = R_E C_E. \tag{5-6}$$

Since the firing and subsequent turn-OFF are both very fast compared to the charging time, the total period of the oscillations is about equal to $R_E C_E$. Frequency equals the reciprocal of period, so Eq. (5-5) is valid.

A UJT's standoff ratio is fairly stable as temperature changes, varying less than 10% over an operating temperature range of $-50°C$ to $+125°C$ for a high-quality UJT. Relaxation oscillators can be made frequency stable to within 1% over the same temperature range by properly adjusting R_2 in Fig. 5-3(a). The standoff ratio tends to *decrease* with increasing temperature, whereas the total internal resistance, r_{BB}, tends to *increase* with increasing temperature. External resistor R_2 is constant as temperature changes, so the voltage between the base terminals, V_{B2B1}, increases with increasing temperature since r_{BB} becomes a larger portion of the total resistance from V_s to ground. Therefore V_{B2B1} gets larger as η gets smaller. These effects can be made to just cancel each other if R_2 is properly chosen. Under these circumstances V_P is held constant. If V_P is constant, oscillation frequency is also constant, because C_E will always have to charge to the same voltage to fire the UJT, no matter what the temperature.

The batch stability, or variation between UJTs of the same type number, is not so good as the temperature stability. Two supposedly identical UJTs may have standoff ratios which differ by 30% or more. For this reason, UJT relaxation oscillators contain some sort of trim adjustment if a precise oscillation frequency is desired. This is easily done by inserting a potentiometer in series with R_E.

Example 5-3

Refer to the relaxation oscillator shown in Fig. 5-3. Assume that the UJT has the following characteristics:

$$\eta = 0.63, \qquad r_{BB} = 9.2 \text{ K}, \qquad V_V = 1.5 \text{ V},$$
$$r_{B1} = 5.8 \text{ K}, \qquad I_P = 5 \text{ } \mu A,$$
$$r_{B2} = 3.4 \text{ K}, \qquad I_V = 3.5 \text{ mA}.$$

a. Find V_P.

b. What is the approximate output frequency?

c. Prove that a 10-K R_E is within the acceptable range. That is, $R_{E\text{min}} < R_E < R_{E\text{max}}$.

d. Describe the waveform that appears across R_1. How tall are the spikes? What voltage appears across R_1 during the time that the UJT is OFF?

Solution

a. From Eq. (5-1),
$$V_P = (0.63)(V_{B2B1}) + 0.6 \text{ V}.$$

The voltage from base 2 to base 1 can be found by the proportion

$$\frac{V_{B2B1}}{V_s} = \frac{r_{BB}}{R_{\text{total}}} = \frac{r_{BB}}{R_2 + r_{BB} + R_1},$$

$$\frac{V_{B2B1}}{24 \text{ V}} = \frac{9200 \text{ } \Omega}{470 \text{ } \Omega + 9200 \text{ } \Omega + 100 \text{ } \Omega},$$

$$V_{B2B1} = 22.6 \text{ V}.$$

Therefore,
$$V_P = (0.63)(22.6 \text{ V}) + 0.6 \text{ V} = \textbf{14.8 V}.$$

b. Since $\eta = 0.63$, Eq. (5-5) will predict the oscillator's output frequency quite accurately:

$$f = \frac{1}{(10 \text{ K})(0.2 \text{ } \mu F)} = \frac{1}{2 \times 10^{-3}} = \textbf{500 Hz}.$$

c. From Eq. (5-3),

$$R_{E\text{max}} = \frac{V_s - V_P}{I_P} = \frac{24 \text{ V} - 14.8 \text{ V}}{5 \text{ } \mu A} = \textbf{1.84 M}.$$

From Eq. (5-4),

$$R_{E_{\min}} = \frac{V_s - V_P}{I_v} = \frac{24 \text{ V} - 1.5 \text{ V}}{3.5 \text{ mA}} = 6.4 \text{ K}.$$

The actual value of R_E, 10 K, is between 6.4 K and 1.84 M, so it is acceptable. It will allow enough emitter current to flow to fire the UJT but not enough to prevent it from turning back OFF.

d. The peak value of the spikes across R_1 is given approximately by

$$V_{R_1} = V_P - V_V = 14.8 \text{ V} - 1.5 \text{ V} = 13.3 \text{ V}.$$

This equation is valid because the capacitor voltage always equals the voltage from emitter to base 1 plus the voltage across R_1. At the instant of firing, the capacitor voltage equals V_P, and the emitter to base 1 voltage is approximately equal to V_v. Naturally the peak value of V_{R_1} occurs at the instant the UJT fires, so it can be calculated as shown in the above equation.

The voltage level to which V_{R_1} returns when the UJT is OFF can be calculated by the series circuit voltage division formula:

$$\frac{V_{R_1}}{R_1} = \frac{V_s}{R_{\text{total}}},$$

$$\frac{V_{R_1}}{100 \text{ } \Omega} = \frac{24 \text{ V}}{470 \text{ } \Omega + 9200 \text{ } \Omega + 100 \text{ } \Omega},$$

$$V_{R_1} = 0.25 \text{ V}.$$

The V_{R_1} waveform could therefore be described as a rest voltage of 0.25 V with fast spikes rising to 13.3 V, occurring at a frequency of 500 Hz.

5-3 UJT TIMING CIRCUITS

5-3-1 UJT relay timer

An example of a UJT timing circuit to provide the time delay in picking a relay is shown in Fig. 5-4. In this circuit, power is applied to the load when relay CR picks up. This will occur a certain time (adjustable) after SW1 is closed. The time delay is adjusted by adjusting R_{E_v}. The circuit works as follows.

When SW1 is closed and 24 V is applied to the top of R_3, a small amount of current starts to flow into the CR relay coil. R_3 is sized so that this current is not large enough to *pick* the coil, but it is large enough to keep the coil energized once it has already been picked up. This is possible because the holding current for a relay coil is usually only about one half of the pickup current. That is, a relay coil which requires a current of 0.5 A to actually move the armature and switch the contacts might require only 0.25 A to *maintain* contact closure.

The 20-μF capacitor C_E charges through R_{E_f} and the 1-M pot R_{E_v}, at a rate determined by the setting of R_{E_v}. When C_E reaches a high enough voltage, the UJT fires, dumping the capacitor charge into relay coil CR. This is sufficient to energize the coil, picking CR. The current pulse into the coil ceases almost immediately, but now the current through R_3 is sufficient to hold the relay energized.

Figure 5-4. A UJT timing circuit. Relay *CR* energizes a certain time after the switch is closed. The time delay can be varied by potentiometer R_{E_v}.

The N.O. *CR* contact closes and applies power to the load. The time delay is given by Eq. (5-6):

$$t = (R_{E_f} + R_{E_v})C_E.$$

5-3-2 An improved one-shot using a UJT

We have encountered one-shots in Sec. 3-10, and we have seen some of their uses in industrial digital circuitry. A method of constructing a one-shot was presented in Fig. 3-18. This design is adequate in most one-shot applications, but it does have two shortcomings:

a. When the output pulse is complete, the one-shot is not ready to be fired again immediately. It has a certain nonzero *recovery time*. Recovery time is the time that must elapse between the finish of an output pulse and the arrival of the next trigger input pulse.
b. Long firing times are difficult to achieve with this design. Output pulses longer than a few seconds cannot be accomplished.

Refer back to Fig. 3-18 and let us see why these problems exist. Here is the reason for problem **a**.

At the instant the output pulse is finished, the voltage across *C* is close to zero. Actually it is about 0.6 V, positive on the right, just enough to overcome the base-emitter junction of T_2. At this instant, T_2 turns ON and T_1 turns OFF. When this happens, *C* starts charging through R_C, through the base-emitter junction of T_2, to ground. Until the capacitor can charge fully, the one-shot is not ready to fire again. That is, C must charge to a voltage equal to $V_s - 0.6$ V, positive on the left, before the one-shot can be triggered again. If a trigger pulse arrives at the one-shot before the capacitor has charged fully, the resulting output pulse will be too short.

To fully charge C, a time equal to five charging time constants must elapse. Therefore recovery time is given by

$$t_{rec} = 5(R_{C_1})C.$$

Now here is the reason for problem **b**. The output pulse duration (firing time) equals the time it takes to *discharge* C when T_1 turns ON. The discharge path is down from V_s, through R_{B_2}, through C, through T_1, to ground. When C is discharged to 0 V and charged just slightly in the opposite direction (about 0.6 V as mentioned) it turns ON T_2. Turning ON T_2 drives the one-shot back into its stable state and terminates the output pulse. Therefore the size of C and the size of R_{B_2} determine the duration of the output pulse.

To obtain long-duration output pulses, either C or R_{B_2} or both must be made large. However, we saw above that the larger C is made, the longer the recovery time becomes. Therefore C must be held to a reasonably small value. As far as R_{B_2} is concerned, it cannot be made very large either because it may prevent T_2 from saturating. To pass sufficient base current to saturate T_2, R_{B_2} must be held to a reasonably small value. Since C and R_{B_2} must both be held small, it is impossible to obtain long firing times.

These two problems can be eliminated by using the improved one-shot shown in Fig. 5-5, which contains a UJT.

Here is how it works. In the rest state, T_2 is ON and T_1 is held OFF. The reason T_2 is ON instead of T_1 is that R_{B_2} is lower than R_{B_1} (10 K compared to 56 K). This ensures that T_2 turns ON and that its 0-V collector holds T_1 OFF. The fact that the T_2 collector is at 0 V means that capacitor C_E is completely discharged.

When a trigger pulse arrives at the trigger terminal, T_1 is driven into the ON state. This forces T_2 to turn OFF because the collector of T_1 falls to 0 V. When T_2 turns OFF, its collector rises quickly and cleanly to almost V_s, thus causing the output pulse to appear at the Q output. When this happens, C_E begins charging. Its charging path is down from V_s, through R_{C_2}, through R_E, and into C_E. When V_{C_E} climbs to the peak voltage of the UJT, the UJT fires. This creates a positive pulse across R_1 in the base 1 lead of the UJT. This positive pulse is fed back to the base of T_2, thus turning it back ON. The one-shot's output pulse terminates at this instant.

Now let us ask ourselves if this circuit has the same drawbacks as the circuit of Fig. 3-18. Is there a recovery time needed before the one-shot can be triggered again? The answer is no, because the only capacitor in the circuit, C_E, is completely discharged and ready to begin charging again whenever required. (C_E discharged all at once through the E to $B1$ circuit of the UJT.)

Is there a limit on the firing time? Again the answer is no, because now the firing time is determined by R_E and C_E. These components can be made very large

Figure 5-5. One-shot built with a UJT. This one-shot is superior to the one shown in Fig. 3-18, because its recovery time is zero and it can deliver very long output pulses.

without any adverse effects on the operation of the rest of the circuit. The R_E and C_E values given in Fig. 5-5 would create a firing time of about 30 sec, since the time to reach V_P is about one time constant, or

$$t_f = (1.5 \text{ M})(20 \ \mu\text{F})$$
$$= (1.5 \times 10^6)(20 \times 10^{-6}) = 30 \text{ sec.}$$

5-4 UJTS IN SCR TRIGGER CIRCUITS

The UJT is almost ideal as a firing device for SCRs. Most of the UJT triggering principles discussed in this chapter in regard to SCRs apply equally well to triacs, as will be seen in Chapter 6.

There are several reasons for the compatibility between UJTs and SCRs:

a. The UJT produces a pulse-type output, which is excellent for accomplishing sure turn-ON of an SCR without straining the SCR's gate power dissipation capability.

b. The UJT firing point is inherently stable over a wide temperature range. It can be made even more stable with very little extra effort, as explained in Sec. 5-3. This nullifies the temperature instability of SCRs.

c. UJT triggering circuits are easily adaptable to feedback control. We will explore this method of control as we proceed.

5-4-1 A line-synchronized UJT trigger circuit for an SCR

The classic method of triggering an SCR with a unijunction transistor is shown in Fig. 5-6(a). In this circuit, zener diode ZD1 clips the V_s waveform at the zener voltage (usually about 20 V for use with a 120-V ac supply) during the positive half cycle of the ac line. During the negative half cycle, ZD1 is forward biased and holds V_s near 0 V. The V_s waveform is shown in Fig. 5-6(b).

Once the dc voltage V_s has been established, which occurs very shortly after the positive-going zero crossing of the ac line, C_E begins charging through R_E. When C_E reaches the peak voltage of the UJT, the UJT fires, creating a voltage pulse across R_1. This fires the SCR, thus allowing current flow through the load for the remainder of the positive half cycle. The V_{R_1} waveform and V_{load} waveform are shown in Figs. 5-6(c) and (d).

This circuit arrangement provides automatic synchronization between the firing pulse of the UJT and the SCR polarity. That is, whenever the UJT delivers a pulse, the SCR is guaranteed to have the correct anode to cathode voltage polarity for turning ON. A simple relaxation oscillator powered by a normal dc supply would not provide such synchronization; the UJT pulses would be as likely to occur during the negative half cycle as during the positive half cycle. Of course, pulses occurring during the negative half cycle would be worthless.

The load power is controlled by potentiometer R_E. When R_E is low, C_E charges quickly, causing early firing of the UJT and SCR. This results in large average current through the load. When R_E is large, C_E charges more slowly, causing late firing and lower average load current.

5-4-2 Component sizes for a UJT trigger circuit

In Fig. 5-6(a), special care must be taken when selecting R_1. The value of R_1 must be held as low as possible while still being able to generate large enough voltage pulses to fire the SCR reliably. There are two reasons for this:

a. Even before the UJT fires, there is some current flow through R_1, because of the connection through the main body of the UJT to V_s. This current can easily be several milliamps because the OFF-state resistance of the UJT, r_{BB}, is only about 10 K. This is shown by the equation

$$I_{R_1} = \frac{V_s}{R_2 + r_{BB} + R_1} = \frac{20\text{ V}}{10\text{ K}} = 2\text{ mA}$$

In this calculation, R_1 and R_2 have been neglected since they are always small compared to r_{BB}. Because of this nonnegligible current, R_1 must be held at a low

Figure 5-6. (a) A UJT used to trigger an SCR. When the UJT fires, it triggers the SCR. The firing delay angle is adjusted by R_E. (b) V_s waveform. It is almost a perfect square wave. (c) V_{R_1} waveform, which is applied to the SCR gate. The V_{R_1} resting voltage (the voltage between spikes) must be less than the gate trigger voltage of the SCR. (d) Load voltage waveform, with a firing delay angle of about 60°.

value so the Ohm's law voltage across its terminals, which is applied to the SCR gate, is also low. Otherwise the SCR may fire inadvertently.

b. With a low value of R_1, there is less chance of an undesirable noise spike falsely triggering the SCR. External sources of noise (dc motor armatures, welders, switchgear, etc.) create unwanted noise signals which might cause this to happen. Small resistors are not as likely to pick up noise signals as large resistors. Specifically, when R_1 is kept small, there is less chance of a noise signal being generated across it which could trigger the SCR.

A method for sizing all the components in Fig. 5-6(a) will be presented now. Assume the UJT to be a 2N4947, which has the following typical characteristics at a supply voltage of 20 V:

$$r_{BB} = 6\text{ K,} \qquad I_V = 4\text{ mA,}$$
$$\eta = 0.60, \qquad V_V = 3\text{ V,}$$
$$I_P = 2\ \mu\text{A.}$$

If ZD1 has a zener breakdown voltage of 20 V, then the current through R_1 before firing is given by

$$I_{R_1} = \frac{20\text{ V}}{R_2 + r_{BB} + R_1}.$$

Again, neglecting R_2 and R_1 because they are quite a bit smaller than r_{BB}, we can say to a fair approximation that

$$I_{R_1} = \frac{20\text{ V}}{r_{BB}} = \frac{20\text{ V}}{6\text{ K}} = 3.3\text{ mA}$$

Since most SCRs fire at a V_{GK} of about 0.7–1.0 V, it is reasonable to allow V_{R_1} to go no higher than about 0.3 V while the UJT is waiting for the signal to fire. This would allow a noise margin of at least 0.4 V (0.7 V − 0.3 V), which is usually adequate. Therefore

$$R_1 = \frac{V_{R_1}}{I_{R_1}} = \frac{0.3\text{ V}}{3.3\text{ mA}} = 100\ \Omega.$$

As explained in Sec. 5-3, R_E must be small enough to allow enough current, I_P, to flow into the emitter to trigger the UJT. Also, R_E must be large enough to prevent the UJT from latching up; that is, R_E must not permit the emitter to carry a current equal to the valley current, I_v, after C_E has discharged. If a current equal to I_v does continue flowing, the UJT cannot turn back OFF and is said to have *latched up.*

From Eq. (5-4),

$$R_{E_{\min}} = \frac{V_s - V_V}{I_V} = \frac{20\text{ V} - 3\text{ V}}{4\text{ mA}} = 4.25\text{ K}$$

which means that R_E must be greater than 4.25 K to allow the UJT to turn OFF.

It should be pointed out that for the circuit of Fig. 5-6(a), UJT latch-up could not persist for longer than one half cycle, because V_s disappears when the ac line reverses. However, even a one-half-cycle latch-up is undesirable because it would result in continuous gate current to the SCR during the entire conduction angle. This causes the gate power dissipation to increase and could conceivably cause thermal damage to the SCR gate.

Proceeding, we note that V_P is given by Eq. (5-1):

$$V_P = \eta V_{B2B1} + V_D = (0.60)(20 \text{ V}) + 0.6 \text{ V} = 12.6 \text{ V},$$

where V_{B2B1} has been taken as 20 V, which is approximately correct due to the small sizes of R_2 and R_1.

From Eq. (5-3),

$$R_{E_{max}} = \frac{V_s - V_P}{I_P} = \frac{20 \text{ V} - 12.6 \text{ V}}{2 \text{ } \mu\text{A}} = 3.7 \text{ M},$$

meaning that R_E must be smaller than 3.7 M in order to deliver enough emitter current to fire the UJT.

To size R_E, there would be nothing wrong with averaging $R_{E_{min}}$ and $R_{E_{max}}$, yielding

$$R_E = \frac{4.25 \text{ K} + 3.7 \text{ M}}{2} = 1.85 \text{ M}.$$

However, in situations like this where it is desired to find a happy medium between two values that differ by a few orders of magnitude, it is customary to take the *geometric mean*, instead of an average (arithmetic mean). Doing this gives

$$R_E = \sqrt{(R_{E_{min}})(R_{E_{max}})} = \sqrt{(4.25 \times 10^3)(3.7 \times 10^6)}$$
$$= 125 \text{ K}.$$

The nearest standard potentiometer value is 100 K, so

$$R_{E_v} = \textbf{100 K}.$$

To calculate the correct size for C_E, recognize that when all the variable resistance is dialed in, the V_P charging time should be almost one half of the ac line period (the time for a half cycle). This will permit a large delay angle adjustment.

The time to charge to V_P is given approximately by Eq. (5-4). For a 60-Hz ac line, the half cycle time is about 8 msec, so

$$R_{E_{total}} C_E = 8 \times 10^{-3},$$

or

$$C_E = \frac{8 \times 10^{-3}}{110 \times 10^3} = 0.073 \text{ } \mu\text{F}.$$

The nearest standard size is $C_E = \textbf{0.068 } \boldsymbol{\mu}\textbf{F}$.

R_2 is difficult to calculate and is usually determined experimentally or by referring to graphs. For most UJTs, the best temperature stability is realized with an R_2 between 500 Ω and 3 K. Detailed manufacturers' data sheets have graphs which enable the user to choose R_2 for the temperature response desired. In most cases, good stability results when $R_2 = \textbf{1 K}$.

One way of sizing ZD1 and R_d is to proceed as follows. Assume that ZD1 must be no larger than a 1-W zener diode. This is a reasonable condition since zener regulating characteristics tend to get sloppier at the higher power ratings, and the cost goes up considerably.

If ZD1 can dissipate an average power of 1 W, it can dissipate almost 2 W during the positive half cycle because the power burned during the negative half cycle is negligible, due to the low voltage drop when the diode is forward biased ($P = VI$). Therefore, the allowable average current through the zener during the positive half cycle is

$$I = \frac{P_{+\text{half}}}{V_s} = \frac{2\text{ W}}{20\text{ V}} = 100\text{ mA}.$$

R_d must be sized to allow no more than 100-mA average current during the positive half cycle. To a rough approximation, the average voltage across R_d during the positive half cycle will be 100 V, because

$$V_{\text{line}} - V_z = 120\text{ V} - 20\text{ V} = 100\text{ V}.$$

Therefore,

$$R_d = \frac{100\text{ V}}{100\text{ mA}} = 1\text{ K}.$$

Naturally R_d should be somewhat larger than this for a safety margin. A power dissipation safety margin of 2 to 1 is considered desirable, so we might choose

$$\mathbf{R_d = 2.2\text{ K}.}$$

The power rating of R_d can be determined by assuming a 100-V rms voltage drop across the resistor.

$$P_{R_d} = \frac{V^2}{R_d} = \frac{(100)^2}{2.2\text{ K}} = 4.5\text{ W}$$

This would call for a 5-W resistor, the nearest standard rating which is greater than 4.5 W. Naturally all such calculations are approximate and would have to be tested experimentally.

5-4-3 A sequential switching circuit using UJTs for gate control

An interesting example of the UJT-SCR combination is the sequential switching circuit shown in Fig. 5-7. In this circuit, the three loads are energized in sequence, and each load is energized for a certain length of time. The times are individually variable. That is, it would be possible to have load 1 energized for 5 sec, after which load 1 would deenergize and load 2 would energize for 10 sec, after which load 2 would deenergize and load 3 would energize for 7 sec. The times of 5, 10, and 7 sec can be independently adjusted.

This circuit works as follows. The sequence begins when a positive pulse is applied at the START terminal in the lower left of Fig. 5-7. This generates a voltage between gate and cathode of SCR1, firing the SCR. When SCR1 fires,

Figure 5-7. Sequential switching circuit using UJT-SCR pairs. When a UJT fires, it causes the next SCR to fire. When that SCR fires, it connects a charged commutating capacitor across the main terminals of the preceding SCR, thereby turning it OFF.

load 1 energizes because the top lead is connected to +48 V and the bottom lead is connected to ground through the SCR.

Also, when SCR1 fires, the left terminal of C_1 is connected to the ground through the SCR. The right terminal of that capacitor is connected through the resistance of load 2 to the +48-V supply line. C_1 quickly charges to 48 V because the load resistance would be fairly low. The charge polarity is plus on the right-hand side and minus on the left-hand side.

While SCR1 and load 1 are carrying current, *pnp* transistor Q_1 also switches ON because of the base current flow path through the 33-K base resistor, through SCR1, to ground. The *RC* network including C_{E_1} will charge up to the peak voltage of UJT1, causing that UJT to deliver a pulse of current into its 68-Ω base 1 resistor. This in turn fires SCR2, energizing load 2. When SCR2 fires, the positive (right) terminal of C_1 is connected to ground through SCR2. C_1 had previously been charged to 48 V, and since a capacitor cannot discharge instantaneously, the −48-V potential on the left side of C_1 is applied to the anode of SCR1. This effectively reverse-biases SCR1 for an instant, shutting it OFF and deenergizing load 1. Transistor Q_1 also switches OFF, so C_{E_1} does not charge up again.

The above action is repeated in the second stage of the switching circuit, with C_{E_2} charging through Q_2 at a rate determined by the 1-M pot in series with C_{E_2}. When the proper time has elapsed, UJT2 fires, which fires SCR3 and connects the C_2 terminals in parallel with SCR2. C_2 had charged plus on the right and minus on the left during the time load 2 was energized, so it now reverse-biases SCR2, turning it OFF.

When the load 3 energization time has elapsed, UJT3 fires, causing SCR4 to fire. The only purpose of SCR4 is to connect C_3 in parallel with SCR3, to turn it OFF. SCR4 turns OFF on its own, after the cessation of the voltage pulse at its gate. This occurs because the 47-K resistor in its anode lead is so large that the current through the SCR4 main terminals is less than the holding current. That is,

$$I_{AK} = \frac{48 \text{ V}}{47 \text{ K}} = 1 \text{ mA},$$

which is below the holding current for a medium-power SCR. I_{HO} for a medium SCR is about 10 mA, as mentioned in Sec. 4-3.

The current of Fig. 5-7 could easily be extended to any number of stages. Such a circuit could be applied in an industrial control situation whenever there are several loads which must be energized in a given sequence.

5-4-4 A logic output amplifier using an SCR-UJT combination

In Sec. 2-8, we discussed output amplifiers used to interface between low-voltage logic circuitry and industrial actuating devices. As mentioned then, up-to-date output amplifiers often contain an SCR with a UJT in its gate control circuit. A popular design of such an output amplifier is shown in Fig. 5-8(a).

Figure 5-8. (a) Schematic of a logic output amplifier using a UJT and an SCR. When the input line goes HI, it causes the relaxation oscillator to start oscillating at a high frequency, delivering a rapid train of gate pulses to the SCR. (b) The train of gate pulses, shown relative to the 115-V ac supply. (c) V_{AK} waveform, showing that the SCR fires very shortly after the start of a half cycle. (d) Load voltage waveform.

Here is how the output amplifier works. Look first at the right-hand side of Fig. 5-8(a). The load, in this case a solenoid coil, is placed in the ac power line in series with a bridge rectifier which is controlled by a single SCR. This method of controlling both half cycles of the ac line was presented in Sec. 4-6-3, Fig. 4-13(a). Recall that during the positive half cycle of the ac line, diodes D_1 and D_3 are forward biased, and the SCR is also forward biased and capable of firing. If the SCR does fire, the ac line voltage will be impressed across the load for the rest of the positive half cycle. During the negative half cycle of the ac line, diodes D_2 and D_4 are forward biased, and the SCR is still forward biased and capable of firing. Therefore if it does fire, the negative ac line voltage will be impressed across the load for the remainder of the negative half cycle.

The SCR gate is controlled by a *pulse transformer*. Pulse transformers are specially designed transformers used for transforming quick voltage pulses. They are frequently seen in SCR gate trigger circuits. The secondary winding of the pulse transformer is connected between the gate and the cathode of the SCR. Therefore if a voltage pulse is produced in the secondary winding, the SCR will turn ON.

The primary winding of the pulse transformer is wired in the base 1 lead of the UJT. Thus when the UJT fires, a burst of current flows through the primary winding of the transformer. This burst creates a voltage pulse in the secondary winding which fires the SCR. The arrangement in Fig. 5-8(a) is an example of a situation in which the power supply for the gate control circuit is *not* the same power supply that drives the load. In fact the gate control circuit is completely isolated from the main terminal circuit. The coupling between the two circuits is by way of the magnetic coupling between the primary and secondary windings of the pulse transformer. This provides the usual benefits of electrical isolation between the noisy power circuit and the low-voltage electronic control circuit.

The firing of the UJT is determined as always by R_E, C_E, and the input voltage signal at the top of R_E. If that input voltage is LO, C_E cannot charge, so the UJT never fires. In that case the SCR never fires either, and the load is deenergized.

However, if the input voltage from the logic circuit goes HI ($+15$ V in this example), the emitter circuit will start charging with a charging time given by Eq. (5-6):

$$t_{charge} = R_E C_E$$
$$= (10 \text{ K})(0.1 \ \mu\text{F}) = 1 \text{ msec.}$$

As long as the input terminal remains HI, the UJT circuit will behave like a relaxation oscillator, producing output pulses spaced about 1 msec apart. These output pulses continually create gate pulses to the SCR, thereby continually ordering the SCR to fire. The continual arrival of gate trigger pulses is illustrated in Fig. 5-8(b).

With this barrage of trigger pulses arriving at the SCR gate, it can be seen that the SCR is bound to turn ON very early in every half cycle. The *latest* it can turn ON is 1 msec into the half cycle; in all probability a trigger pulse will arrive before

1 msec has elapsed. There is no synchronization between V_{AK} and the gate control circuit in this case, but none is needed.

The SCR main terminal voltage waveform is drawn in Fig. 5-8(c), assuming about a 1-msec delay between the zero crossover and the firing. The resulting load waveform is shown in Fig. 5-8(d). The overall result is that the load is energized when the input signal goes HI.

QUESTIONS AND PROBLEMS

1. Is a unijunction transistor a continuously variable device or a switching device? Explain.

2. Roughly speaking, what range of values does η fall into?

3. For the relaxation oscillator of Fig. 5-3, what would be the effect on the oscillation frequency of doubling C_E? Of doubling R_1?

4. Explain why there is a maximum limit on the size of the emitter resistor in a UJT circuit.

5. Explain why there is a minimum limit on the size of the emitter resistor in a UJT circuit.

6. Why is it that inserting a resistor in the $B2$ lead of a UJT relaxation oscillator tends to stabilize the oscillation frequency in the face of temperature changes? Explain the two temperature effects which tend to cancel each other.

7. Equation (5-5) for a relaxation oscillator is only approximate. Identify two reasons it is not exact.

8. In Fig. 5-4, the UJT relay timer, how would the load be deenergized?

9. When the UJT is being used to trigger a thyristor, as in Fig. 5-6, why is there a limit on the size of R_1?

10. In the circuit of Fig. 5-6, does C_E begin charging again immediately after the UJT has fired during a positive half cycle? Explain.

11. In Fig. 5-7, C_1, C_2, and C_3 are all marked nonpolarized (NP). Why must they be nonpolarized?

12. In Fig. 5-7, SCRs 1–3 are turned OFF by connecting a negatively charged capacitor across their anode to cathode terminals. What is it that turns OFF SCR4?

13. Explain in detail why the one-shot of Fig. 5-5 can be retriggered immediately after it has fired, whereas the one-shot of Fig. 3-18 cannot be retriggered until a certain recovery time has elapsed. Focus your explanation on C of Fig. 3-18 and C_E of Fig. 5-5.

14. Why is it necessary for the relaxation oscillator in the gate control circuit in Fig. 5-8 to have such a high frequency?

15. Would the phase relationship between primary and secondary windings of the pulse transformer be important in Fig. 5-8(a)? Explain what would happen if the phasing were reversed.

16. For the circuit of Fig. 5-1(b), assume that the standoff ratio is 0.70. Calculate the peak point voltage, V_P.

17. For the relaxation oscillator of Fig. 5-3, if $R_E = 10$ K, $C_E = 0.005$ μF, and $\eta = 0.63$, what is the approximate oscillation frequency? What would be the effect on the frequency if η were larger than 0.63? What would be the effect if it were smaller than 0.63?

18. What is the longest possible time delay in Fig. 5-4? Assume that $\eta = 0.63$.

19. For the one-shot of Fig. 5-5, calculate R_E and C_E to give an output pulse duration of 5 sec. Assume $\eta = 0.63$. Is the size of R_{C_2} important in determining the firing duration?

20. In Fig. 5-7, suppose that the SCRs have a gate trigger voltage of $V_{GT} = 0.7$ V. What is the absolute minimum allowable r_{BB} for the UJTs, so that the SCRs don't fire until ordered to fire? Neglect noise margin considerations.

SUGGESTED LABORATORY PROJECTS

Project 5-1: A UJT RELAXATION OSCILLATOR

Purpose

a. To determine the intrinsic standoff ratio of a UJT
b. To observe and graph the output waveforms of a UJT relaxation oscillator
c. To observe the temperature stability of a UJT relaxation oscillator

Procedure

a. Find the UJT's standoff ratio.
 1. With an ohmmeter, measure the interbase resistance of a UJT, r_{BB}.
 2. Find the two individual emitter to base resistances of the UJT, r_{B1} and r_{B2}. This cannot be done accurately with an ohmmeter. Instead, connect a variable dc supply between the emitter and base 1 with a 10-mA ammeter in the emitter lead. Adjust the dc supply voltage until the ammeter reads 5 mA. Measure V_{EB1}, and subtract 0.6 V for the pn junction. The remainder is the voltage actually applied to r_{B1}. Use Ohm's law to calculate r_{B1}:

$$r_{B1} = \frac{V_{EB1} - 0.6\text{ V}}{5\text{ mA}}.$$

Why would an ohmmeter give false readings for r_{B1} and r_{B2} but give correct readings for r_{BB}?
 3. Repeat step 2 above for the emitter to base 2 circuit to find r_{B2}.
 4. Calculate the intrinsic standoff ratio of the UJT from Eq. (5-2).
b. Find the peak voltage of the UJT.
 1. Build the circuit of Fig. 5-3, with $R_E = 100$ K, $C_E = 100$ μF, and $V_s = 15$ V dc. Let R_1 and R_2 have the values given in that drawing. Install a switch in the V_s supply line. Place a 50-V voltmeter across C_E and a 10-V voltmeter across R_1. Discharge C_E completely; then close the switch and watch the voltmeters. What is the V_P for this circuit? Does it agree with what you would expect from Eq. (5-1)?
 2. Repeat step 1 above with a V_s of 10 V.
 3. Measure the time delay before firing. Does it agree with what you would expect from Eq. (5-6)? Does the time delay depend on V_s? Explain this.
c. Open the power switch and change R_E to 22 K and C_E to 0.5 μF. Remove the voltmeters and connect the channel 1 and channel 2 vertical inputs of a dual-trace oscilloscope

across C_E and R_1 to observe the waveforms of V_{C_E} and V_{R_1}. If a dual-trace scope is not available, use a single-trace scope to observe V_{C_E} first; then later move the input to see V_{R_1}.

1. Using the methods of Example 5-3, predict the appearance of the V_{C_E} and V_{R_1} waveforms. Sketch what you expect to see for each waveform.
2. Close the power switch and adjust the scope controls to display several cycles of the waveforms. Sketch the actual waveforms. Do they agree with your predictions? Try to explain any discrepancies.

d. Heat the UJT with a soldering iron. Does the oscillation frequency change as the temperature changes? Experiment with different values of R_2 to get the best temperature stability. A frequency counter will come in handy here, but if none is available, you can detect frequency changes by carefully watching the oscillation period on the oscilloscope screen.

Project 5-2: UJT CHARACTERISTIC CURVES

Purpose

a. To observe and graph the voltage-current characteristic curves for a UJT at several different interbase voltages
b. To observe the batch spread among several UJTs of the same type

Procedure

a. Build the circuit of Fig. 5-9. This circuit will display the V_{EB1} versus I_E curve of the UJT.

As with Project 4-1, the ac supply should be isolated from the earth. If this is not possible, then the scope chassis must be isolated from earth ground by a two-prong adapter and then must not be touched.

Figure 5-9. Circuit for displaying the voltage vs. current curve of a UJT. V_{EB1} is applied to the vertical input of the scope. The signal across the 1000-Ω resistor, which represents I_E, is applied to the horizontal input.

The vertical amplifier of the oscilloscope is connected between the emitter terminal and the base 1 terminal in Fig. 5-9, so it will display V_{EB1}.

With the scope's horizontal sweep disabled and the horizontal display switched to EXTERNAL, the scope horizontal amplifier will be connected across the 1000-Ω resistor in the base 1 lead. The only way for current to flow through that resistor is to come from the emitter, because base 2 current will return to the negative terminal of the variable dc supply before passing through the 1000-Ω resistor. Therefore the signal developed across the 1000-Ω resistor *represents* emitter current by this equation:

$$I_E = \frac{V_{\text{horiz}}}{1000 \ \Omega}.$$

Set the horizontal amplifier gain to 1 volt/cm. With this setting, each centimeter of horizontal deflection represents 1 mA of emitter current, since

$$1 \ \text{mA/cm} = \frac{1 \ \text{volt/cm}}{1000 \ \Omega}.$$

Note that Fig. 5-9 shows the I_E signal applied to the *negative* horizontal input terminal, because the signal developed across the 1000-Ω resistor will be "backwards" or 180° out of phase with the actual emitter current. The negative horizontal input causes this signal to be reinverted, or put back into phase. If your scope does not have a negative horizontal input, use the positive input; then the characteristic curve will be reversed from that shown in Fig. 5-2(b).

b. With the scope's vertical amplifier gain set at 2 V/cm, adjust the variable dc supply to 10 V.

1. Close the switch and observe the *V-I* curve of the UJT. Make an accurate graph of the curve, paying special attention to V_P, V_V, I_P, and I_V. You will have to increase the horizontal sensitivity to see I_P accurately. A sensitivity of about 5 mv/cm will probably work best if your scope can go that low. At a sensitivity of 5 mV/cm, each centimeter represents 5 μA of emitter current.
2. Repeat step 1 above with the variable supply set at 15 V ($V_{B2B1} = 15$ V).
3. Repeat for $V_{B2B1} = 20$ V, and again for 30 V. Explain what you are seeing.

c. Substitute several different UJTs of that same type, and compare their characteristics for $V_{B2B1} = 20$ V. How much batch spread is there in V_P? How much in V_v? How much in I_v? How much in η?

Project 5-3: A UJT GATE CONTROL CIRCUIT FOR AN SCR

Purpose

a. To build and observe a UJT gate control circuit for use with an SCR

Procedure

a. Build the UJT-SCR circuit of Fig. 5-6. Use the component sizes calculated in Sec. 5-4-2. If you cannot obtain a 2.2-K, 5-W resistor for R_d, a 6.8-K, 2-W resistor will also work. ZD1 can then be reduced to a $\frac{1}{2}$-W rating.

As usual the ac supply should be isolated from the earth, or the scope chassis must be isolated and then not touched. Use any good UJT (2N4947, for example) and any good medium-sized SCR (TIC106B, for example).

1. Use your oscilloscope to study the waveforms of V_{CE}, V_{R_1}, V_{AK}, and V_{load}. Graph all these waveforms to the same time reference for a firing delay angle of 90°. Do they all look like what you expected?
2. What is the minimum possible firing delay angle?
3. What is the maximum firing delay angle?

b. Readjust the firing delay angle to about 90°. Heat up the SCR with a soldering iron. What happens to the firing delay angle? Does this make sense?

6
Triacs and Other Thyristors

Triacs behave roughly like SCRs, except that they can carry current in either direction. Both triacs and SCRs are members of the *thyristor* family. The term thyristor includes all the semiconductor devices which show *inherent* ON-OFF behavior, as opposed to allowing gradual change in conduction. All thyristors are regenerative switching devices, and they cannot operate in a linear manner. Thus a transistor is not a thyristor because although it *can* operate ON-OFF, that is not its inherent nature; it is possible for a transistor to operate linearly.

Some thyristors can be *gated* into the ON state, as we saw in Chapter 4 for SCRs. Triacs are the same in this respect. Other thyristors cannot be gated ON, but they turn ON when the applied voltage reaches a certain breakover value. Four-layer diodes and diacs are examples of this type of thyristor. Small thyristors which do not switch the main load current are usually referred to as *breakover devices*, and we will use that term in this book. They are useful in the gate triggering circuit of a larger load power switching thyristor, such as a triac. In this chapter we will deal with the smaller breakover device thyristors as well as with triacs.

OBJECTIVES

After completing this chapter and performing the suggested laboratory projects, you will be able to:

1. Explain the operation of a triac in controlling both alternations of an ac supply driving a resistive load
2. Define and discuss the important electrical parameters of triacs, such as gate trigger current, holding current, etc.
3. Explain the operation of breakover-type devices in the triggering circuits of triacs and discuss the advantages of using these devices
4. Describe the current-voltage behavior of the following breakover devices: diacs, four-layer diodes, silicon bilateral switches (SBSs), and silicon unilateral switches (SUSs)
5. Explain the flash-on effect seen with triacs and why it occurs, and explain how it can be eliminated with an SBS trigger circuit
6. Explain in detail the operation of resistance feedback for firing a UJT in a triac trigger circuit
7. Explain in detail the operation of voltage feedback for firing a UJT in a triac trigger circuit
8. Calculate resistor and capacitor sizes for a triac's UJT trigger circuit employing either resistive feedback or voltage feedback
9. Construct a triac control circuit for controlling a resistive load, and measure some of the triac's electrical parameters
10. Construct a circuit which provides an oscilloscope display of the current-voltage characteristic curve of a thyristor
11. Interpret the characteristic curve of a thyristor, reading the breakover voltages, breakback voltages, and holding currents

6-1 THEORY AND OPERATION OF TRIACS

A *triac* is a three-terminal device used to control the average current flow to a load. A triac is different from an SCR in that it can conduct current in *either* direction when it is turned ON. The schematic symbol of a triac is shown in Fig. 6-1(a), along with the names and letter abbreviations of its terminals.

When the triac is turned OFF, no current can flow between the main terminals no matter what the polarity of the externally applied voltage. The triac therefore acts like an open switch.

When the triac is turned ON, there is a very low-resistance current flow path from one main terminal to the other, with the direction of flow depending on the polarity of the externally applied voltage. When the voltage is more positive on $MT2$, the current flows from $MT2$ to $MT1$. When the voltage is more positive on $MT1$, the current flows from $MT1$ to $MT2$. In either case the triac is acting like a closed switch.

Figure 6-1. (a) Schematic symbol and terminal names of a triac. (b) Triac circuit showing how the supply voltage, the load, and the triac are connected.

The circuit relationship among the supply voltage, the triac, and the load is illustrated in Fig. 6-1(b). A triac is placed in series with the load just like an SCR, as this figure shows. The average current delivered to the load can be varied by varying the amount of time per cycle that the triac spends in its ON state. If a small portion of the time is spent in the ON state, the average current flow over many cycles will be low. If a large portion of the cycle time is spent in the ON state, then the average current will be high.

A triac is not limited to 180° of conduction per cycle. With the proper triggering arrangement, it can conduct for a full 360° per cycle. It thus furnishes full-wave power control instead of the half-wave power control possible with an SCR.

Triacs have the same advantages that SCRs and transistors have over mechanical switches. They have no contact bounce, no arcing across partially opened contacts, and they operate much faster than mechanical switches, therefore yielding more precise control of current.

6-2 TRIAC WAVEFORMS

Triac waveforms are very much like SCR waveforms except that they can fire on the negative half cycle. Figure 6-2 shows waveforms of both load voltage and triac voltage (across the main terminals) for three different conditions.

The Fig. 6-2(a) waveforms show the triac OFF during the first 30° of each half cycle; during this 30° the triac is acting like an open switch. During this time the entire line voltage is dropped across the main terminals of the triac, with no voltage applied to the load. Thus there is no current flow through the triac or the load. The portion of the half cycle during which this situation exists is called the *firing delay angle*, just as it was for an SCR.

Continuing with Fig. 6-2(a), after 30° has elapsed, the triac fires or turns ON,

(a)

(b)

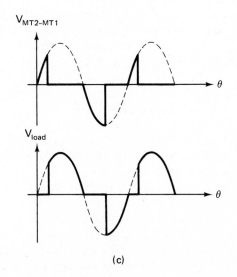

(c)

Figure 6-2. Waveforms of triac main terminal voltage and load voltage for three different conditions. (a) Firing delay equals 30° for both the positive half cycle and the negative half cycle. (b) Firing delay equals 120° for both half cycles. (c) Unequal firing delay angles for the positive and negative half cycles. This is usually undesirable.

and it becomes like a closed switch. At this instant the triac begins conducting current through its main terminals and through the load, and it continues to carry load current for the remainder of the half cycle. The portion of the half cycle during which the triac is turned ON is called its *conduction angle*. The conduction angle in Fig. 6-2(a) is 150°. The waveforms show that during the conduction angle the entire line voltage is applied to the load, with zero voltage appearing across the triac's main terminals.

Figure 6-2(b) shows the same waveforms with a larger firing delay angle. The delay angle is 120° and the conduction angle is 60° in Fig. 6-2(b). Since the current is flowing during a smaller portion of the total cycle in this case, the average current is less than it was for the condition in Fig. 6-2(a). Therefore less power is transferred from source to load.

Triacs, like SCRs and most other semiconductor devices, show notoriously wide variations in their electrical characteristics. This problem is especially evident with triacs because it usually happens that the triggering requirements are different for the two different supply voltage polarities. Figure 6-2(c) shows waveforms which illustrate this problem. The triac waveform in Fig. 6-2(c) shows a smaller delay angle on the positive half cycle than on the negative half cycle, due to the tendency of the triac to trigger more easily on the positive half cycle. Another triac of the same type might have a tendency to trigger more easily on the negative half cycle; in that case the negative delay angle would be smaller. Sometimes such inconsistent firing behavior cannot be tolerated. Methods for eliminating unequal firing delays will be discussed in Sec. 6-3.

6-3 ELECTRICAL CHARACTERISTICS OF TRIACS

When a triac is biased with an external voltage more positive on $MT2$ (called *forward* or *positive* main terminal bias), it is usually triggered by current flow from gate to $MT1$. The polarities of the voltages and the directions of the currents in this case are shown in Fig. 6-3(a).

When biased as shown in Fig. 6-3(a), the triggering of a triac is identical to the triggering of an SCR. The G terminal is positive with respect to $MT1$, which causes the trigger current to flow into the device on the gate lead and out of the device on the $MT1$ lead. The gate voltage necessary to trigger a triac is symbolized V_{GT}; the gate current necessary to trigger is symbolized I_{GT}. Most medium-sized triacs have a V_{GT} of about 0.6–2.0 V and an I_{GT} of 0.1–20 mA. As usual, these characteristics vary quite a bit as temperature changes. Typical variations in electrical characteristics with temperature are graphed on the manufacturer's specification sheets.

When the triac is biased more positive on $MT1$ (called *reverse* or *negative* main terminal bias), as shown in Fig. 6-3(b), triggering is usually accomplished by sending gate current into the triac on the $MT1$ lead and out of the triac on the G

Figure 6-3. (a) The situation when a triac has forward main terminal bias. Normally, the gate current and gate voltage would have the polarities indicated. (b) The situation at a different point in time when the triac is reverse biased. Normally the gate current and voltage are also reversed.

lead. The gate voltage will be negative with respect to $MT1$ to accomplish this. The voltage polarities and current directions for reverse main terminal bias are illustrated in Fig. 6-3(b).

For a particular individual triac, I_{GT} for the forward main terminal polarity may be quite a bit different from I_{GT} for the reverse main terminal polarity, as mentioned in Sec. 6-2. However, if many triacs of the same type are considered, the I_{GT} for the forward main terminal bias will be equal to the I_{GT} for the reverse main terminal bias.

A triac, like an SCR, does not require continuous gate current once it has been fired. It will remain in the ON state until the main terminal polarity changes or until the main terminal current drops below the holding current, I_{HO}. Most medium-sized triacs have an I_{HO} rating of 100 mA or less.

Other important electrical characteristics which apply to triacs are (a) the maximum allowable main terminal rms current, $I_{T(RMS)}$ and (b) the breakover voltage, V_{DROM}, which is the highest main terminal peak voltage the triac can block in either direction. If the instantaneous voltage applied from $MT2$ to $MT1$ should exceed V_{DROM}, the triac breaks over and begins passing main terminal current. This does not damage the triac, but it does represent a loss of gate control. To prevent breakover, a triac must have a V_{DROM} rating greater than the peak value of the ac voltage driving the circuit. The most popular V_{DROM} ratings for triacs are 100, 200, 400, and 600 V.

For many manufacturers, the sequence of $I_{T(RMS)}$ ratings available is 1, 3, 6, 10, 15, and 25 A; other similar sequences are also used by triac manufacturers.

One other important electrical rating which is given on manufacturers' specification sheets is V_{TM}, the ON-state voltage across the main terminals. Ideally, the

ON-state voltage should be 0 V, but V_{TM} usually falls between 1 and 2 V for real triacs, the same as for SCRs. A low V_{TM} rating is desirable because it means the triac closely duplicates the action of a mechanical switch, applying full supply voltage to the load. It also means that the triac itself burns very little power. The power burned in a triac is given by the product of main terminal current and main terminal voltage. Large power dissipation is undesirable from the point of view of protecting the triac from high temperatures and also from the point of view of economical energy transfer from source to load.

6-4 TRIGGERING METHODS FOR TRIACS

6-4-1 RC gate control circuits

The simplest triac triggering circuit is shown in Fig. 6-4(a). In Fig. 6-4(a), capacitor C charges through R_1 and R_2 during the delay angle portion of each half cycle. During a positive half cycle, $MT2$ is positive with respect to $MT1$, and C charges positive on its top plate. When the voltage at C builds up to a value large enough to deliver sufficient gate current (I_{GT}) through R_3 to trigger the triac, the triac fires.

(a) (b)

Figure 6-4. (a) Simple gate control circuit (triggering circuit) for a triac. Firing delay is adjusted by potentiometer R_2. (b) An improved gate control circuit, which allows a wider range of adjustment of firing delay.

During a negative half cycle, C charges negative on its top plate. Again, when the voltage across the capacitor gets large enough to deliver sufficient gate current in the reverse direction through R_3 to trigger the triac, the triac fires.

The charging rate of capacitor C is set by the resistance of R_2. For large R_2, the charging rate is slow, causing a long firing delay and small average load current. For small R_2, the charging rate is fast, the firing delay angle is small, and the load current is high.

As was true with SCR triggering circuits, a single *RC* network cannot delay triac firing much past 90°. To establish a wider range of delay angle adjustment, the double *RC* network of Fig. 6-4(b) is often used. Typical component sizes are shown for use with a medium-sized triac.

6-4-2 Breakover devices in gate control circuits of triacs

The gate control circuits of Fig. 6-4 can be improved by the addition of a breakover device in the gate lead, as shown in Fig. 6-5(a). The breakover device pictured in Fig. 6-5(a) is a *diac*, but there are several other breakover devices which also work well. Use of a breakover device in the gate triggering circuit of a triac offers some important advantages over simple *RC* gate control circuits. These advantages stem from the fact that breakover devices deliver a *pulse* of gate current rather than a sinusoidal gate current.

The ability of a breakover device to provide a current pulse can be understood by studying Fig. 6-5(b), which shows a typical current-voltage characteristic curve for a diac. (A diac is also known by the names *bidirectional trigger diode* and *symmetrical trigger diode*.)

Let us interpret the diac's characteristic curve now. The curve shows that for applied forward voltages less than the *forward breakover voltage* (symbolized $+V_{BO}$) the diac permits virtually no current to flow. Once the forward breakover voltage is reached, however, the diac switches into conduction and the current surges up as the voltage across the terminals declines. Refer to Fig. 6-5(b) to see this. This surge of current on the characteristic curve accounts for the pulsing ability of the diac.

In the negative voltage region, the behavior is identical. When the applied reverse voltage is smaller than the *reverse breakover* voltage (symbolized $-V_{BO}$) the diac permits no current to flow. When the applied voltage reaches $-V_{BO}$, the diac switches into conduction in the opposite direction. This is graphed as negative current in Fig. 6-5(b). Diacs are manufactured to be relatively temperature stable and to have fairly close tolerances on breakover voltages. There is very little difference in magnitude between forward breakover voltage and reverse breakover voltage for a diac. The difference is typically less than 1 V. This enables the trigger circuit to maintain nearly equal firing delay angles for both half cycles of the ac supply.

The operation of the circuit in Fig. 6-5(a) is the same as that of the circuit in Fig. 6-4(a) except that the capacitor voltage must build up to the breakover voltage of the diac in order to deliver gate current to the triac. For a diac, the breakover voltage would be quite a bit higher than the voltage which would be necessary in Fig. 6-4(a). The most popular breakover voltage for diacs is 32 V $(+V_{BO} = +32 \text{ V}, -V_{BO} = -32 \text{ V})$. This value is convenient for use with a 115-V ac supply. Therefore when the capacitor voltage reaches 32 V, in either polarity, the diac breaks over, delivering the turn-ON pulse of current to the gate of the triac.

(a)

(b)

(c)

Figure 6-5. (a) Triac gate control circuit containing a diac (bidirectional trigger diode). This triggering method has several advantages over the methods shown in Fig. 6-4. (b) Current vs. voltage characteristic curve of a diac. (c) Another schematic symbol for a diac.

Because the capacitor voltage must reach higher values when a diac is used, the charging time constant must be reduced. This means that Fig. 6-5(a) would have smaller component values (resistor and capacitor values) than Fig. 6-4(a).

A second schematic symbol for the diac is presented in Fig. 6-5(c) This symbol is less frequently used, and the diac symbol in Fig. 6-5(a) is preferred.

Example 6-1

Suppose the circuit of Fig. 6-5(a) contains a diac with $V_{BO} = \pm 32$ V. Suppose also that the resistor and capacitor sizes are such that the firing delay angle $= 75°$. Now if the 32-V diac is removed and replaced by a 28-V diac but nothing else is changed, what will happen to the firing delay angle? Why?

Solution. If the 32-V diac is replaced by a 28-V diac, it means that the capacitor will only have to charge to ± 28 V in order to fire the triac, instead of ± 32 V. With given sized components, C can certainly charge to 28 V in less time than it can charge to 32 V. Therefore it causes the diac to break over earlier in the half cycle, and the firing delay angle is **reduced**.

6-5 SILICON BILATERAL SWITCHES

6-5-1 Theory and operation of an SBS

There is another breakover device which is capable of triggering triacs. Its name is the *silicon bilateral switch* (SBS), and it is popular in low-voltage trigger control circuits. SBSs have lower breakover voltages than diacs, ± 8 V being the most popular rating. The current-voltage characteristic curve of an SBS is similar to that of the diac, but the SBS has a more pronounced "negative resistance" region. That is, its decline in voltage is more drastic after it enters the conducting state. An SBS is shown schematically in Fig. 6-6(a). Its current-voltage characteristic curve is shown in Fig. 6-6(b). Notice that when the SBS switches into its conducting state the voltage across its anode terminals drops almost to zero (to about 1 V). The SBS is said to have a *breakback voltage* of 7 V, because the voltage between $A2$ and $A1$ decreases by about 7 V when it turns ON.

The characteristic curve of Fig. 6-6(b) is for the gate terminal of the SBS disconnected. The gate terminal can be used to alter the basic current-voltage behavior of an SBS, as we will see shortly. However, the SBS is quite useful even without its gate terminal, just by virtue of the snap-action breakover from $A2$ to $A1$.

To make use of an SBS without its gate terminal, it could be installed in place of the diac in Fig. 6-5(a). Because of the lower V_{BO} of the SBS, the RC timing components would have to be increased in value. You may wonder why we would want to use an SBS in this control circuit instead of a diac. Well, generally speaking, the SBS is a superior device compared to the diac. Not only does the SBS show a more vigorous switching characteristic, as Fig. 6-6(b) indicates, but an SBS is more temperature stable and more symmetrical and has less batch spread than a diac.

To say it with numbers, a modern SBS has a temperature coefficient of about $+0.02\%/°C$. This means that its V_{BO} increases by only 0.02% per degree of temperature change, which figures out to only 0.16 V/100°C, which is very temperature stable indeed.

SBSs are symmetrical to within about 0.3 V. That is, the difference in magnitude between $+V_{BO}$ and $-V_{BO}$ is less than 0.3 V. This yields virtually identical firing delays for positive and negative half cycles.

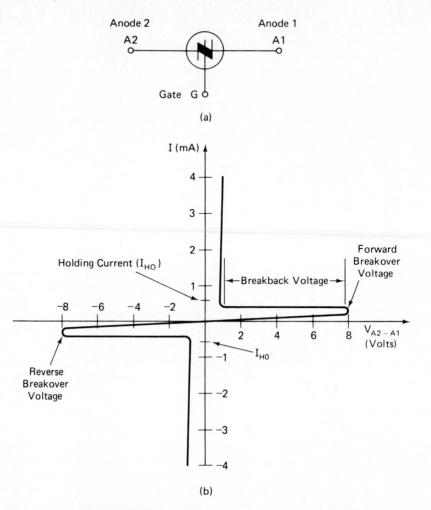

Figure 6-6. (a) Schematic symbol and terminal names of an SBS
(silicon bilateral switch). (b) Current-voltage characteristic curve of
an SBS, with important points indicated.

The batch spread of SBSs is less than 0.1 V. This means that the difference in
V_{BO} among all the SBSs in a batch is less than 0.1 V. The batch spread among diacs
is about 4 V, by contrast.

6-5-2 Using the gate terminal of an SBS

As mentioned in Sec. 6-5-1, the gate terminal of an SBS can be used to alter
its basic breakover behavior. For example, if a zener diode is connected between
G and $A1$, as shown in Fig. 6-7(a), the forward breakover voltage ($+V_{BO}$) changes
to approximately the V_z of the zener diode. With a 3.3-V zener diode connected,

Figure 6-7. (a) An SBS combined with a zener diode to alter the breakover point in the forward direction. (b) Characteristic curve of the SBS-zener diode combination. The forward breakover voltage is lower, but the reverse breakover voltage is unchanged.

$+V_{BO}$ would equal 3.3 V + 0.6 V (there is an internal *pn* junction). This would yield

$$+V_{BO} = 3.9 \text{ V}.$$

The reverse breakover voltage would be unaffected and would remain at −8 V. The new current-voltage behavior would be as drawn in Fig. 6-7(b). This behavior would be useful if it were desired to have different firing delay angles for the positive and negative half cycles (which would be unusual).

6-5-3 Eliminating triac flash-on
(hysteresis) with a gated SBS

One of the nicest things about using a gated SBS for the trigger control of a triac is that it can eliminate the *hysteresis* or *flash-on* effect. Let us first come to an understanding of the flash-on problem. The explanation of flash-on is rather involved, so be forewarned.

Refer back to Fig. 6-5(a). Suppose that R_2 is adjusted so that C cannot quite charge up to 32 V in either direction. In this case, the diac would never fire, and the load would be completely deenergized. If the load were a lighting load, it would give no light at all. Since C never discharges any of its built-up charge, it always starts a new half cycle with a residual charge of the opposite polarity. That is, when a positive half cycle of the ac supply starts, the initial charge on C is minus on the top and plus on the bottom; this charge is left over from the preceding negative half cycle. Likewise, when a negative half cycle of the ac supply line begins, the initial charge on C is plus on the top and minus on the bottom, left over from the preceding positive half cycle. The effect of this initial charge is to make it more difficult for the capacitor to charge up to the breakover voltage of the diac.

Now suppose that we slowly decrease R_2 until the capacitor can *just barely* charge to V_{BO} of the diac. Assume that the first breakover occurs on the positive half cycle (it is just as likely to occur on the negative half cycle as the positive). When the diac breaks over, it discharges part of the $+$ charge which has built up on the top plate of C. The discharge path is through the G to MT_1 circuit of the triac. During the remainder of the positive half cycle, no further charging of C takes place, because the triac shorts out the entire trigger circuit when it turns ON. Therefore when that positive half cycle ends, the next negative half cycle begins, and the initial $+$ charge on the top of C is *less* than it was for all the prior negative half cycles. The capacitor has a "head start" this time, as it attempts to charge to $-V_{BO}$.

Because of this head start, C will reach $-V_{BO}$ much *earlier* in the negative half cycle than it reached $+V_{BO}$ in the preceding positive half cycle. Furthermore, since C will lose some of the $-$ charge on its top plate when the diac breaks over during the negative half cycle, it will begin the next positive half cycle with less initial charge than ever before. Therefore it will fire much earlier in the *next* positive half cycle than it fired in the *first* positive half cycle.

The overall result of this phenomenon is this: You can adjust R_2 to just barely fire the triac, expecting to get very dim light from the lamps, but as soon as the first firing takes place, all subsequent firings take place much earlier in the half cycle. It is impossible to smoothly adjust from the completely off condition to the glowing-dimly condition. Instead, the lamps "flash on."

What you *can* do, once the lamps have come on, is to adjust the R_2 resistance back up to a higher value to delay the diac breakover until later in the half cycle. In other words, you must turn the potentiometer shaft back in the direction it just came from in order to create very dim light. You can demonstrate this with almost any commercial lamp dimmer in your home. Unless it is a very good one, it will exhibit flash-on and subsequent reduction in light output as the knob is turned back.

What we have here is a situation in which a single given resistance value of R_2 can cause two completely different circuit results, depending on the *direction* in which R_2 is changing. This phenomenon occurs quite often in electronics and, in fact, in all of nature. Its general name is *hysteresis*. The flash-on of a triac is a specific example of hysteresis.

Example 6-2

Suppose that it requires an R_2 resistance of 5000 Ω to just barely cause the diac to break over in Fig. 6-5(a).

a. If the R_2 resistance is 6000 Ω and we reduce it to 5025 Ω, will any light be created?

b. If the R_2 resistance is 4700 Ω and we increase it to 5025 Ω, will any light be created?

c. What word would you use to sum up this behavior?

Solution

a. When $R_2 = 6000$ Ω, the diac will not be breaking over because R_2 must decline all the way to 5000 Ω to just barely cause breakover. If we then reduce R_2 to 5025 Ω, the resistance is still a little too high to allow diac breakover, so the triac is not firing, and no light is created.

b. If R_2 is 4700 Ω, this is less than the resistance which just barely causes breakover, so the diac will be breaking over and firing the triac and the lamps will be glowing. If we raise the resistance to 5025 Ω, the diac will still continue breaking over because now the capacitor always starts charging with a smaller opposite charge on its plates than it had in part a. The smaller charge results from the fact that the capacitor partly discharged on the preceding half cycle. With the diac breaking over, the triac is firing, and the lamps are giving off some light.

c. The fact that 5025 Ω coming from *above* (from 6000 Ω) caused no light, but 5025 Ω coming from *below* (from 4700 Ω) caused some light means that a given resistance value causes two completely different results, depending on the direction of approach. Therefore we can describe the behavior as showing hysteresis.

Triac hysteresis can be almost completely eliminated with the circuit of Fig. 6-8(a). To understand how it works, we must investigate the action of an SBS when a small amount of current flows in its gate lead. Refer to Figs. 6-8(b) and (c).

Figure 6-8(b) shows a resistor R inserted in the gate lead of an SBS and a certain amount of current, I_G, flowing from A2 to G. This implies that the voltage applied to the gate resistor is negative relative to A2.

If a small gate current flows between A2 and G, the forward breakover characteristic is changed drastically. The $+V_{BO}$ voltage drops to about 1 V, as shown in Fig. 6-8(c). This means that the SBS will break over as soon as the A2 to A1 voltage reaches 1 V. As the curve shows, $-V_{BO}$ is not affected by gate current from A2 to G.

Now look at Fig. 6-8(a). Suppose that R_2 is set so that the capacitor voltage cannot reach ± 8 V to make the SBS break over. The triac will not fire, and the lamp will be extinguished. During the positive half cycle, C will charge plus on the

(a)

(b)   (c)

Figure 6-8. (a) A more complex triac triggering circuit. Triac flash-on can be eliminated with this circuit. (b) Direction of gate current through the SBS as the ac supply approaches its zero crossover. (c) The very low forward breakover voltage when gate current is flowing in the SBS.

top and minus on the bottom. Now look what happens as the ac supply completes the positive half cycle and approaches 0 V. When the top supply line gets near zero relative to the bottom line, it means that the top of R_3 is near zero volts relative to the bottom of C. However, the top of C is positive relative to the bottom of C at this time, because of the capacitor charge. Therefore there is a voltage impressed between $A2$ and the top of R_3; this voltage is plus at $A2$ and minus at the top of

R_3. It forward-biases diode D_1 and causes a small amount of SBS gate current to flow. The flow path is into the SBS at $A2$, out of the SBS at G, through D_1, and through R_3. With this small I_G, even a very low forward voltage from $A2$ to $A1$ will break over the SBS, as Fig. 6-8(c) shows. There is a small forward voltage between $A2$ and $A1$ at this time, namely the capacitor voltage. As long as it is greater than about 1 volt, the SBS will break over. When the SBS breaks over, it dumps the capacitor charge through R_4. The negative half cycle of the ac supply therefore starts with the capacitor almost completely discharged. The result is that the capacitor starts charging with the same initial charge (about zero) no matter whether the triac is firing or is not firing. Therefore the triac hysteresis is eliminated.

It has been left as a question at the end of the chapter to explain the purpose of diode D_2.

6-6 UNILATERAL BREAKOVER DEVICES

The diac and the SBS are classed as *bilateral* or *bidirectional* breakover devices because they can break over in either direction. There are also breakover devices which break over in only one direction; they are in the class of *unilateral* or *unidirectional* breakover devices. We have already seen one unilateral breakover device, the four-layer diode, in Sec. 4-5. We will now take a detailed look at two unidirectional breakover devices, the four-layer diode and the *silicon unilateral switch* (SUS). Although unilateral breakover devices are more frequently seen in SCR triggering circuits, they can also be used in triac triggering circuits if they have some extra supporting circuitry.

The schematic symbols and terminal names of a four-layer diode and an SUS are shown in Figs. 6-9(a) and (b). Their characteristic current-voltage behavior is illustrated in Fig. 6-9(c).

As can be seen, the behavior of four-layer diodes and SUSs is similar to the behavior of an SBS except that only forward breakover is possible. Reverse *breakdown* can happen, but only at a much greater voltage level than $+V_{BO}$. Reverse breakdown is destructive for the device.

The SUS, like the SBS, has a gate terminal which can alter the basic breakover behavior shown in Fig. 6-9(c). By connecting a zener diode between the gate and cathode of an SUS, the breakover voltage can be reduced to

$$V_{BO} = V_z + 0.6 \text{ V}.$$

When this is done, the cathode of the zener diode must be connected to the gate of the SUS, and the anode of the zener must be connected to the cathode of the SUS.

The SUS can be fired at a very low anode to cathode voltage (about 1 volt) if gate current flows from anode to gate. This is the same kind of control as illustrated in Fig. 6-8 for an SBS.

The breakover behavior of a four-layer diode cannot be altered.

Four-Layer Diode

Figure 6-9. (a) Schematic symbol of a four-layer diode. (b) Symbol and terminal names of an SUS (silicon unilateral switch). (c) Current-voltage characteristic curve of an SUS.

Four-layer diodes are available with breakover voltages ranging from 10 to 400 V. They can carry large pulsed currents if the pulses are of short duration. Some four-layer diodes can carry 100-A current pulses.

Silicon unilateral switches are low-voltage, low-current devices. Most SUSs have a breakover voltage of 8 V and a current limit of less than 1 A.

6-7 A FOUR-LAYER DIODE USED TO TRIGGER A TRIAC

An example of a four-layer diode and its supporting circuitry in a triac gate control circuit is shown in Fig. 6-10(a).

Here is how it works. As the ac supply goes through its positive and negative alternations, the bridge rectifier delivers a full-wave rectified voltage to the *RC* timing network. This is called V_{bridge} in Fig. 6-10(c). The voltage across the capacitor, V_C, tends to follow V_{bridge}, with the amount of lag determined by the setting of R_2. At some point in the half cycle, V_C will reach the breakover voltage of the four-layer diode, shown in Fig. 6-10(d) as 20 V. When that point is reached, the four-layer diode breaks over and allows the capacitor to discharge through the primary winding of the pulse transformer.

The capacitor discharge creates a burst of current in the transformer primary, shown in Fig. 6-10(e). The current burst continues until the capacitor discharges to the point where it cannot supply current equal to the holding current of the four-layer diode.

The pulse transformer couples this current pulse into the *G-MT*1 circuit of the triac, thereby firing the triac. The gate current waveform is graphed in Fig. 6-10(f). The pulse transformer is necessary because the *RC* timing circuit must be electrically isolated from the *G-MT*1 circuit, since the *RC* circuit is driven by a bridge rectifier which is itself connected to *MT*1. That is, if the transformer were removed from Fig. 6-10(a) and the trigger circuit were connected directly to *G* and *MT*1, there would be a short circuit across the diode in the lower right-hand corner of the bridge. You should verify this for yourself.

Notice that the I_{sec} pulses all flow in the same direction regardless of the main terminal polarity. Although it has not been mentioned until now, the direction of a triac's gate current does not *have* to agree with its main terminal polarity. That is, a triac with a positive main terminal polarity can be fired by a negative gate current. Likewise, when the triac sees negative main terminal polarity, it can be fired by a positive gate current. SCRs are not like this; reverse gate current will damage an SCR.

To make this clear, the four possible triggering modes of a triac are summarized below:

a. Positive main terminal voltage, positive gate current
b. Positive main terminal voltage, negative gate current

Figure 6-10. (a) A complete schematic diagram of a triac control circuit containing a four-layer diode and a pulse transformer. (b) ac supply voltage waveform. (c) Full-wave rectified voltage (V_{bridge}) which is applied to the triggering circuit. (d) Capacitor voltage waveform, shown reaching the V_{BO} of the four-layer diode. (e) Pulse transformer primary current. (f) Inverted pulses of secondary current from the pulse transformer. (g) Load voltage waveform.

(a)

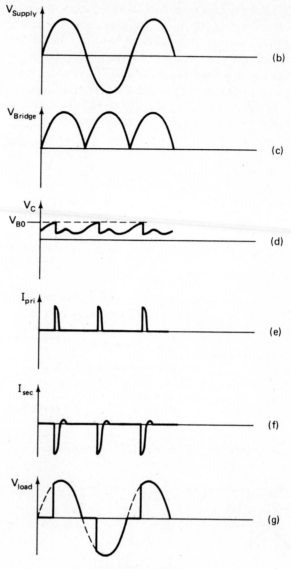

Figure 6-10. (Cont.)

    c. Negative main terminal voltage, negative gate current
    d. Negative main terminal voltage, positive gate current

The positive main terminal voltage means $MT2$ is more positive than $MT1$; the negative main terminal voltage is the opposite. Positive gate current means conventional current flows into the gate and out of main terminal 1; negative gate current is the opposite.

    Modern triacs trigger very well in modes **a**, **b**, and **c**, but they are more difficult to trigger in mode **d**. Therefore mode **d** is avoided. For this reason, whenever both main terminal polarities must be triggered by just one direction of gate current, the direction is *negative*.

A negative gate current pulse can be supplied readily just by inverting the pulse transformer secondary. This has been done in Fig. 6-10(a). Notice the transformer phasing dots indicating that when primary current flows *in* on the top lead, secondary current flows *out* on the bottom lead of the winding. This is the negative direction for gate current, as stated above. I_{sec} is shown negative in Fig. 6-10(f).

At the instant the I_{sec} pulse occurs in the gate circuit, the triac fires and causes the supply line voltage to appear across the load. This is drawn in the waveform of Fig. 6-10(g). The delay angle could be reduced or increased by adjusting R_2 to be smaller or larger.

6-8 CRITICAL RATE OF RISE OF OFF-STATE VOLTAGE (dv/dt)

In Fig. 6-10, notice the *RC* circuit connected in parallel with the triac. Such *RC* circuits are sometimes installed across triacs in industrial settings. The circuit's purpose is to prevent fast-rising transient signals from appearing across the main terminals of the triac. The reason it is necessary to eliminate fast-rising voltage surges is that all triacs have a certain *dv/dt rating*, which is the maximum rate of rise of the main terminal voltage they can withstand. If this rate of rise is exceeded, the triac may inadvertently turn ON, even when no gate signal occurs.

For most medium-sized triacs, the dv/dt rating is around 100 V/μsec. As long as any transient surges appearing across the main terminals have slopes less than the dv/dt rating, the triac will not turn ON without a gate signal. If a transient surge appears having a voltage vs. time slope in excess of the dv/dt rating, the triac may turn ON.

It should be emphasized that this is not the same thing as V_{DROM} breakover. The V_{DROM} rating of a triac is the maximum peak voltage the triac can withstand without breaking over, if that voltage is approached *slowly*. The dv/dt rating refers to *fast* voltage surges, which may have a peak value far less than V_{DROM}. The point is that even though the transient surges may be small in magnitude, their steep *slope* may cause a triac to fire.

If the ac supply line is guaranteed to be free of transient surges, the *RC* suppression circuit of Fig. 6-10 is not needed. However, in industrial settings the ac lines are usually bristling with transient surges due to switchgear operation, etc. Therefore the *RC* suppression circuit is almost always included.

Surge suppression is actually provided by the *C* part of the *RC* circuit. The 0.1-μF capacitor in Fig. 6-10 tends to short out the triac for high-frequency signals. Thus any fast transients on the ac lines are dropped across the load resistance, since *C* presents a very low impedance to anything fast.

The reason for including *R* is to limit the large capacitive discharge current when the triac turns ON under normal operating conditions. Resistor R is of no use in actually suppressing the fast transients; its only purpose is to limit the discharge of capacitor C through the main terminals of the triac.

The R and C sizes shown in Fig. 6-10 are typical. Such RC suppression circuits are also used for SCRs, because SCRs have the same problem with fast-rising transients.

6-9 UJTS AS TRIGGER DEVICES FOR TRIACS

So far, the firing delay angle of all our triac and SCR circuits has been set by a potentiometer resistance adjustment. In industrial power control, there are times when the firing point is set by a feedback voltage signal. A feedback voltage signal is a voltage which somehow represents actual conditions at the load. For example, for a lighting load, a voltage proportional to light intensity could be used as a feedback signal to automatically control the triac's firing delay angle and therefore the light which is produced; for a motor load, a voltage proportional to motor shaft speed could be used as a feedback signal to control the firing delay angle and consequently the motor speed. Whenever triac (or SCR) firing control is accomplished via a feedback voltage signal, the UJT is a popular triggering device.

Sometimes a feedback signal takes the form of a varying resistance instead of a varying voltage. In these cases too, the UJT is compatible with the feedback situation.

Figure 6-11 illustrates a frequently seen UJT triggering setup for use with feedback. In Fig. 6-11(a), which shows the complete power control circuit, resistive feedback has been depicted. Resistor R_F is a variable resistance which varies as the load conditions change. The same circuit can be adapted to voltage feedback by removing R_F and inserting in its place the voltage feedback network drawn in Fig. 6-11(b). The resistance feedback circuit is a little simpler, so we will begin with it.

6-9-1 A UJT trigger circuit with resistive feedback

Transformer T_1 is an *isolation transformer*. An isolation transformer has a 1:1 turns ratio, and its purpose is to electrically isolate the secondary and primary circuits. In this case the isolation transformer is isolating the ac power circuit from the triggering circuit. Many isolation transformers contain transient suppression components. When they do contain such components, high-frequency transient signals appearing at the primary are *not* coupled into the secondary winding, thereby helping to keep the secondary circuit noise-free.

The 115-V ac sine wave from the T_1 secondary is applied to a bridge rectifier. The full-wave rectified output of the bridge is applied to a resistor-zener diode combination, which then delivers a 24-V waveform synchronized with the ac line. This waveform is drawn in Fig. 6-11(c).

(a)

(b)

Figure 6-11. (a) A complete schematic diagram of a triac control circuit. The triggering circuit uses a UJT and a constant-current source, which is controlled by resistance feedback. (b) The same triggering circuit as part (a) except that the current source is controlled by voltage feedback. (c) The zener-clipped full-wave voltage which drives the triggering circuit. (d) The capacitor voltage waveform. It rises at a constant slope until it hits V_P of the UJT (15 V in this case). (e) Secondary current pulses from the pulse transformer. (f) Load voltage waveform.

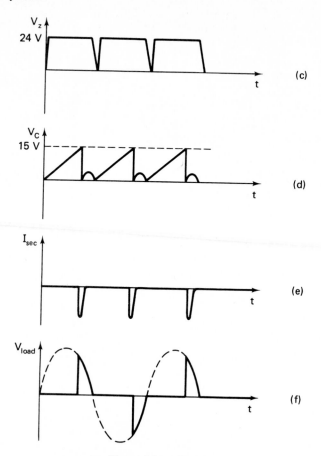

Figure 6-11. (Cont.)

When the 24-V source is established, C_1 begins charging. When it charges up to V_P of the UJT, the UJT fires and creates a current pulse in the primary winding of pulse transformer T_2. This is coupled into the secondary winding, and the secondary pulse is delivered to the gate of the triac, turning it ON for the rest of the half cycle. The waveforms of capacitor voltage, V_{C_1}, and T_2 secondary current, I_{sec}, and load voltage, V_{load}, are drawn in Figs. 6-11(d), (e), and (f). The firing delay angle is about 135° in these waveforms.

The rate at which C_1 charges is determined by the ratio of R_F to R_1. R_F and R_1 form a voltage divider. Between them they split up the 24-V dc source which supplies the trigger circuit. If R_F is small relative to R_1, then R_1 will receive a large share of the 24-V supply. This will cause *pnp* transistor Q_1 to conduct hard, because the R_1 voltage is applied to the base-emitter circuit of Q_1. With Q_1 conducting hard, C_1 charges rapidly, because C_1 is charged by the Q_1 collector current. Under these conditions the UJT fires early, and the average load current is high.

On the other hand, if R_F is large relative to R_1, then the voltage across R_1 will be smaller than before because of the voltage division effect. This causes a smaller voltage to appear across the Q_1 base-emitter circuit, reducing the Q_1 drive. With Q_1 conducting less, the charging rate of C_1 is reduced, and it takes longer to build up to the V_P of the UJT. Therefore the UJT and triac fire later in the half cycle, and the average load current is lower than before.

The C_1 charging circuit shown produces a *constant* rate of voltage rise across the capacitor, as Fig. 6-11(d) makes clear. The slope of the voltage waveform is constant because the capacitor charging current is constant. Let us now quantitatively analyze this *constant-current supply* circuit.

First, considering R_1 and R_F to be a series circuit, we can say that

$$V_{R_1} = (24 \text{ V}) \frac{R_1}{R_1 + R_F}, \tag{6-1}$$

which expresses the proportionality between voltage and resistance for a series circuit. Of course, strictly speaking, R_1 and R_F are not really in series. The base lead of Q_1 ties into the point between these two resistors; because of that, R_F carries a slight bit of extra current that R_1 does not carry. However, if these resistors are correctly sized, their current draw will far exceed the transistor's base current. With the base current very small by comparison, the percent difference between the R_F current and the R_1 current is negligible. That being so, it is alright to consider R_F and R_1 as a series circuit, and Eq. (6-1) is justified.

V_{R_1} appears across resistor R_2 and the base-emitter junction of Q_1. Since R_2 is in the emitter lead of Q_1, we can say that

$$V_{R_1} = (I_{E_1})R_2 + 0.6 \text{ V},$$

where I_{E_1} stands for the Q_1 emitter current. If Q_1 is a high-beta transistor, its collector current is virtually equal to its emitter current, so to a good approximation,

$$V_{R_1} = (I_{C_1})R_2 + 0.6 \text{ V},$$

where I_{C_1} is the collector current in the transistor and also the charging current of the capacitor C_1. Rewriting this equation for I_{C_1} and combining with Eq. (6-1), we obtain

$$I_{C_1} = \frac{V_{R_1} - 0.6 \text{ V}}{R_2} = \frac{1}{R_2}\left[\frac{(24 \text{ V})R_1}{R_1 + R_F} - 0.6 \text{ V.}\right] \tag{6-2}$$

Equation (6-2) expresses the fact that the capacitor charging current goes up as R_F goes down, and it also shows that for a given R_F the charging current is constant as time goes by.

Intuitively speaking, this circuit is able to maintain a constant current flow because it reduces the Q_1 collector-to-emitter voltage as the capacitor voltage rises. That is, for every volt that V_{C_1} rises, V_{CE} of Q_1 decreases by 1 volt. In this way, the continually rising capacitor voltage cannot retard the flow of current as it normally does in a simple *RC* circuit.

With I_{C_1} a constant, the rate of voltage buildup is constant, because for any capacitor,

$$\frac{\Delta v}{\Delta t} = \frac{I_C}{C},$$ (6-3)

where $\Delta v/\Delta t$ is the time rate of change of capacitor voltage. Therefore as long as current is constant, the voltage buildup rate is constant, as drawn in Fig. 6-11(d).

Example 6-3

For the circuit of Fig. 6-11(a), assume the following conditions: $R_1 = 5$ K, $R_F = 8$ K, $R_2 = 2.5$ K, $\beta_1 = 150$, $C_1 = 0.5$ μF, $\eta = 0.70$.

a. Find V_{R_1}.

b. Find I_{C_1}.

c. What is the rate of voltage buildup across C_1?

d. How much time elapses between the beginning of a half cycle and the firing of the triac?

e. What is the firing delay angle?

f. What value of R_F would cause a firing delay angle of 120°?

Solution

a. Assuming that R_1 and R_F are in series, we can use Eq. (6-1):

$$V_{R_1} = (24 \text{ V})\frac{5 \text{ K}}{5 \text{ K} + 8 \text{ K}} = \textbf{9.2 V.}$$

b. From Eq. (6-2),

$$I_{C_1} = \frac{1}{2.5 \text{ K}}\left[\frac{(24 \text{ V})(5 \text{ K})}{5 \text{ K} + 8 \text{ K}} - 0.6 \text{ V}\right] = \textbf{3.45 mA.}$$

c. From Eq. (6-3),

$$\frac{\Delta V_C}{\Delta t} = \frac{I_C}{C} = \frac{3.45 \times 10^{-3}}{0.5 \times 10^{-6}} = 6.9 \times 10^3 \text{ V/sec} = \textbf{6.9 V/msec.}$$

d. The capacitor must charge to V_P of the UJT, which is given by

$$V_P = \eta V_{B2B1} + 0.6 \text{ V} = (0.70)(24 \text{ V}) + 0.6 \text{ V} = 17.4 \text{ V.}$$

The time required to charge that far can be found from

$$t = \frac{V_P}{\Delta v/\Delta t} = \frac{17.4 \text{ V}}{6.9 \text{ V/msec}} = \textbf{2.5 msec.}$$

e. Let θ stand for the firing delay angle. Since 360° represents one cycle period and the period of a 60-Hz supply is 16.67 msec, we can set up the proportion

$$\frac{\theta}{2.5 \text{ msec}} = \frac{360°}{16.67 \text{ msec}},$$

$$\theta = \textbf{54°.}$$

f. For a firing delay angle of 120°, the time between zero crossover and firing is given by the proportion

$$\frac{t}{120°} = \frac{16.67 \text{ msec}}{360°},$$

$$t = 5.55 \text{ msec.}$$

The peak point of the UJT is still 17.4 V, so to delay the firing for 5.55 msec, the rate of voltage buildup must be

$$\frac{\Delta v}{\Delta t} = \frac{17.4 \text{ V}}{5.55 \text{ msec}} = 3.14 \text{ V/msec.}$$

Applying Eq. (6-3), we get a charging current of

$$\frac{I_{C_1}}{C_1} = \frac{\Delta v}{\Delta t},$$

$$I_{C_1} = \frac{3.14 \text{ V}}{1 \times 10^{-3} \text{ sec}} (0.5 \times 10^{-6}) = 1.57 \text{ mA.}$$

From Eq. (6-2), we can find R_F:

$$I_{C_1} = 1.57 \text{ mA} = \frac{1}{2.5 \text{ K}} \left[\frac{(24 \text{ V})(5 \text{ K})}{(5 \text{ K} + R_F)} - 0.6 \text{ V} \right]$$

Manipulating this equation and solving for R_F, we obtain

$$R_F = \mathbf{21.4 \text{ K}.}$$

Therefore if the feedback resistance is increased to 21.4 K, the firing delay angle would be increased to 120°, and the load current would be reduced accordingly.

6-9-2 A UJT trigger circuit with voltage feedback

As stated earlier, UJTs are also compatible with voltage feedback circuits. Mentally remove R_F from Fig. 6-11(a) and replace it with the *npn* transistor circuit shown in Fig. 6-11(b). Now the variable feedback voltage V_F controls the firing delay angle of the triac. Quantitatively, here is how it works. Applying Ohm's law to the base-emitter circuit of Q_2, we obtain

$$V_F = (I_{E_2})R_3 + 0.6 \text{ V}$$

in which I_{E_2} stands for the emitter current in transistor Q_2. Since the collector current is almost equal to the emitter current for a high-gain transistor, this equation can be written as

$$I_{C_2} = \frac{V_F - 0.6 \text{ V}}{R_3}.$$

I_{C_2}, the Q_2 collector current, is the same as the current through R_1 if we neglect the Q_1 base current. Therefore V_{R_1}, which drives Q_1, is determined by I_{C_2}. That is,

$$V_{R_1} = (I_{C_2})R_1$$

$$V_{R_1} = \frac{R_1}{R_3}(V_F - 0.6 \text{ V}). \tag{6-4}$$

From this point on, the circuit action is identical to that in the resistive feedback circuit. The greater V_{R_1}, the faster the capacitor charging rate and the earlier the UJT and triac fire. The smaller V_{R_1}, the slower the charging rate and the later the UJT and triac fire.

Notice that Q_2 has a rather large resistance in its emitter lead. This provides a high input impedance to the V_F source, resulting in easy loading on the feedback voltage source. Notice also that the feedback voltage source is electrically isolated from the main ac supply lines by virtue of transformers T_1 and T_2, which completely isolate the trigger circuitry.

The Q_2 transistor beta and temperature are made unimportant by normal emitter follower action. Namely, if Q_2 tries to conduct too hard, the voltage developed across R_3 will rise a little and will throttle back the Q_2 base current. This offsets any tendency of the transistor itself to carry a greater than normal collector current. Conversely, if Q_2 starts loafing (not conducting hard enough), the voltage developed across R_3 will drop a little and admit extra base current to offset the tendency of Q_2 to loaf. In the end, the transistor will behave in such a way that Ohm's law, Eq. (6-4), is obeyed.

Example 6-4

For a voltage feedback situation as shown in Fig. 6-11(b), assume that $\beta_2 = 200$ and $R_3 = 2$ K. All the other trigger circuit component sizes are the same as in Example 6-3.

a. If the range of feedback voltage is from 2 to 7 V dc, what is the control range of the firing delay angle?

b. What is the maximum current drawn from the feedback source?

Solution

a. For $V_F = 2$ V, Eq. (6-4) yields

$$V_{R_1} = \frac{5\text{ K}}{2\text{ K}}(2\text{ V} - 0.6\text{ V}) = 3.5\text{ V}.$$

Equation (6-2) still applies in the voltage feedback situation, so

$$I_{C_1} = \frac{V_{R_1} - 0.6\text{ V}}{R_2} = \frac{3.5\text{ V} - 0.6\text{ V}}{2.5\text{ K}},$$

$$I_{C_1} = 1.16\text{ mA}.$$

Equation (6-3) gives

$$\frac{\Delta v}{\Delta t} = \frac{I_{C_1}}{C_1} = \frac{1.16\text{ mA}}{0.5\ \mu\text{F}} = 2.32\text{ V/msec}.$$

The time to charge to V_P is

$$t = \frac{17.4\text{ V}}{2.32\text{ V/msec}} = 7.5\text{ msec}.$$

A 7.5-msec delay figures out to a firing delay angle of

$$\frac{\theta}{7.5\text{ msec}} = \frac{360°}{16.67\text{ msec}},$$

$$\theta = 162°.$$

This is the latest the triac can fire.

When $V_F = 7$ V,

$$V_{R_1} = \frac{5\text{ K}}{2\text{ K}}(7\text{ V} - 0.6\text{ V}) = 16.0\text{ V},$$

$$I_{C_1} = \frac{V_{R_1} - 0.6\text{ V}}{R_2} = \frac{16.0\text{ V} - 0.6\text{ V}}{2.5\text{ K}} = 6.16\text{ mA}.$$

The slope of the voltage ramp is then

$$\frac{\Delta v}{\Delta t} = \frac{I_{C_1}}{C_1} = \frac{6.16 \text{ mA}}{0.5 \ \mu\text{F}} = 12.32 \text{ V/msec.}$$

The time to fire is

$$t = \frac{V_P}{\Delta v/\Delta t} = \frac{17.4 \text{ V}}{12.32 \text{ V/msec}} = 1.4 \text{ msec.}$$

This figures out to a delay angle of

$$\frac{\theta}{1.4 \text{ msec}} = \frac{360°}{16.67 \text{ msec}},$$

$$\theta = 30°.$$

The control range of firing delay angle of the UJT and triac is thus **30° to 162°**.

b. The maximum current draw from the feedback source will occur when $V_F = 7$ V. It can be found by applying Ohm's law to the input of Q_2. First we must find the input impedance (resistance) of Q_2.

Recall from electronic fundamentals that the input resistance of an emitter follower is given approximately by

$$R_{\text{in}} = \beta R_E,$$

where β is the current gain of the transistor and R_E is the resistance in the emitter lead. In this case,

$$R_{\text{in}} = (200)(2 \text{ K}) = 400 \text{ K.}$$

Therefore the maximum current draw from the V_F source is given by

$$I_{\text{max}} = \frac{V_{E_{\text{max}}} - 0.6 \text{ V}}{R_{\text{in}}} = \frac{7 \text{ V} - 0.6 \text{ V}}{400 \text{ K}},$$

$$I_{\text{max}} = 16 \ \mu\text{A.}$$

QUESTIONS AND PROBLEMS

1. Draw the schematic symbol of a triac and identify all three terminals.
2. What polarity of main terminal voltage is called the *positive* polarity? What is the *negative* polarity?
3. Does the firing delay angle of a triac during the positive half cycle necessarily equal the firing delay angle during the negative half cycle?
4. Define V_{GT}. What range of values does it have for medium-sized triacs?
5. Repeat Question 4 for I_{GT}.
6. Repeat Question 4 for I_{HO}.
7. Define V_{DROM}. All else being equal, which would cost more, a triac with high V_{DROM} or a triac with low V_{DROM}?
8. Define V_{TM}. Which is considered better, a high V_{TM} or a low V_{TM}? Why?
9. A triac controls the average ac current through a load. A rheostat can do the same thing. In what ways is a triac superior to a rheostat? Is there any way in which a rheostat is superior to a triac?

10. Are triacs inherently temperature stable? What happens to firing delay angle as temperature increases, assuming everything else is constant?

11. Draw the schematic symbol and sketch the characteristic curve of a diac. What other names are used for a diac?

12. Roughly, what is the maximum difference between the magnitudes of $+V_{BO}$ and $-V_{BO}$ for a diac?

13. Compare the two situations of using a 30-V diac in the gate lead of a triac versus triggering the gate directly from the charging circuit. Which of these two situations requires a shorter time constant in the charging circuit? Explain.

14. Define *breakback voltage* of a thyristor.

15. How large is the breakback voltage of an SBS or an SUS, approximately?

16. Roughly speaking, how symmetrical are SBSs? When is symmetry important?

17. If it were desired to alter the basic characteristic of an 8-V SUS so that the breakover voltage was 2.8 V, how could this be done?

18. If it were desired to alter the basic characteristic of a 40-V four-layer diode so that the breakover voltage was 16 V, how could this be done?

19. What essential feature sets thyristors apart from other semiconductors?

20. Elaborate on this statement, explaining in your own words why it is correct: "Although a triac cannot create continuous changes in *instantaneous* current, it can create continuous changes in *average* current."

21. With the tendency of circuit designers to use low-voltage, low-power circuits whenever possible, why are high-power triacs and SCRs even necessary? What is the justification for their existence?

22. In Fig. 6-5(a), if the triac is not firing at all, explain why the capacitor starts every half cycle with a residual charge of the "wrong" polarity.

23. For the same circuit as in Question 22, if the triac is firing, does the capacitor still start every half cycle with a residual charge of the wrong polarity? Explain.

24. Try to explain the flash-on effect seen with the circuit of Fig. 6-5(a). Be sure to explain why the load power can be reduced by turning R_2 backwards once the triac has just barely started firing.

25. In general, what is hysteresis? Can you think of any good examples of hysteresis in other areas of electricity and magnetism? Can you think of any examples in other nonelectrical areas?

26. What is the purpose of diode D_2 in Fig. 6-8(a)? It will require several sentences to explain this; there is no simple terse explanation.

27. What is the difference between *breakover* and *breakdown* for an SUS? Is breakdown healthy for an SUS?

28. Which word, *breakover* or *breakdown*, would be appropriate to describe the action of a zener diode?

29. Is breakdown healthy for a zener diode?

30. List the four triggering modes of a triac. Which mode is usually avoided?

31. Why is it that we always deliver *negative* gate current to a triac whenever both main terminal polarities must be triggered by a single gate current polarity?

32. Explain the meaning of the *dv/dt* rating of a triac. Do SCRs have a *dv/dt* rating, too?

33. What simple precaution do we take to protect power thyristors from dv/dt triggering? Give some typical component sizes.

34. In Fig. 6-11(a), suppose $\eta = 0.55$, $R_1 = 10$ K, $R_2 = 1.5$ K, and $C_1 = 1$ μF. What value of R_F will cause a firing delay angle of 90°? Make a graph of V_{C_1} versus time for this situation.

35. If everything were to stay the same as in Question 34 except that a different UJT, having an $\eta = 0.75$, is substituted, what would be the new firing delay angle?

36. For the circuit of Fig. 6-11(a), suppose $R_1 = 50$ K, $R_F = 100$ K, $R_2 = 1.2$ K, $\beta_1 = 100$, and $C_1 = 1$ μF. (a) Prove that this is not a good design because the base current of Q_1 is not negligible compared to the voltage divider current through R_1 and R_F. (b) Change the values of R_1 and R_F so that the firing delay is the same but the transistor base current is less than one tenth of the voltage divider current.

37. For the circuit of Fig. 6-11(b), suppose $R_1 = 10$ K, $R_2 = 1$ K, $R_3 = 15$ K, $C_1 = 0.7$ μF, and $\eta = 0.60$.

 a. What is the firing delay angle if $V_F = 3.2$ V?
 b. What is the firing delay angle if $V_F = 8.8$ V?

38. For the circuit of Fig. 6-11(b), $R_1 = 18$ K, $R_2 = 2.2$ K, $R_3 = 25$ K, and $\eta = 0.68$. The range of feedback voltages is from 3 to 12 V. Select a size of C_1 to give a firing delay angle range of about 120°, approximately centered on the 90° mark. In other words, the firing delay angle is to vary from about 30° to about 150°. It will be impossible to get this *exact* range. Try to come as close as you can.

SUGGESTED LABORATORY PROJECTS

PROJECT 6-1: CONTROLLING AC CURRENT WITH A TRIAC

Procedure

a. Construct the circuit of Fig. 6-4(b). Use an isolated ac supply if possible. If it is not possible, take the steps described in Project 4-1.

 Use the following component sizes and ratings: $R_{load} = 100$ Ω, 100-W resistor or 100-W light bulb; triac T2302B (RCA), or any triac with a V_{DROM} of at least 200 V, an $I_{T(RMS)}$ of at least 3 A, and with similar gate characteristics. If the gate characteristics (V_{GT} and I_{GT}) are not similar to the T2302B, the triggering circuit components must be changed.

$$R_1 = 10 \text{ K}; \qquad R_4 = 1 \text{ K}\Omega;$$
$$R_2 = 250\text{-K pot}; \qquad C_1 = 0.22 \ \mu\text{F};$$
$$R_3 = 33 \text{ K}; \qquad C_2 = 0.22 \ \mu\text{F}.$$

1. What is the range of adjustment of the firing delay angle? Are the firing delays equal for both half cycles?

2. Make a sketch showing the waveforms of V_{load}, $V_{MT2\text{-}MT1}$, and V_G, all to the same time reference.

3. Measure I_{GT}, the current necessary to trigger the triac, for both main terminal polarities. This must be done by measuring the voltage across R_4 at the instant of firing and then applying Ohm's law to R_4. The scope must be connected to display

the V_{R_4} waveform. Compare the measured I_{GT} to the manufacturer's specifications, if you have them.

4. Measure V_{TM}, the voltage across the triac after firing. Compare to the manufacturer's specs.

5. Heat the triac with a soldering iron and note the effect on the firing delay angle. Does this make sense?

6. Investigate the effect of substituting different triacs of the same type number. Explain your results.

b. Insert a diac in the gate lead of the triac. Use a TI43A diac or any equivalent diac with about the same breakover voltage rating (32 V). Change the following component sizes:

$$R_2 = 200\text{-K or } 250\text{-K pot}, \qquad C_1 = 0.1\ \mu\text{F},$$
$$R_3 = 4.7\ \text{K}, \qquad\qquad\qquad C_2 = 0.02\ \mu\text{F}.$$

Leave everything else the same.

1. Are the firing delay angles the same for both half cycles? Why?

2. Investigate the effects of heating the triac. Explain your results.

3. Investigate the effects of substituting different triacs of the same type number. Explain your results.

4. Investigate the effects of substituting different diacs of the same type number. Comment on your results.

PROJECT 6-2: DIAC AND SBS CHARACTERISTICS

Procedure

Construct the circuit shown in Fig. 6-12. It will furnish a display of the current-voltage characteristic curve of the devices being tested. Adjust the variable ac supply to zero before turning on the power every time a new test is made.

The horizontal signal applied to the scope represents the voltage applied to the breakover device. The vertical signal represents the device's current, because this signal is developed across the 100-Ω resistor in series with the device, which carries the same current as the device. The current sensitivity of the oscilloscope is then given by Ohm's law,

$$\text{amps/cm} = \frac{\text{volts/cm}}{1000}.$$

a. Insert a diac and adjust the ac supply until the diac's breakover points can be seen.

1. Measure $+V_{BO}$ and $-V_{BO}$. Are they nearly equal in magnitude?

2. What is the breakback voltage?

3. What is the holding current (the minimum current necessary to maintain conduction once it has begun)?

4. Investigate the temperature stability of the diac. Make comments on your findings.

5. Investigate the batch spread between different diacs of the same type. Comment on your findings.

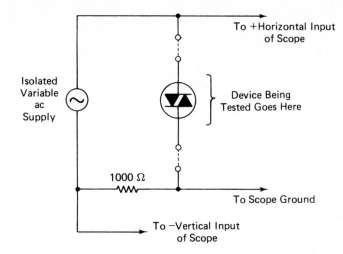

Figure 6-12. Circuit for displaying on an oscilloscope the current-voltage characteristic curve of a breakover device.

b. Insert an SBS (MBS 4991 or equivalent) in place of the diac. Adjust the ac supply until the breakover is visible on the scope screen. Repeat steps 1–5 of part a. Make a comparison between the results you obtained for the two devices.

c. Investigate the alteration in SBS characteristics which can be brought about by using the gate.

  1. Insert a low-voltage zener diode (V_z less than 6 V) as described in Sec. 6-5-2 to alter $+V_{BO}$. Sketch the current-voltage curve obtained. Does this agree with what you expected?

  2. Repeat step 1 to alter $-V_{BO}$.

  3. Connect two zeners having different zener voltages so that both $+V_{BO}$ and $-V_{BO}$ are altered. Sketch the curve and explain.

  4. With the zeners removed, insert a 2.2-K resistor between G and $A1$. Observe and sketch the new characteristic curve. Does it bear out what you know about SBSs? Repeat with the gate resistor connected to $A2$.

PROJECT 6-3: A 100-V rms VOLTAGE REGULATOR

Note: A true rms voltmeter is needed to perform this project. Without such a voltmeter, the regulation of the circuit cannot be observed.

The circuit of Fig. 6-13 is an rms voltage regulator using resistive feedback. It will maintain the lamp voltage at 100 V rms in the face of the line voltage varying anywhere from 110 to 250 V rms.

The amount of light given off by the lamp depends on the true rms value of the voltage across it, since the rms value tells the power-transferring capability of the voltage supply. Therefore if the amount of light given off by the lamp is held constant, the rms

The photocell should have a
Resistance of about 10 K at
2 foot candles of Illumination.
Its Light to Dark Ratio should
be about 100:1

Figure 6-13. A circuit which will regulate the rms voltage applied to
the lamp. The photocell provides resistive feedback to the triggering
circuitry. (*Courtesy of Motorola, Inc.*)

voltage across the lamp terminals is being held constant. The circuit of Fig. 6-13 tries to
do just that, namely hold the light output constant.

The lamp can be a 100-W incandescent bulb, mounted in a box lined on the inside with
aluminum foil. A hole is punched in the box, and a hollow cardboard tube is inserted in
the hole. The tube can be the kind that kitchen paper towels are wrapped on. A photocell
is mounted at the other end of the tube in such a way that it cannot detect ambient light.
It responds only to the light reflected from the aluminum foil down the cardboard tube.

The photocell resistance is the feedback resistance in this circuit, R_F. Its position is
different from the position shown in Fig. 6-11(a), which had R_F on the bottom. In this
circuit R_F is on top. The proper location depends on whether R_F increases as load power
increases or whether R_F decreases as load power increases. In this example, the resistance
of a photocell *decreases* as load power increases (as light intensity increases), so R_F must
be on top. If the opposite had been true, R_F would have been placed in the bottom position.

Here is how the circuit attempts to maintain a constant light output. If the line voltage
increases, tending to increase the light output, more light impinges on the photocell,
thereby lowering its resistance. As R_F decreases, the voltage developed across the photocell
becomes a smaller portion of the 24-V dc supply voltage. This reduces the drive on transistor Q_1, thereby lowering the C_1 charging rate and causing the triac to fire later. The

later firing compensates for the higher line voltage, and the rms lamp voltage increases by only a very small amount.

On the other hand, if the ac line voltage decreases, tending to lower the rms lamp voltage, the reduced light output will cause R_F to increase. This allows R_F to receive a larger share of the 24-V dc supply voltage. The Q_1 drive voltage therefore rises, causing C_1 to charge faster and the triac to fire earlier. The earlier firing cancels the decrease in ac line voltage, and the rms lamp voltage is held almost constant.

To test this circuit, connect a *true rms voltmeter* across the lamp. Adjust potentiometer R_2 so that the load voltage is 100 V rms when the ac line voltage is 110 V rms (line voltage can be measured on a standard peak-detecting voltmeter, since it is a sine wave). Then increase the line voltage to 250 V rms (if possible) and adjust R_3 so that the lamp voltage is still 100 V rms. You may have to go back and forth between these adjustments a few times, since the pots will interact with each other. If the ac line voltage cannot be raised all the way to 250 V rms, take it as high as it can go and then make the R_3 adjustment.

When the adjustments are complete, you should be able to vary the ac line voltage anywhere from 110 to 250 V rms with the lamp voltage maintained at 100 V rms $\pm$ 2 V.

Do this in equal increments of line voltage and make a table showing ac line voltage, true rms lamp voltage, and firing delay angle (measured on an oscilloscope).

Note that when the firing delay angle is in the 90° neighborhood, it takes only a small change in firing delay to compensate for a given change in line voltage. Yet when the delay angle is far from 90°, either above or below, the circuit will produce a greater change in firing delay to compensate for the same change in line voltage. Can you explain this?

If you are mathematically inclined, you might find it interesting to integrate some of the load voltage waveforms. Look in a good engineering calculus book to see how to integrate these waveforms to solve for the rms values. The integration is not easy because you are not integrating the load voltage itself, but the *square* of the load voltage. The "s" in "rms" stands for *square*.

7

An Industrial Automatic

Welding System

with Digital Control

In this chapter we will explain the operation of an automatic welding system. The system presented is a slightly simplified version of a real automobile wheel-welding system capable of a production rate of 600 wheels per hour. Although system operation is explained in terms of *wheel* welding, the design of the system has much in common with other welding operations utilizing the basic industrial automatic welding sequence of (1) **Squeeze**, (2) **Weld**, (3) **Hold**, (4) **Release**, (5) **Standby**. In this chapter, those five words will be written in boldface whenever they refer to the specific intervals within the automatic welding sequence.

OBJECTIVES

After completing this chapter, you will be able to:

1. Explain in detail how a solid-state logic system receives information from its signal converters about conditions in the apparatus being controlled
2. Explain how a solid-state logic system exerts control over automated machinery through its output amplifiers

3. Explain how the system operator can set automatic cycle specifications on selector switches and how those specifications are entered into the memory devices of the logic system

4. Explain how the logic system keeps track of the progress of the automated cycle, knowing which steps have been completed and which step is to come next

5. Discuss the *block diagram* approach to complicated industrial systems, and explain the advantages of breaking up a complicated system into small subcircuits

6. Interpret a block diagram, identifying which blocks interact with each other and which *direction* the interaction takes

7. Explain in detail the often-seen industrial practice of shifting, or presetting, selector switch settings into a down-counter

8. Discuss the common practice of using the ac power line oscillations to "time" the occurrence of events

9. Explain in detail the action of a decoder circuit in converting a sequence of binary bits into useful form

10. Explain in detail the action of a diode encoding matrix in converting selector switch settings into a form compatible with down-counters

11. Show why germanium diodes are preferred to silicon diodes in an encoding matrix

12. Explain the use of buffers (drivers) to prevent digital signal degeneration

13. Name the four variables which are adjusted to produce the best possible weld from an automatic welder, and explain with waveform drawings the meaning of each variable

14. Describe an ignitron, and point out its advantage over SCRs

15. Show how an ignitron can be fired by an SCR

16. Describe how welding current conduction angle is controlled by an ignitron-SCR-UJT control circuit

17. Discuss the necessity of discharging a UJT's emitter capacitor during the negative half cycle in the control circuit above

18. Describe the saturation problem of welding transformers. Using waveform drawings, show how the saturation problem can be solved

19. Describe the operation of a delta-connected three-phase welding transformer, and explain why the conduction angle per phase must be limited to 120°

7-1 A PHYSICAL DESCRIPTION OF THE WHEEL-WELDING SYSTEM

The arrangement of the wheel handling and lifting mechanism, and the welding electrodes and their associated hydraulic controls is shown in Fig. 7-1.

The relationship between the wheel *rim* and the wheel *spider* is shown in Fig. 7-1(a), which is a top view of a spider resting inside a rim. The *rim* of an automobile wheel is the circular outside part that the tire mounts on. The *spider* is the flange-like middle part which contains the holes for the wheel studs and the center hole for the axle-bearing cap. The spider is welded to the rim to form a complete wheel.

Wheel Spider

Ears of
Lifting Cradle

Wheel Rim

Welds are
Made Here

Welds are
Made Here

(a)

X: Welding Electrodes
Y: Electrode Insulator
Z: Electrode Power Leads

DISENGAGE
ELECTRODES

PS1

ENGAGE
ELECTRODES

Z   Z   Z   Z

X X   X X

Y Y   Y Y

Wheel Spider
(Inside Rim)

Wheel Rim

Lifting Cradle

Lower Wheel
Out of Welding
Position

Raise Wheel
into Welding
Position

Lift Cylinder

(b)

Figure 7-1. (a) Top view of a wheel spider resting on the inner lip of a wheel rim. The ears of the lifting cradle restrain the wheel rim from sliding horizontally. (b) Side view showing the physical layout of the welding mechanism. The notes indicate the functions of the various hydraulic lines.

The spider is so named because it is welded to the rim in *eight* places. Figure 7-1(a) shows that the spider has four flaps, which are the protuberances that rest on the inside lip of the rim. Each flap is spot-welded to the rim in two places, making eight welds in all.

Figure 7-1(b) shows a side view of a spider-rim combination resting in the *lifting cradle*. The spider could not really be seen through the solid metal of the wheel rim, even though the spider is seen in this figure. The lifting cradle is positioned underneath the hydraulic cylinders which engage and disengage the welding electrodes. The lifting cradle is raised and lowered by the *lift cylinder*. When the lift cylinder extends, it raises the lifting cradle, thus positioning the wheel rim and spider so that the welding electrodes can engage and produce a weld. When the weld is complete and the electrodes have disengaged, the lift cylinder retracts, lowering the lift cradle and wheel.

There is a roof-surface above the welding electrode cylinders which holds the wheel in perfect alignment for the electrode cylinders, but for the sake of simplicity this is not shown in Fig. 7-1(b). Likewise, the two limit switches which detect when the wheel rim is perfectly positioned against the roof-surface have been deleted from Fig. 7-1(b) to keep the drawing uncluttered.

When the lifting cradle has been raised and the wheel is in position, the hydraulic line marked ENGAGE ELECTRODES is pressured up, causing all four electrode cylinders to extend. This brings the electrodes into electrical contact with the wheel rim on the outside and with the spider on the inside. Hydraulic pressure switch PS1 is set to detect when the motion of the cylinders has stopped, which means that the electrodes are pressing against the metal surfaces to be welded. Figure 7-1(b) shows only two pairs of welding electrodes. As stated before, the spider is welded to the rim in eight separate places, so there are really eight pairs of electrodes. Only two pairs are shown in order to keep the drawing simple.

After the welds have been made, the ENGAGE ELECTRODES hydraulic line is relieved, and the DISENGAGE ELECTRODES hydraulic line is pressured up. This causes the electrode cylinders to retract, disengaging the electrodes. The lift cylinder then retracts, lowering the finished wheel.

7-2 THE SEQUENCE OF OPERATIONS IN MAKING A WELD

When the wheel is in position to be welded, the welding electrodes come forward to engage the metal, as explained in Sec. 7-1. Once the electrodes have engaged the metal, they are allowed to press against the surfaces for a short time before the welding current is turned on. This is done to allow the electrodes to conform to the curvature of the surfaces and to make perfect electrical contact. This portion of the overall welding sequence is called the **Squeeze** interval. The time allotted for this interval in the welding sequence is called *squeeze time*, and it can be adjusted by the system operator.

Squeeze time begins when the electrode cylinder hydraulic pressure is up to its normal value, as detected by PS1 in Fig. 7-1(b). Squeeze time is usually about 1 second. When the squeeze time has elapsed, the **Squeeze** interval is complete and the **Weld** interval begins.

During the **Weld** interval, the *welding transformer* [not shown in Fig. 7-1(b)] is energized. Current flows down the electrode power leads to the electrodes and through the metal-to-metal contact between wheel rim and spider, thereby creating the weld. The **Weld** interval usually takes from 2 to 10 seconds.

The welding current does not flow continuously during the **Weld** interval. It is turned on and off in short bursts, called *pulsations*. The operator sets the number of pulsations which are used to create the weld.

Besides the number of pulsations, the number of cycles of current which flow during a single pulsation is also adjusted by the system operator, as is the number of cycles "missed" between pulsations. Figure 7-2(a) shows a graph of current versus time during the **Weld** interval, assuming that the welding current flows during the entire 180° of an ac half cycle.

In Fig. 7-2(a), it can be seen that welding current flows for three ac cycles. This is followed by an absence of current for two cycles. At the end of these two cycles, the current is turned on for another three cycles. Each time three cycles of current are completed, the system is said to have completed one current *pulsation*.

The example given here shows three cycles of current flow followed by two

Figure 7-2. Waveforms of welding current pulsations. (a) Three cycles of rectified current flow, followed by two cycles of current absence. This means that the **Heat** subinterval is set to three cycles and the **Cool** subinterval is set to two cycles. During the **Heat** subinterval, current flows for the entire 180° of a positive half cycle. (b) The same situation, except that current flows for only 90° of a positive half cycle.

Current

(a)

Current

(b)

cycles of current absence. It should be understood that these numbers are adjustable. The operator could select five cycles of current flow followed by three cycles of current absence, or eight cycles of flow followed by two cycles of absence, etc.

The portions of the **Weld** interval during which welding current is flowing are called the **Heat** subintervals. The portions of the **Weld** interval during which the current is absent are called the **Cool** subintervals. The number of cycles in the **Heat** and **Cool** subintervals are adjusted to suit the alloy type and the gauge of the metal.

We have seen that the operator adjusts the number of pulsations in the **Weld** interval, the number of ac cycles in the **Heat** subinterval, and the number of ac cycles in the **Cool** subinterval. In addition to these variables, the number of degrees per half cycle that welding current flows is also adjustable. This number of degrees per half cycle during which current actually flows is called the *conduction angle*. Figure 7-2(b) shows a welding current waveform in which the conduction angle is approximately 90°. Alloy type and gauge of the metal and the type of electrode material are taken into account when setting the conduction angle.

Much research has been done to determine the best combination of these four variables for each different welding situation. Each variable has some effect on the final quality of the weld. For every welding situation encountered, the system operators refer to researched tables of settings for each of these four variables (pulsations per **Weld** interval, cycles per **Heat** subinterval, cycles per **Cool** subinterval, and conduction angle). In this way the system produces welds of the highest possible quality. Because all the variables are held perfectly consistent from one weld to another, the weld strength is also perfectly consistent. Of course, consistency is the benefit that arises from any automated machinery.

When the proper number of current pulsations have been delivered, the system leaves the **Weld** interval and enters the **Hold** interval. During the **Hold** interval, electrode pressure is maintained on the metal surfaces, but the welding current is turned off. The purpose of the **Hold** interval is to allow the fused metal of the weld to harden before the mechanical force exerted by the electrodes is removed from the wheel. This prevents any distortion of the wheel while the welded metal is in its molten state.

At the end of the **Hold** interval, which usually lasts about 1 second, the system enters the **Release** interval. During the **Release** interval, the welding electrode cylinders are retracted, releasing the wheel from the electrodes.

When the **Release** interval is complete, the system enters the **Standby** interval, during which the lift cylinder is retracted, lowering the finished wheel from the welding location. Once down, the wheel is removed from the lifting cradle. The system remains in **Standby** until a new wheel rim and spider are loaded onto the lifting cradle, and the lift cylinder is once again given the signal to extend.

To recap and summarize, the entire welding sequence consists of five intervals. In order of occurrence, they are **Standby**, **Squeeze**, **Weld**, **Hold**, **Release**, return to **Standby**.

Once entered, the **Standby** interval is maintained until a new wheel is loaded onto the lifting cradle and the signal is given to lift it up.

After the **Standby** interval comes the **Squeeze** interval, during which the welding electrodes make contact with the wheel and press against it. The amount of time spent in this interval is determined by how long it takes the welding electrode cylinders to extend and make firm contact, followed by how much squeeze time the system operator has set. The squeeze time is set by positioning two 10-position selector switches. The circuit details of these selector switches and associated circuits are explained in Sec. 7-6. At this point it is only necessary to realize that the settings of the selector switches determine how many time increments must elapse to complete the **Squeeze** interval. One time increment in this system is the period of the ac line, namely $\frac{1}{60}$ sec (16.67 msec). Therefore the number set on the 10-position selector switches can be thought of as being the number of cycles the ac line must make, in order to complete the **Squeeze** interval.

The **Squeeze** interval is followed by the **Weld** interval. The system stays in the **Weld** interval until the proper number of welding current pulsations are delivered to the electrodes. This number of current pulsations is set by positioning two other 10-position selector switches. The number of cycles *per* pulsation (the number of cycles of the **Heat** subinterval) is also set on yet another pair of 10-position selector switches. The same is true for the number of cycles *between* pulsations (the number of cycles of the **Cool** subinterval); this number is also set on a pair of 10-position selector switches.

When the desired number of welding current pulsations has been delivered, the system leaves **Weld** and enters the **Hold** interval. The time spent in the **Hold** interval is again set on a pair of 10-position selector switches. The counting of the ac cycles begins immediately when the **Hold** interval is entered. In the **Squeeze** interval, by contrast, there is a considerable delay before the count of ac cycles begins; this delay occurs as the system waits for the welding electrode cylinders to stroke and for the hydraulic pressure on the ENGAGE ELECTRODES line to reach normal. There is no delay at all in the **Hold** interval.

When the **Hold** interval has counted out (interval has been completed), the system enters the **Release** interval. Timing of the **Release** interval again is set by two 10-position selector switches. The number is set to allow adequate time for the electrode cylinders to retract, freeing the wheel from contact with the welding electrodes. When the **Release** interval has counted out, the system enters **Standby**, which causes the cradle containing the finished wheel to be lowered. This completes one welding sequence. An entire sequence takes from 6 to 15 seconds.

7-3 BLOCK DIAGRAM OF THE SEQUENCE CONTROL CIRCUIT

Figure 7-12 at the end of Chapter 7 shows the complete logic circuit for controlling the wheel-welding sequence. Such large diagrams are difficult to understand all in one chunk. Instead, with complex circuits of this type, it is better to break them up into several small parts (subcircuits). We can then concentrate on these small subcircuits one at a time without being overwhelmed by the entire circuit.

7-3-1 A complex system broken up into small subcircuits or blocks— explanation of the block diagram approach

In Fig. 7-3, the complete circuitry for controlling the welding sequence has been broken down into subcircuits, with each subcircuit identified by a block. The circuitry contained within a single block has a certain single purpose which contributes to the overall operation of the system. Before studying the specific circuitry of any individual block, it is helpful to understand how each block fits into the overall control scheme. In this section we will try to gain an understanding of the purpose of each block and an understanding of how each block interacts with the other blocks. In the subsequent sections of this chapter we will make a careful study of the specific circuitry of each block.

In further discussion of the welding system, a distinction is made between the terms *welding sequence* and **Weld** *interval*. In fact we have already started making a distinction between them. *Welding sequence* refers to the complete sequence of actions necessary to weld a wheel, from **Standby**, through each interval of the sequence, back into **Standby** again. *Weld interval*, on the other hand, refers to that portion (interval) of the sequence during which welding current pulsations are being delivered to the wheel surfaces.

In Fig. 7-3 there are nine blocks. Each block has been assigned a short name which roughly describes its function. Each block has also been assigned a letter, A–I.

The lines between blocks show that there is direct interaction between those blocks or, more precisely, that there is wiring running between the circuitry of those blocks. The arrowhead denotes the *direction of flow* of information, from signal *sender* to signal *receiver*. In this way, the line running out of block A into block B shows that there is interaction between block A and block B and furthermore that block A is sending the information and block B is receiving the information.

The fact that Fig. 7-3 shows only one *line* running from one block to another should not be interpreted to mean that there is only one *wire* running between the circuits in the actual wiring. There may be many wires running between those circuits. Figure 7-3 symbolizes the flow of information only; it is not an exact wiring diagram.

In the discussion which follows, a block connection line will be identified by two letters which show the two blocks being connected. The first letter denotes the sending block and the second letter the receiving block. For example, line AB would be the line that goes from block A to block B. Line BE is the line that goes

Figure 7-3. Block diagram of the entire wheel-welding system. Each block contains a subcircuit which has a certain specific purpose within the overall control scheme.

from block B to block E. Line EB is the line that goes from block E to block B. Note that a given block may *send* information *to* another block as well as *receive* information *from* that block.

7-3-2 How the sequence initiation circuit (block A) fits into the overall system

Block A, the sequence initiation circuit, has the function of lifting a new wheel into position for welding, and detecting when the wheel is properly positioned. It signals this to the interval triggering and gating circuit on line AB, allowing block B to step the system out of **Standby** and into **Squeeze**. Just how block B performs this action is explained in the description of the interval triggering and gating circuit in Sec. 7-3-3. When block A receives the signal via line CA that the system has entered the **Squeeze** interval, it causes the welding electrodes to come forward and engage the wheel rim and spider. When the electrode cylinder hydraulic oil pressure is high enough, meaning that the squeeze time can begin, this condition is detected and signaled to block B. Thereafter block A drops out of the picture until the system enters the **Release** interval.

As the system enters the **Release** interval, that action is signaled to block A on line CA, causing block A to retract the welding electrode cylinders and to release the wheel. After the **Release** interval is over and the system reenters the **Standby** interval, block A receives that information on line CA. At that time it lowers the cradle containing the finished wheel.

In summary, the sequence initiation circuit has the responsibility to raise and lower the wheel and to engage and disengage the welding electrodes at the proper times. It also sends out signals telling the interval triggering and gating circuit when the system is to enter the **Squeeze** interval and when the squeeze time is to begin.

7-3-3 How the interval triggering and gating circuit (block B) fits into the overall system

The interval triggering and gating circuit, block B, has the job of receiving the information that an interval is complete from the interval time counter. It receives this information on line EB and then acts on that information. The actions it provides are

a. It triggers the interval stepping circuit on line BC, causing that circuit to step into the next interval of the welding sequence.

b. It signals the interval time counter preset circuit on line BD, telling it to preset the interval time counter to the proper numbers. These numbers have been selected on 10-position selector switches, as mentioned in Sec. 7-2.

c. After the first two actions have been completed, it *gates* 60-Hz pulses into the interval time counter to begin the actual timing of the new interval. This is done on line BE. The term *gates* means "allows to pass." The gating is done by a logic gate of the type studied in Chapter 2.

The **Weld** interval is the only exception to action c. During the **Weld** interval the count pulses which are passed to the interval time counter do not occur at 60 Hz; instead one count pulse is delivered each time a pulsation of welding current is finished. Note that there is a line running from block F to block B. During the **Weld** interval, line FB delivers a count pulse each time a welding current pulsation is finished. Block B then gates *these* count pulses into the interval time counter, instead of the usual 60-Hz pulses.

7-3-4 How the interval stepping circuit (block C) fits into the overall system

The interval stepping circuit receives a pulse on line BC each time block B learns that an interval is complete. This pulse causes the circuitry of block C to advance one step. This advancing is done by triggering flip-flops, as will be described in detail in Sec. 7-5. Thus the interval stepping circuit is composed of flip-flops, and the states of the flip-flops indicate which interval the system is in at any point in time. The information about which interval the system is currently in is important to several other subcircuits in the system, as can be seen from Fig. 7-3. The block diagram shows lines leading from block C to blocks A, B, D, F, and G, indicating that information regarding the system's interval is sent to all those blocks. The reason each subcircuit needs to know which interval the system is currently in will become clear when we study the specific circuitry of each block.

The interval stepping circuit is accompanied by a decoder, as the label in block B says. The decoder converts the states of the flip-flops into a useful signal for routing around to the various subcircuit blocks. The decoder has one output line for each interval, five output lines altogether.

For example, if the flip-flops in the interval stepping circuit are indicating that the system is currently in the **Hold** interval, then the **Hold** output line of the decoder would go HI, while the **Standby**, **Squeeze**, **Weld**, and **Release** output lines would all be LO. By detecting which one of the five decoder output lines is HI, the other subcircuits can then tell which of the five intervals the system is currently in.

7-3-5 How the interval time counter preset circuit (block D) fits into the overall system

Block D, the interval time counter preset circuit, has the job of setting the correct two-digit number into the interval time counter. It does this immediately after the system enters a new interval. As mentioned in Sec. 7-2, each interval except **Standby** has two 10-position selector switches associated with it, on which the system operator selects the desired number of time increments (ac line cycles) for that interval. The **Weld** interval is different in this regard, as stated before.

The interval time counter preset circuit decides which pair of 10-position selector switches to read, depending on which sequence interval the system has just entered. It knows which interval has been entered by way of connection line CD. The interval time counter preset circuit then shifts the numbers from those selector switches into the interval time counter. It does this via line DE.

7-3-6 How the interval time counter (block E) fits into the overall system

The interval time counter is the circuit which actually counts the time increments during **Squeeze**, **Hold**, and **Release** and the welding current pulsations during **Weld**. It is a *down*-counter, counting backwards from its preset number to zero. When it reaches zero, it sends a signal on line EB to the interval triggering and gating circuit that the interval is complete and that therefore the system is ready to step into the next interval.

For example, if the preset number shifted into the interval time counter for the **Hold** interval is 45, the counter will count down one unit for each ac line cycle that occurs. After 45 ac line cycles, taking $\frac{45}{60}$ sec, the counter will reach zero. When this happens, it signals block B that the system is ready to be triggered into the next interval, the **Release** interval in this example.

7-3-7 How the heat-cool stepping and gating circuit (block F) fits into the overall system

The heat-cool stepping and gating circuit has several functions:

a. When the system is in the **Weld** interval, this circuit steps the system back and forth between the **Heat** and **Cool** subintervals.

b. Whenever a new subinterval is entered, it sends a signal on line FG to the heat-cool counter preset circuit, telling that circuit to preset the proper number into the heat-cool counter.

c. Once the presetting has been accomplished, the heat-cool stepping and gating circuit gates 60-Hz pulses down line FH into the heat-cool counter so it can count the number of ac cycles in the subinterval. When a **Heat** or **Cool** subinterval is completed, the heat-cool stepping and gating circuit receives this information from the heat-cool counter via line HF. This is how it knows when to step to the next subinterval [when to perform function a].

To summarize the purposes of the heat-cool stepping and gating circuit, it takes care of routing the proper signals to those blocks which have to do with the **Heat** and **Cool** subintervals, blocks G, H, and I. It also sends out a signal whenever a welding current pulsation is finished, so that other subcircuits (blocks B and E) can keep track of the progress of the **Weld** interval itself.

7-3-8 How the heat-cool counter preset circuit (block G) fits into the overall system

The heat-cool preset circuit is identical in concept to the interval time counter preset circuit. The heat-cool counter preset circuit shifts a two-digit number from a pair of selector switches into the heat-cool counter. This takes place on line GH. There is a pair of 10-position selector switches which determines the number of ac line cycles in the **Heat** subinterval and another pair of switches which determines the number of cycles in the **Cool** subinterval. The heat-cool counter preset circuit reads the correct pair of switches, depending on which subinterval the system is in. It has access to this information by way of line FG.

7-3-9 How the heat-cool counter (block H) fits into the overall system

Likewise, the heat-cool counter is identical in concept to the interval time counter. It counts backwards from the preset number down to zero, making one backward count for each cycle of the ac line. When it reaches zero, it sends out the signal that the subinterval is finished. This signal is sent to block F on line HF, informing block F that it is time to step into the next subinterval.

7-3-10 How the weld power circuit (block I) fits into the overall system

The weld power circuit, block I, receives a signal from block G whenever the system is in the **Heat** subinterval of the **Weld** interval. When the **Heat** signal is received, the weld power circuit energizes the welding transformer and thus delivers current to the welding electrodes. How this is done and how the weld power circuit adjusts the conduction angle will be discussed in detail in Sec. 7-9.

7-4 A DETAILED DESCRIPTION OF THE SEQUENCE INITIATION CIRCUIT AND THE INTERVAL TRIGGERING AND GATING CIRCUIT

Starting in this section, and through Sec. 7-9, we will look closely at the details of operation for each of the subcircuits of Fig. 7-3. Before we can do this effectively, we must decide on certain rules of notation. The rules we will use are explained in Sec. 7-4-1.

7-4-1 Notation used on schematic drawings and in the written text

Figure 7-4 is a schematic diagram showing the sequence initiation circuit and the interval triggering and gating circuit. Notice that some wires are labeled with capital letters. Each of these letters refers to a note at the bottom of the drawing,

Figure 7-4. Schematic diagram of the Sequence Initiation Circuit (block A) and the Interval Triggering and Gating Circuit (block B). The circles represent input and output terminals, which connect to other subcircuits (other blocks).

which explains the significance of that wire, telling what that wire accomplishes in the operation of the circuit. In the explanatory text, these wires will be identified by their letter labels.

If there is a set of parentheses following the explanatory note, the condition identified in the parentheses is the condition necessary for the wire to accomplish its purpose. For example, note Y reads "enable the count gate (LO)." This means that the wire labeled Y is the wire that enables the count gate to pass count pulses and that wire Y allows these count pulses to be passed only when it is LO.

When a wire comes into a subcircuit from another subcircuit, its function is identified by a circular terminal having a label which is a word or short phrase. For example, the circular terminal labeled "release" in Fig. 7-4 indicates that the wire attached to that terminal originally comes from a **Release** terminal in some other subcircuit and that the terminal goes HI when the system enters the **Release** interval.

As another example, the wire attached to the terminal "interval time counter is counted out (LO)" comes from another subcircuit. When that terminal goes LO, it means that the interval time counter has counted out (counted down to zero). A precise explanation of the action of that terminal will be given in the written text.

Furthermore, the particular subcircuit being illustrated in a schematic drawing will have *outputs going to* other subcircuits as well as *inputs coming from* other subcircuits. Labeled circular terminals are used to indicate this situation too. When this is done, you can expect to come across that same label on an input terminal in some other subcircuit schematic. For example, the terminal labeled "preset the interval time counter (HI)" has a wire attached to it which goes off to some other subcircuit. The schematic diagram of that other subcircuit will show an input terminal with exactly the same label.

You cannot confuse input terminals with output terminals because input terminals are always wired to the *inputs* of solid-state gates, flip-flops, etc., while output terminals are always wired to the *outputs* of the solid-state circuit devices.

Regarding the identification of circuit parts in the explanatory text, here is the legend we shall follow.

Specific intervals and subintervals of the welding sequence are capitalized and written in boldface; examples: **Release, Standby, Weld, Heat, Cool**.

Particular circuit devices (gates, flip-flops, etc.) appearing in the drawings having specific names are written in all-capital letters; examples: INTERVAL STEPPING ONE-SHOT, LIFT WHEEL PUSH BUTTON, SIGNAL CONVERTER 2, NOR3.

Specific subcircuits which are defined in the block diagram of Fig. 7-3 are written capitalized; examples: Interval Time Counter, Interval Stepping Circuit, Heat-Cool Counter Preset Circuit.

Terminal labels and wire descriptions (notes) are enclosed in quotation marks; examples: "welding current pulsation count pulses," "step the Interval Stepping

Circuit (neg. edge)," "60-Hz pulses." An exception is if the terminal label is one of the specific intervals or subintervals of the system. Then it is written in boldface. Example: **Squeeze**.

7-4-2 The circuit operation

The welding sequence begins when the operator presses and holds the LIFT WHEEL PUSH BUTTON on the left side of Fig. 7-4. In the production process, this would be done when the operator sees that a wheel rim and spider have been properly loaded onto the lifting cradle, as described in Sec. 7-1. If the process were completely automated, there would be a solid-state contact or a relay contact instead of the push-button switch. At any rate, the application of 115 V ac to the input of SIGNAL CONVERTER 3 will cause a voltage of $+15$ V dc, a HI, at the input of OR2. The bottom input of OR2 is LO at this time, due to the fact that the **Standby** terminal is HI while the system is in **Standby**. The output of OR2 goes HI, causing output amplifier OA1 to energize relay RLW. The normally open contact of relay RLW in the lower left of Fig. 7-4 closes, applying 115 V ac to the LIFT WHEEL SOLENOID. The energization of this solenoid shifts the hydraulic valve which causes the lift cylinder to extend. When the lifting cradle has lifted the wheel rim and spider into proper position to be welded, the limit switches in the upper left of Fig. 7-4 close their contacts. This applies 115 V ac to SIGNAL CONVERTER 1, causing a HI at the input of inverter I1. When this HI appears, the output of I1 goes LO, causing a negative edge on wire X. This negative edge appears at the T_1 terminal of the INTERVAL STEPPING ONE-SHOT.

The INTERVAL STEPPING ONE-SHOT has two trigger terminals, T_1 and T_2. It will fire when a negative edge appears at either one of its trigger terminals. Therefore the negative edge at T_1 causes the one-shot to fire, delivering an output pulse 100 μsec in duration. As it fires, the $\bar{Q}$ output goes LO, creating a negative edge at the terminal labeled "step the Interval Stepping Circuit (neg. edge)." This causes the Interval Stepping Circuit to step out of **Standby** and into the **Squeeze** interval. This stepping action is discussed in detail in Sec. 7-5.

Meanwhile, the Q output of the INTERVAL STEPPING ONE-SHOT remains HI for 100 μsec, far longer than is necessary for the system to step into **Squeeze**. During this 100 μsec, the "preset the Interval Time Counter" terminal is HI. The HI level on this terminal shifts the digits set on the **Squeeze** 10-position selector switches into the Interval Time Counter. This shifting is discussed further in Sec. 7-6.

When the system steps out of **Standby** into **Squeeze**, the **Standby** terminal goes LO, and the **Squeeze** terminal goes HI on the left of Fig. 7-4. Because **Standby** is LO, the output of I2 goes HI. Since the bottom input of AND1 is also HI, the output of AND1 goes HI. This means that the LIFT WHEEL PUSH BUTTON can be released, because the bottom input of OR2 is now HI, eliminating the need for the top input of that gate to be HI. The OR2 gate has sealed itself up, as long as **Standby** remains LO. This keeps RLW energized, keeping the wheel lifted up into the welding position. The OR2-AND1 combination just described is the familiar sealing circuit seen several times in Chapter 2.

As stated above, the **Squeeze** terminal on the left of Fig. 7-4 goes HI as the system steps into the **Squeeze** interval. This causes the output of OR3 to go HI. The output feeds back into AND2, sealing up OR3 as long as the **Release** terminal is LO. OR3 drives OA2, which in turn drives relay REWE. This relay energizes the ENGAGE ELECTRODES SOLENOID at the bottom of Fig. 7-4, causing the welding electrode cylinders to extend, bringing the electrodes into contact with the wheel rim and spider. The OR3-AND2 combination is another sealing circuit.

When the pressure on the electrode cylinders is high enough, meaning that the welding electrodes have made firm contact with the metal, the ELECTRODE PRESSURE UP PR. SW. contact closes at the upper left of Fig. 7-4. When the output of SIGNAL CONVERTER 2 goes HI, the output of NOR1 goes LO. At this time both inputs of OR1 are LO, causing wire Y to go LO. This enables the COUNT GATE, OR4, to pass any pulses which show up on its number 2 input. There *are* 60-Hz pulses existing on input 2 of OR4 at this time, so they feed through OR4 and show up on the "count pulses to Interval Time Counter" terminal. The squeeze time has begun, and the Interval Time Counter starts counting down from its preset number.

Let us pause at this point to discuss the action of OR4 in gating the count pulses through to the Interval Time Counter. As long as wire Y was HI, OR4 could not pass count pulses because its output was locked in the HI state by the HI on its number 1 input. Under this condition, pulses appearing at its number 2 input could not get through to the output. Now that input 1 is LO, the OR4 output is able to respond to pulses applied to its number 2 input (assuming that input 3 is LO). This is an example of *gating* pulses to a counter. The gate either passes or blocks the count pulses, in response to the command signal on its number 1 input. Of course, input 3 of OR4 has the same control ability to tell the COUNT GATE to pass or block count pulses.

The number 3 input of OR4 (the COUNT GATE) is LO at this time due to the **Weld** terminal being LO. The I4 output is HI, causing the NOR4 output to go LO, bringing the number 3 input of OR4 to a LO logic level.

It was stated above that 60-Hz count pulses do in fact exist at input 2 of the COUNT GATE at the instant the squeeze time begins. Figure 7-4 shows that such pulses must come from the output of NOR3. Inspection of NOR3 reveals that the pulses appearing on its top input from the "60-Hz pulses" terminal will be passed to the output of NOR3 only if the bottom input is LO. When they *are* passed, the count pulses arrive at the output of NOR3 inverted in phase, but that is unimportant in this application. This situation is almost the same as for OR4. If the bottom input were HI, the NOR3 output would be locked in the LO state, and NOR3 would not pass the pulses on its top input. However, its bottom input is LO at this time because the **Squeeze** terminal feeding into NOR2 is HI.

The final result of all this circuit action is that the Interval Time Counter is able to start counting down at the rate of one count per ac line cycle. The **Squeeze** interval is being timed out. When the Interval Time Counter reaches zero, the terminal labeled "Interval Time Counter is counted out" goes LO. This supplies a negative edge to terminal T_2 of the INTERVAL STEPPING ONE-SHOT, causing it to fire once again. As before, a negative edge appears at the "step the Interval Stepping Circuit" terminal. The Interval Stepping Circuit steps into the **Weld** interval. The HI signal on the Q output of the one-shot repeats its function of shifting numbers into the Interval Time Counter. This time it shifts the numbers set on the **Weld** 10-position selector switches.

Since the system is now in the **Weld** interval, all three inputs of NOR2 are LO, causing its output to go HI. This HI is applied to the bottom input of NOR3, disabling that gate by locking its output LO. Thus the 60-Hz pulses are prevented from passing through NOR3 during the **Weld** interval, and they cannot be counted by the Interval Time Counter. However, the **Weld** terminal driving the input of I4 is now HI, which causes a LO level to the top input of NOR4. This LO enables NOR4 to pass any pulses which appear on the "welding current pulsation count pulses" terminal.

Remember that the **Weld** interval differs from the **Squeeze**, **Hold**, and **Release** intervals in that the preset number represents how many welding current pulsations are required to complete the interval rather than how many ac line cycles. Every time a current pulsation is completed, the Heat-Cool Stepping and Gating Circuit delivers a count pulse to the "welding current pulsation count pulses" terminal. From there the pulse is passed through NOR4, through OR4, and eventually into the Interval Time Counter.*

As before, the Interval Time Counter counts backward one bit for each pulse it receives. When it reaches zero, it once again supplies a negative edge to T_2 of the INTERVAL STEPPING ONE-SHOT. The one-shot triggers the Interval Stepping Circuit by means of the negative edge appearing at the "step the Interval Stepping Circuit" terminal. The system leaves **Weld** and enters **Hold**, and the **Hold** 10-position selector switch settings are preset into the Interval Time Counter. The Interval Stepping Circuit takes away the **Weld** signal and sends out the **Hold** interval signal. Therefore the **Weld** terminal in Fig. 7-4 goes LO, disabling NOR4. The **Hold** terminal feeding NOR2 goes HI, causing the bottom input of NOR3 to go back LO. Once again the 60-Hz pulses are routed through NOR3, through OR4, to the Interval Time Counter. The **Hold** interval begins counting out.

When **Hold** is complete, the INTERVAL STEPPING ONE-SHOT receives another negative edge on its T_2 input, and its $\bar{Q}$ output delivers a negative edge to the "step the Interval Stepping Circuit (neg. edge)" terminal. That negative edge steps the system into **Release**. The same actions occur again, resulting in the **Release** selector switch settings being shifted into the Interval Time Counter.

---

*The word *time* is a little misleading during the **Weld** interval, because the counter is not actually counting time increments but welding current pulsations. During all the other intervals the Interval Time Counter really does count time increments ($\frac{1}{60}$-sec increments).

NOR3 immediately begins passing the 60-Hz pulses, and the **Release** interval begins counting out.

The **Release** terminal on the lower left of Fig. 7-4 goes HI at this time. This causes the top input of AND2 to go LO, breaking the seal on OR3 for the first time since the system entered the **Squeeze** interval. Output amplifier OA2 goes to a LO level, deenergizing relay REWE. When the ENGAGE ELECTRODES SOLENOID deenergizes, the welding electrode cylinders retract, releasing the wheel. Although the ELECTRODE PRESSURE UP PR. SW. contact opens, allowing SIGNAL CONVERTER 2 to go LO, the output of NOR1 remains LO because its bottom input is now held HI by the **Release** terminal. It is necessary to keep the NOR1 output LO in order to keep wire Y at a LO level, allowing the COUNT GATE, OR4, to continue passing count pulses. When these pulses have driven the Interval Time Counter to zero, the INTERVAL STEPPING ONE-SHOT is triggered once again by the "Interval Time Counter is counted out" terminal.

When the system's Interval Stepping Circuit leaves the **Release** interval, it steps into the **Standby** condition. On the far left of Fig. 7-4, the **Standby** terminal goes HI, causing the I2 output to go LO. This LO is applied to the top input of AND1, breaking the seal on OR2. Output amplifier OA1 goes LO, which causes relay RLW to deenergize. This deenergizes the LIFT WHEEL SOLENOID, lowering the finished wheel. The WHEEL PROPERLY POSITIONED LIMIT SWITCHES open, causing SIGNAL CONVERTER 1 to go LO. The output of I1 goes HI, returning wire X to its initial HI state. Wire Y is also HI at this time.

This completes the discussion of circuit action for the Sequence Initiation Circuit and the Interval Triggering and Gating Circuit. In the next section we will deal with the Interval Stepping Circuit and Decoder.

7-5 A DETAILED DESCRIPTION OF THE INTERVAL STEPPING CIRCUIT AND DECODER

Figure 7-5 is a schematic diagram of the Interval Stepping Circuit and Decoder. These circuits are not extensive. The Interval Stepping Circuit itself consists of three flip-flops and an AND gate. The Decoder is a diode decoding matrix having six input lines and five output lines. The Decoder also has five output drivers.

7-5-1 The interval stepping circuit

In Fig. 7-5, the flip-flop outputs have been identified by the letter name of the individual flip-flop. That is, the outputs of flip-flop A are labeled A and $\bar{A}$ instead of Q and $\bar{Q}$, and the same for flip-flops B and C. Table 7-1 shows the sequence of the flip-flops as pulses are delivered. The digit 1 in Table 7-1 means that the flip-flop is ON, while a 0 means that the flip-flop is OFF.

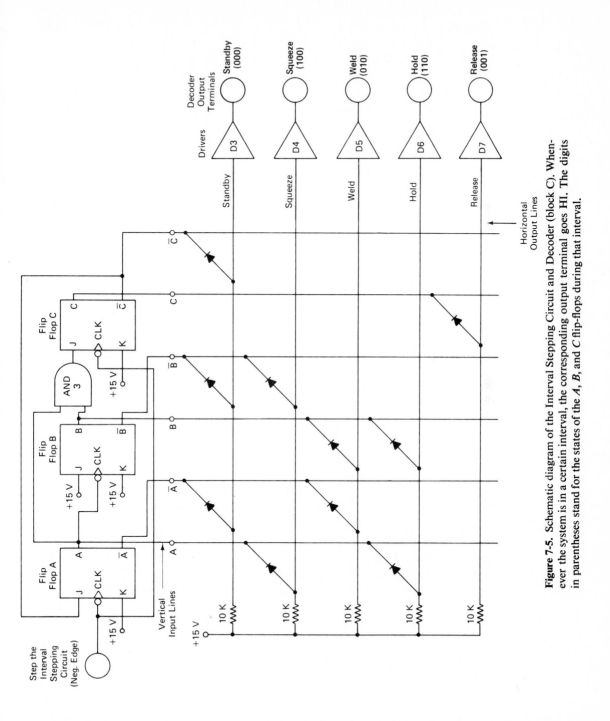

Figure 7-5. Schematic diagram of the Interval Stepping Circuit and Decoder (block C). Whenever the system is in a certain interval, the corresponding output terminal goes HI. The digits in parentheses stand for the states of the A, B, and C flip-flops during that interval.

233

TABLE 7-1

		Flip-flops			
		A	*B*	*C*	System Interval
	Start	0	0	0	**Standby**
	1	1	0	0	**Squeeze**
Number of	2	0	1	0	**Weld**
Stepping					
Pulses	3	1	1	0	**Hold**
Delivered					
	4	0	0	1	**Release**
	5	0	0	0	**Standby**

All three flip-flops are negative edge-triggered *JK* flip-flops, as discussed in Sec. 3-5. To understand the operation of the Interval Stepping Circuit, refer to Table 7-1 and Fig. 7-5.

In the **Standby** condition all flip-flops are OFF. When the first negative edge arrives at the circuit's input terminal, the "step the Interval Stepping Circuit (neg. edge)" terminal on the left of Fig. 7-5, FF*A* toggles into the ON state because both *J* and *K* are HI. *J* of FF*A* is HI because $\bar{C}$ is HI with FF*C* in the OFF state.

The negative edge at CLK of FF*A* also appears at CLK of FF*C*. At this time, though, the output of AND3 is LO, holding *J* of FF*C* LO. FF*C* therefore stays OFF. AND3 is LO because both *A* and *B*, the inputs to AND3, are LO at the instant the negative edge appears. Therefore after the first pulse, the states of the flip-flops are $ABC = 100$.

When the second negative edge appears at the input terminal, *J* of FF*A* is still HI, so FF*A* toggles to the OFF state. The negative edge is delivered to CLK of FF*C* also, but the AND3 output is still LO because *B* is LO at this instant. Therefore FF*C* again stays OFF. When the *A* output goes LO it delivers a negative edge to CLK of FF*B*. This causes FF*B* to toggle into the ON state. The state of the circuit is $ABC = 010$ after the second pulse.

When the third stepping pulse negative edge arrives at the input terminal, FF*A* toggles into the ON state as before, since *J* of FF*A* is HI. The negative edge appears at CLK of FF*C*, but once again *J* of FF*C* is LO. It is LO because the top input of AND3 is LO at the instant the negative edge arrives. After the third pulse, the state of the flip-flop circuit is $ABC = 110$.

On the fourth negative edge, FF*A* toggles into the OFF state because its *J* input is still HI. The negative edge appears at CLK of FF*C* also; this time the output of AND3 is HI, so FF*C* toggles into the ON state. The output of AND3 is HI when the edge arrives because both FF*A* and FF*B* are ON at that instant.

Remember from Sec. 3-7-5 that the thing that matters is the condition at the J input at the *instant* the edge arrives. FFB also receives a negative edge at its CLK input when A goes LO. It therefore toggles into the OFF state. The state of the flip-flop circuit is $ABC = 001$ after the fourth pulse.

When the fifth step pulse negative edge arrives at the input terminal, J of FFA is LO, because $\bar{C}$ is now LO. Therefore FFA remains in the OFF state. The negative edge appears at CLK of FFC as usual. This time the J input of that flip-flop is LO because both AND gate inputs are LO. FFC therefore turns OFF, making the state of the circuit $ABC = 000$. After five step pulses the Interval Stepping Circuit has returned to its original state.

It can be seen that the Interval Stepping Circuit steps through five separate states, never varying the order. It remains in any given state until it gets a step signal to step into a new state. These features make it an ideal circuit for keeping track of which interval the system is currently in. All that is necessary is to convert the states of the flip-flops, expressed as a sequence of binary bits, into a useful form for the other subcircuits of the system. This is the function of the Decoder.

7-5-2 The Decoder

The Decoder of Fig. 7-5 has the same basic purpose as the BCD-to-decimal decoder discussed in Sec. 3-8. It takes in coded information and puts out a logic HI on one of its output terminals. All the other output terminals are held LO while the proper one goes HI.

The way the Decoder does this is by looking at a *portion* of the binary sequence that represents the complete state of the Interval Stepping Circuit. It focuses on that portion that makes one particular state unique. For example, Table 7-1 shows that when the Interval Stepping Circuit is in the **Squeeze** interval, the state is 100. A search of the other entries in Table 7-1 reveals that no other row has $A = 1$ and $B = 0$. Therefore the $AB = 10$ combination makes that state unique, different from all other states. In Fig. 7-5, the horizontal squeeze line has two diodes wired to it, one pointing into the A vertical input line and the other pointing into the $\bar{B}$ input line. If either of these inputs is LO, the squeeze line will be pulled down LO by one of the two diodes. But if both these inputs are HI, the squeeze line will be allowed to go HI. An output line will go HI if no diode attached to that line points to a LO potential (O V). With no diode pointing into a LO, there is no route for current to flow to ground, and therefore there is no voltage drop across the 10-K lead-in resistor. With no voltage drop across the resistor, the output line is at the same potential as the supply voltage, namely $+15$ V. Thus the squeeze line goes HI whenever both A and $\bar{B}$ are HI. Of course, $\bar{B}$ being HI is equivalent to B being LO (0). Therefore the squeeze line will go HI whenever $A = 1$ and $B = 0$. This will cause driver D4 to bring the **Squeeze** terminal HI, which signals to the other subcircuits that the system is in the **Squeeze** interval.

If the system is in **Squeeze**, all four of the other horizontal output lines will be LO, because at least one diode pulls each line down LO. For example, the weld output line is being pulled down by the diode pointing into $\bar{A}$. (It is also being pulled down by the diode pointing into B, but one diode is sufficient.)

An output line being pulled down LO means that current is flowing through the 10-K lead-in resistor on the left of that line and then through a diode to ground. The 15 V of the supply is dropped across that 10-K lead-in resistor, leaving only a small voltage on the line itself. Germanium diodes are used in this diode matrix because of their lower forward voltage drop across the *pn* junction (about 0.2 V for germanium versus 0.6 V for a silicon diode).

You should verify for yourself that the other three horizontal output lines, standby, hold, and release, are all pulled down LO when $AB = 10$.

Another example may help to clarify the working of the Decoder. Consider the **Release** interval in Table 7-1. The state of the Interval Stepping Circuit is $ABC = 001$. A search of the rest of the table reveals that no other interval has $C = 1$. Therefore the single bit $C = 1$ distinguishes the **Release** interval from all four other intervals and makes it unique. The Decoder takes advantage of this fact in that it has a single diode pointing from the release output line to the C vertical input line. If C is HI, as it would be during the **Release** interval, the release output line goes HI. If C goes LO, as it would be during any *other* interval, the release output line is pulled down LO. Therefore the release output line goes HI when the C flip-flop turns ON, and only then. Again, the Decoder is looking at the *unique* portion of the state of the flip-flop circuit and using that portion to control the output line.

The driver attached to each output line has the function of isolating the line from the other subcircuits, so the subcircuits cannot degenerate the quality of the signal level on the output line. Degeneration of signal level (HIs not high enough or LOs not low enough) could occur if the subcircuits were to draw too much current away from the line when it was HI (a current-sourcing logic family) or if they were to dump too much current into the line when it was LO (a current-sinking family).

It will be instructive for you to verify that the Decoder does in fact identify the unique portion of the state of the Interval Stepping Circuit for the other three intervals. Check for yourself that for each interval it brings the proper output terminal HI, leaving all others LO.

7-6 THE INTERVAL TIME COUNTER AND THE INTERVAL TIME COUNTER PRESET CIRCUIT

The Interval Time Counter consists of a pair of decade down-counters, one for units and one for tens, and a simple gating circuit to detect when the counter reads zero. It is shown schematically in Fig. 7-6, along with the Interval Time Counter Preset Circuit.

Figure 7-6. Schematic of the Interval Time Counter (block E) and the Interval Time Counter Preset Circuit (block D). The length of each interval (except **Standby**) is set on a pair of 10-position selector switches.

7-6-1 The interval time counter

The decade down-counters count down one digit each time a negative clock edge appears at the CLK input terminal. The contents of a decade down-counter appear in BCD form at the *DCBA output* terminals ($D = 8$, $C = 4$, $B = 2$, $A = 1$). The *DCBA input* terminals are used for presetting a number into the counters before delivering the count pulses. The binary number appearing at the *DCBA* input terminals is shifted into the down-counter when the LOAD terminal goes LO. When the LOAD terminal is HI, the logic levels at the input terminals are ignored by the counter. Down-counter action was discussed in Sec. 3-13.

For example, if the binary data at the input terminals are *DCBA* = 0111, the decimal number 7 would be preset into the counter when the LOAD terminal went LO. The 7 would appear at the output terminals as *DCBA* = 0111. Once the LOAD terminal returns to the HI level, count pulses at the CLK terminal can be received by the counter. Any count pulses appearing at CLK while the LOAD terminal is LO would be ignored.

As count pulses are received, the counter counts down one digit per pulse. As usual, the actual count transition occurs at the instant the negative-going edge arrives. When the counter gets to zero, all output terminals are LO (*DCBA* = 0000). On the next pulse, the count goes to decimal 9, with *DCBA* = 1001 at the output terminals.

It can be seen from Fig. 7-6 that the units decade counter receives count pulses from the terminal labeled "count pulses to Interval Time Counter." This terminal originates in the Interval Triggering and Gating Circuit, as discussed in Sec. 7-4. The tens decade counter receives its count pulses from the units decade counter, since the two counters are cascaded. When the units decade goes from the 0 state into the 9 state, the *D* output line goes HI. This HI is applied to the input of I5, which delivers a negative edge to CLK of the tens decade counter. The tens decade moves down one digit at that time. For example, if the number contained in the Interval Time Counter is 40, the units decade has a 0 (*DCBA* = 0000), and the tens decade has a 4 (*DCBA* = 0100). On the next pulse, the units decade goes to a 9 (1001), and the negative edge delivered from the output of I5 steps the tens decade down to a 3 (0011). The number contained in the Interval Time Counter is then 39.

As explained in Sec. 7-2, pulses continue to arrive at the Interval Time Counter until it reaches zero, at which time a negative edge appears at the terminal labeled "Interval Time Counter is counted out." OR5, OR6, and OR7 are the gates which detect the zero condition and supply the negative edge to that terminal. The negative edge then triggers the INTERVAL STEPPING ONE-SHOT.

Specifically, when the tens decade contains a 0, the output of OR5 will be LO because every one of its inputs will be LO. When the units decade contains a 0, the OR6 output will be LO because every one of *its* inputs will be LO. When both decades contain Os, the Interval Time Counter has counted out. When that happens, OR7 will see two LOs on its inputs, and its output will go LO. This supplies the negative edge at the "Interval Time Counter is counted out" terminal.

7-6-2 The operation of the preset circuitry

The Interval Time Counter Preset Circuit consists of the following:

a. Four pairs of 10-position selector switches, one pair for each of the intervals

Squeeze, Weld, Hold, and **Release**: Each pair of switches has one switch for the units digit and a second switch for the tens digit.

b. An encoder for each selector switch—eight encoders altogether.

c. Four NAND gates, one for each of the four intervals which has a pair of switches.

When the "preset the Interval Time Counter" terminal gets a HI signal from the INTERVAL STEPPING ONE-SHOT in the Interval Triggering and Gating Circuit, two things happen. First, the LOAD terminals of the two decade down-counters are driven LO by I8, permitting the counters to receive their preset numbers. Second, the gates NAND1, NAND2, NAND3, and NAND4 are partially enabled because all their top inputs go HI. Then, depending on which interval has just been entered, one of these four gate outputs will go LO, applying a LO signal to the common terminals of the appropriate pair of selector switches.

For example, if the system has just entered **Hold**, the NAND3 output will go LO, applying 0 V to the common terminals of the **Hold** selector switches. The other three pairs of selector switches will continue to have +15 V on their common terminals. Recall that we said we would expect our encoders to respond to a *LO* decimal input rather than a *HI* decimal input. This point was covered in Sec. 3-13-2. Therefore only the selector switches which receive 0 V are able to drive their preset encoders, so the other three pairs drop out of the picture.

In this example, with the system having just entered the **Hold** interval, the HOLD UNITS SWITCH applies its number setting to the units preset lines, and the HOLD TENS SWITCH applies its number setting to the tens preset lines. Thus, BCD data are available on the units preset lines and tens preset lines at the same time that the LOAD terminals of the units and tens decade counters are held LO. In this way the proper selector switch settings are preset into the Interval Time Counter.

Now let us turn our attention to the preset encoders themselves. Decimal-to-BCD encoders are available in packaged ICs, but Fig. 7-7(a) shows how one could be built from scratch. All eight of the preset encoders in Fig. 7-6 are identical and could be built as shown in Fig. 7-7(a).

In Fig. 7-7(a), whichever vertical input line is selected on the switch has 0 V (ground voltage) applied to it from the switch's common terminal. Lowering the potential of an input line to ground level causes the cathode of each diode attached to that line to be at ground potential. When the cathode goes to ground, the horizontal output line that the anode attaches to is also pulled down LO. Current flows through the 1.5-K lead-in resistor at the left end of the output line, down the horizontal line, through the diode, and into the ground. The current flow through the lead-in resistor at the left is sufficient to drop almost the full 15-V supply voltage. Only a few tenths of a volt remain on the vertical output line itself (0.2 V for a germanium diode). The logic LO on an output line is applied to the inverting buffer at the right-hand side of that line, causing the output terminal from that line to go HI. The result is that every output line which is connected through a diode to the LO input line goes LO itself. Those output terminals of the encoder

Figure 7-7. (a) Detailed schematic of one of the preset encoders. Altogether, there are eight of these preset encoders, as shown in Fig. 7-6. (b) The way in which four output inverting buffers are tied together. During any interval, the three inverting buffers which are not in use all have LO outputs, and they are all isolated by their three transistors. The one inverting buffer which *is* in use may have either a LO output or a HI output. If it has a LO output, its transistor turns OFF, causing Q_5 to turn ON; this causes the preset line to go LO. On the other hand, if that inverting buffer has a HI output, its transistor turns ON, causing Q_5 to turn OFF; this causes the preset line to go HI. The preset line thus follows the state of the one inverting buffer which is in use during that interval.

are then driven HI by their inverting buffers. The inverting buffers serve to isolate, or buffer, the diodes in the encoder from the output terminals and their attached subcircuits. They also serve to invert the internal output lines so that the encoded number appears at the output terminals in *positive* logic.

Those output lines which are *not* connected through diodes to the 0-V input line have no current flow through their 1.5-K lead-in resistors. Therefore there is no voltage drop across their 1.5-K resistors, leaving these output lines at essentially $+15$ V (HI). A HI signal existing on a horizontal output line is applied to the inverting buffer at the right-hand end of the line, causing the output terminal to be LO.

As an example, consider that the selector switch is set to 3. This applies ground potential to the number 3 vertical input line. Current will flow through the two diodes that connect the number 3 input line to the A output line and to the B output line. This current comes through the 1.5-K resistors at the left end of these two horizontal output lines, causing these two lines to drop almost to 0 V. The C output line and the D output line have no current flow, so their potential stays essentially at $+15$ V. Therefore the A and B inverting buffers receive LO inputs and the C and D inverting buffers receive HI inputs. The A and B output terminals thus go HI, and the C and D output terminals go LO. The result is that the decimal number 3 has been encoded to $DCBA = 0011$, which is correct. You should analyze the action of the encoder for several other switch settings and prove to yourself that it does correctly encode the decimal number into BCD.

7-6-3 Connecting the buffers together

Notice in Fig. 7-6 that the output terminals of the four UNITS ENCODERS are all shown tied together to drive the units preset lines. That is, output terminal D of the SQUEEZE UNITS ENCODER ties to output terminal D of the WELD UNITS ENCODER and also to output terminal D of the HOLD UNITS ENCODER and output terminal D of the RELEASE UNITS ENCODER.* Under these circumstances, "disputes" will arise between the different inverting buffers, with one inverting buffer trying to bring the preset line HI and the other three inverting buffers which attach to that preset line trying to bring it LO. Because of these possible disputes, the inverting buffers cannot be connected *directly* together. Buffer connection circuits are necessary. The buffer connection circuits are designed so that the inverting buffer trying to bring the line HI "wins the dispute."

Figure 7-7(b) shows how this is accomplished. The buffer connection circuit of Fig. 7-7(b) is drawn for the D outputs of the four TENS ENCODERS. However, the entire circuit shown in that figure is repeated eight times. It is repeated for all

---

*The same is true for all four of the C output terminals of the four UNITS ENCODERS and also for all four of the B and A output terminals of the four UNITS ENCODERS. The same situation is also true for the four TENS ENCODERS. Figure 7-6 shows all this.

the C, B, and A outputs of the TENS ENCODERS and again for all the D, C, B, and A outputs of the UNITS ENCODERS.

Here is how the buffer connection circuit works. Transistor switches Q_1, Q_2, Q_3, and Q_4 are tied in parallel, with all four collectors connected together. Therefore if the D output of the encoder which has been enabled happens to be HI, the collector tie point is pulled down LO by that transistor. The LO on the collector tie point is applied to the base of Q_5, causing the final D output to go HI.

On the other hand, if the D output of the encoder which has been enabled happens to be LO, then the collector tie point will go HI. That will occur because none of the transistors Q_1–Q_4 will turn ON. The HI on the collector tie point is applied to the base of Q_5, causing the final D output to go LO.

For example, suppose that the HOLD ENCODER has been enabled and that the SQUEEZE, WELD, and RELEASE ENCODERS are disabled. That is, in Fig. 7-6, the NAND3 output is LO, applying 0 V to the common terminals of the **Hold** selector switches and the NAND1, NAND2, and NAND4 outputs are all HI, applying $+15$ V to the common terminals of all their selector switches. In this example, the only D output terminal in Fig. 7-7(b) which *can possibly go HI* is the D output terminal of the HOLD TENS ENCODER. The other three D outputs are guaranteed to be LO because of the HIs at the common terminals of their selector switches. In other words, their encoders are disabled.

If the D output of the HOLD TENS ENCODER happens to be HI, Q_3 will turn ON. This will put a LO on the base of Q_5, causing that transistor to turn OFF. Therefore the final D output (the D line of the tens preset lines) goes HI.

Now consider what would happen if the D output terminal of the HOLD TENS ENCODER happens to be LO. In that case Q_3 will turn OFF. Q_1, Q_2, and Q_4 are guaranteed to be OFF at this time because the SQUEEZE, WELD, and RELEASE ENCODERS are all disabled. Therefore the collector tie point will go HI and will turn ON Q_5. The collector of Q_5 goes LO, applying a LO to the D line of the tens preset lines.

The overall result is that the D line of the tens preset lines obeys the D inverting buffer output of the HOLD TENS ENCODER if that encoder is the one that is enabled. Of course, if some other encoder had been enabled, the D line would have obeyed *that* encoder's inverting buffer.

7-7 THE HEAT-COOL STEPPING AND GATING CIRCUIT

The Heat-Cool Stepping and Gating Circuit goes into action only during the **Weld** interval. During that interval, its functions are to keep track of the **Heat** and **Cool** subintervals and to control the stepping from one subinterval to the next. The circuit is shown schematically in Fig. 7-8.

The circuit begins operation when the **Weld** terminal goes HI on the far left of Fig. 7-8. This terminal goes HI when the system enters the **Weld** interval. At

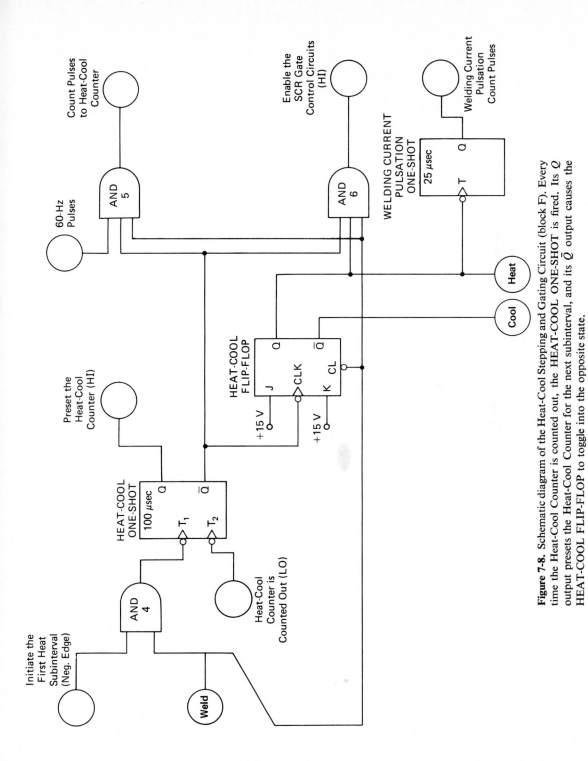

Figure 7-8. Schematic diagram of the Heat-Cool Stepping and Gating Circuit (block F). Every time the Heat-Cool Counter is counted out, the HEAT-COOL ONE-SHOT is fired. Its Q output presets the Heat-Cool Counter for the next subinterval, and its $\bar{Q}$ output causes the HEAT-COOL FLIP-FLOP to toggle into the opposite state.

243

that time the "initiate the first **Heat** subinterval (neg. edge)" terminal also goes HI, coming from the Q output of the INTERVAL STEPPING ONE-SHOT. Therefore the output of AND4 goes HI because both of its inputs are HI at this time.

When the INTERVAL STEPPING ONE-SHOT output pulse ends after 100 μsec, the "initiate the first **Heat** subinterval (neg. edge)" terminal goes back LO, driving the top input of AND4 back LO, causing a negative edge to appear at trigger terminal T_1 of the HEAT-COOL ONE-SHOT. The firing of the HEAT-COOL ONE-SHOT causes its $\bar{Q}$ output to go LO, which causes a negative edge to appear at the clock terminal of the HEAT-COOL FLIP-FLOP. This makes the flip-flop toggle into the ON state. The HEAT-COOL FLIP-FLOP was in the OFF state prior to the system entering the **Weld** interval because its clear input was held LO by the **Weld** terminal. A LO on the CL terminal of a *JK* flip-flop clears the flip-flop, as covered in Sec. 3-5.

While the HEAT-COOL ONE-SHOT is still firing, the terminal labeled "preset the Heat-Cool Counter" is HI. The HI on this terminal is fed to the Heat-Cool Counter Preset Circuit along with the HI signal on the **Heat** terminal. Together, these signals cause the HEAT SELECTOR SWITCH settings to be shifted into the Heat-Cool Counter. This action will be discussed fully in Sec. 7-8.

When the HEAT-COOL ONE-SHOT pulse ends, the $\bar{Q}$ output of the one-shot goes back to HI. This HI signal is routed to AND5 and to AND6. The AND6 gate now has all inputs HI, so it brings the "enable the SCR gate control circuits" terminal HI, allowing the welding SCRs to commence firing. The welding transformer therefore starts delivering welding current to the wheel and spider. This is explained fully in Sec. 7-9.

Meanwhile, AND5 has been enabled to pass 60-Hz pulses to the terminal labeled "count pulses to Heat-Cool Counter." Since the welding transformer is carrying 60-Hz current, one pulse is delivered to the Heat-Cool Counter for every cycle of welding current. The Heat-Cool Counter counts backwards to zero, just like the Interval Time Counter. When the preset number of cycles of welding current have occurred, the "Heat-Cool Counter is counted out" terminal on the left of Fig. 7-8 goes LO, triggering the HEAT-COOL ONE-SHOT again, this time from terminal T_2. The one-shot fires, putting a HI level on the "preset the Heat-Cool Counter" terminal once again and causing the HEAT-COOL FLIP-FLOP to toggle into the OFF state. This makes the **Heat** terminal go LO and the **Cool** terminal go HI. The **Weld** interval is now in the **Cool** subinterval.

The negative edge appearing on the Q terminal of the HEAT-COOL FLIP-FLOP as it turns OFF is applied to trigger terminal T_1 of the WELDING CURRENT PULSATION ONE-SHOT. This one-shot delivers a 25-μsec pulse to the "welding current pulsation count pulses" terminal, indicating that a welding current pulsation has been completed (a **Heat** subinterval has been completed). This pulse is routed to NOR4 in Fig. 7-4. It is passed through NOR4 and OR4 to the Interval Time Counter as described in Sec. 7-4. Therefore the welding current pulsation that has just finished causes the Interval Time Counter to count down by one digit.

Since the **Weld** interval has just now entered the **Cool** subinterval, the **Cool** terminal is HI and the **Heat** terminal is LO. The output of AND6 goes LO, causing a LO to appear on the terminal labeled "enable the SCR gate control circuit." This results in shutting off the welding transformer by disabling the SCR gate control circuit. Meanwhile, the "preset the Heat-Cool Counter" terminal is still HI (it stays HI for 100 μsec as the system steps from **Heat** to **Cool**), so the preset numbers on the COOL SELECTOR SWITCHES are shifted into the Heat-Cool Counter. When the HEAT-COOL ONE-SHOT output pulse ends after 100 μsec, AND5 is enabled once again. The Heat-Cool Counter again starts counting down as it receives the 60-Hz pulses.

When the Heat-Cool Counter reaches zero, indicating the **Cool** subinterval is complete, it sends another negative edge to the HEAT-COOL ONE-SHOT by way of the "Heat-Cool Counter is counted out" terminal. The one-shot repeats its previous actions, namely toggling the HEAT-COOL FLIP-FLOP into the ON (**Heat**) state, presetting the HEAT SELECTOR SWITCH settings into the Heat-Cool Counter and then reenabling AND5 and AND6 when it finishes firing. Notice that as the system goes from the **Cool** subinterval to the **Heat** subinterval, a positive edge appears at T_1 of the WELDING CURRENT PULSATION ONE-SHOT. This one-shot does not fire on a positive edge, and no pulse appears at the "welding current pulsation count pulse" terminal. This is proper because the Interval Time Counter is supposed to count only when a welding current pulsation is complete. No current pulsation has just taken place, so no count pulse is delivered.

The above cycle is repeated over and over until the proper number of welding current pulsations have been counted by the Interval Time Counter. At that point, the system will step out of **Weld** and into **Hold**. The **Weld** terminal in Fig. 7-8 will go LO and the entire Heat-Cool Stepping and Gating Circuit will be disabled.

7-8 THE HEAT-COOL COUNTER AND THE HEAT-COOL COUNTER PRESET CIRCUIT

The Heat-Cool Counter is identical to the Interval Time Counter. It consists of two decade down-counters cascaded together, and it has the same zero detection circuitry. The arrangement for presetting the Heat-Cool Counter is also similar to the arrangement for the Interval Time Counter. Figure 7-9 shows the Heat-Cool Counter along with the Heat-Cool Counter Preset Circuit.

To preset the Heat-Cool Counter, the LOAD terminals must be driven LO at the same time that LOs are delivered to the common terminals of a pair of selector switches. If the HEAT SELECTOR SWITCH common terminals go LO, the digits selected on those switches are preset into the decade counters. This is what is done at the start of a **Heat** subinterval. If the COOL SELECTOR SWITCH common terminals go LO, digits selected on those two switches are preset into the decade counters. This is what is done at the start of a **Cool** subinterval.

When the HEAT-COOL ONE-SHOT in Fig. 7-8 fires, it temporarily raises the

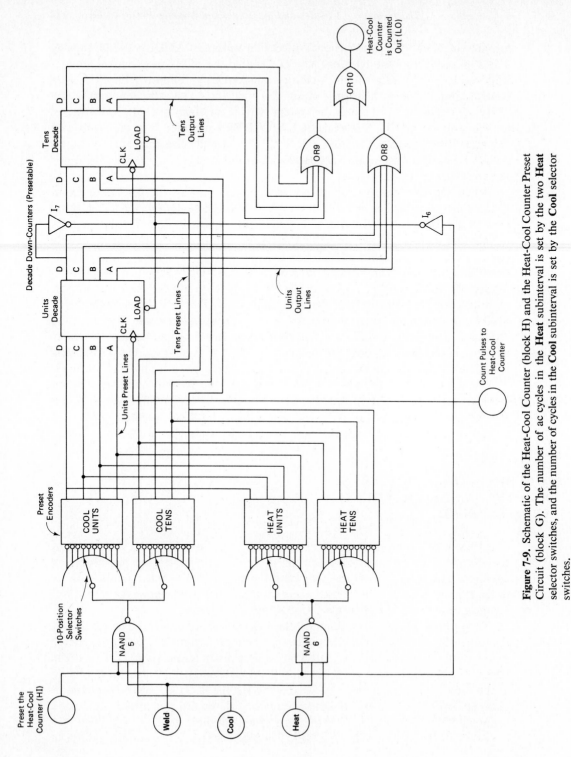

Figure 7-9. Schematic of the Heat-Cool Counter (block H) and the Heat-Cool Counter Preset Circuit (block G). The number of ac cycles in the **Heat** subinterval is set by the two **Heat** selector switches, and the number of cycles in the **Cool** subinterval is set by the **Cool** selector switches.

"preset the Heat-Cool Counter" terminal HI on the left of Fig. 7-9. This terminal applies a HI to I6, which drives the LOAD terminals of both decade counters LO, permitting them to accept the data on their preset inputs. Meanwhile, the 100-μsec pulse on the "preset the Heat-Cool Counter" terminal also goes up and partially enables NAND5 and NAND6. If the system has just entered the **Heat** subinterval of the **Weld** interval at this time, then NAND6 will be fully enabled because the **Weld** and **Heat** terminals will both be HI. The LO output of NAND6 will enable the HEAT ENCODERS, thus presetting the number of **Heat** cycles into the Heat-Cool Counter.

On the other hand, if the system has just entered the **Cool** subinterval, the NAND5 output goes LO. This enables the COOL ENCODERS, loading the number of **Cool** cycles into the Heat-Cool Counter.

When the HEAT-COOL ONE-SHOT pulse ends after 100 μsec, the presetting operation is complete, and the LOAD terminals go back HI. The Heat-Cool Counter is now ready to begin receiving 60-Hz count pulses at the terminal labeled "count pulses to the Heat-Cool Counter." This terminal originates in Fig. 7-8, at the HEAT-COOL COUNT GATE, AND5. This gate starts passing 60-Hz pulses as soon as the HEAT-COOL ONE-SHOT output pulse is finished.

When both decade counters in Fig. 7-9 reach zero, meaning that the Heat-Cool Counter has counted out, the outputs of OR8 and OR9 both go LO. This causes the output of OR10 to go LO, creating a negative edge at the "Heat-Cool Counter is counted out" terminal. This negative edge is fed back to the HEAT-COOL ONE-SHOT in Fig. 7-8, where it causes the Heat-Cool Stepping and Gating Circuit to step into the next subinterval. This action was described in Sec. 7-7.

7-9 THE WELD POWER CIRCUIT

The Weld Power Circuit controls the flow of current to the welding electrodes. Its job is to respond to the signal arriving from the Heat-Cool Triggering and Gating Circuit on the "enable the SCR gate control circuits" terminal. When that signal is LO, the Weld Power Circuit prevents the flow of welding current. When that signal is HI, it permits the flow of welding current. In addition, the Weld Power Circuit maintains the desired conduction angle during the period that welding current is allowed to flow.

7-9-1 A simplified view of the weld power circuit

A simplified schematic of the Weld Power Circuit is shown in Fig. 7-10(a). It shows that current can flow from the 460-V ac incoming power lines, and through the primary winding of the welding transformer, only if the *ignitron* has been fired, because the ignitron is in series with the primary winding.

An ignitron (pronounced *igg-nyé-tron*) is a large mercury-arc rectifying tube. The circuit behavior of an ignitron is very much like the circuit behavior of an

(a)

Figure 7-10. (a) Simplified schematic of the Weld Power Circuit (block I). When the input terminal goes HI, Q_1 turns ON and Q_2 turns OFF. This removes the short circuit across ZD1 and allows C_1 to begin charging when the ac line crosses into its positive half cycle. The pulse transformer triggers an SCR, which in turn fires an ignitron. The ignitron actually carries current to the welding transformer. (b) Waveform of T_1 secondary voltage. (c) Clipped sine wave which powers the UJT timing circuit. (d) Voltage across C_1. When it reaches the V_p of the UJT, the UJT fires; this dumps almost all the charge which is on the top plate of C_1. (e) Pulses of current into the primary side of the pulse transformer. (f) The rectified welding current.

SCR. It acts either like an open switch in series with the load, or like a closed switch in series with the load. It passes current in only one direction, from anode to cathode. It does not automatically turn ON when the anode to cathode voltage polarity goes positive, but it must be turned ON, or fired, by a third control terminal, called the *ignitor*.

A burst of current into the ignitor lead and out the cathode lead will fire the ignitron, after which it will remain ON until the anode to cathode voltage changes polarity.

The advantage of the ignitron over the SCR is a very simple one: current capacity. In situations where tremendous surges of current must be supplied to a load, the ignitron is often the only device equal to the task. Ignitrons are available which can deliver regular current surges as large as 10,000 amperes. No SCR can come close to that current capacity.

The amount of ignitor current needed to fire an ignitron is fairly large, usually about 25 A. Therefore the ignitor circuit alone is worthy of an SCR. This situation is shown in Fig. 7-10(a), where an SCR is connected between the anode terminal and the ignitor terminal of the ignitron. When the SCR fires, it establishes a current flow path as follows: from the L_1 incoming power line, through the SCR, into the ignitor which leads into a mercury pool inside the ignitron, through the pool of liquid mercury, and out the cathode lead. Thus the firing of the ignitron coincides with the firing of the SCR.

The SCR itself is fired when a pulse appears on the secondary winding of pulse transformer T_2 in its gate circuit. A secondary pulse will appear when the UJT delivers a burst of current into the primary side of T_2, as we have seen before.

The UJT triggering circuitry in Fig. 7-10(a) is a fairly standard circuit. When the ac power line goes positive, diodes D_1 and D_2 become forward biased, applying a positive half cycle of ac voltage to the R_3-ZD1 combination. Zener diode ZD1 clips the waveform at $+15$ V shortly after the positive half cycle begins and maintains a constant dc voltage to the UJT triggering circuit for the rest of the positive half cycle. This relationship is illustrated in Fig. 7-10(b) and (c).

Notice transistor Q_2, however. Q_2 is a transistor switch which can short out ZD1 and prevent the UJT from ever triggering. Q_2 is driven by Q_1, which is controlled by the "enable the SCR gate control circuit" terminal in the lower left of Fig. 7-10(a). This terminal originates in Fig. 7-8, the Heat-Cool Stepping and Gating Circuit. When this terminal is HI, Q_1 is ON, causing a LO to the base of Q_2. Q_2 turns OFF, thus permitting the dc voltage to be established across the UJT trigger circuit.

If this control terminal is LO, however, Q_1 turns OFF, causing Q_2 to turn ON. With Q_2 ON, the zener diode is shorted out, and no dc voltage can appear across the UJT. In this case the entire T_1 secondary voltage is dropped across R_3.

In this way, the "enable the SCR gate control circuit" terminal is able to either permit the flow of welding current or prevent the flow of welding current.

If welding current is being permitted, the ZD1 voltage causes capacitor C_1 to start charging when the positive half cycle begins. The charging rate is set by variable resistance R_5. If C_1 charges fast, its voltage reaches the peak voltage of the UJT quickly, and the UJT, SCR, and ignitron fire early in the half cycle. This results in a large conduction angle. If C_1 charges slowly, the UJT, SCR, and ignitron fire late, resulting in a small conduction angle and lower average welding current.

The waveforms of current in the pulse transformer and current through the welding transformer are shown in Figs. 7-10(e) and (f). The welding current waveform is somewhat idealized. It would not really look that clean, because of the inductive properties of the transformer windings.

7-9-2 The real weld power circuit

In the foregoing discussion of the Weld Power Circuit in Sec. 7-9-1, two simplifying changes have been made:

a. We have shown a *single-phase* welding transformer instead of the *three-phase* transformer which is actually used.
b. We have shown only one ignitron-SCR pair, which results in unvarying direction of current flow through the welding transformer and through the wheel metal itself.

In the actual system, there are two ignitron-SCR pairs per phase, which allows the direction of welding current to *reverse* from one current pulsation to the next. This prevents saturation of the welding transformer core. Transformer core saturation can occur due to buildup of residual magnetism if the current always flows in the same direction through the windings.

Figure 7-11(a) shows a three-phase welding transformer, with each phase having *two* ignitron-SCR pairs. The ignitron-SCR pairs point in opposite directions, which allows the welding current reversal spoken of above.

When the three-phase welding transformer is used, only *one phase may be energized at any instant*. To understand this, refer to Fig. 7-11(a). If it is desired to energize the A phase of the transformer, this can be done by firing the A_x ignitron-SCR pair, allowing current to pass through the A primary winding from power line L_1 to L_2; or phase A can be energized by firing the A_y ignitron-SCR pair, allowing current to pass through the A primary winding in the opposite direction, from L_2 to L_1. In either case, voltage will be induced in the phase A *secondary* winding, which then delivers welding current to the electrodes. If phase A is delivering the welding current, phases B and C must not interfere. Note that the B and C secondary windings are connected in series with the phase A secondary winding in Fig. 7-11(a). There must be no voltage created in the B and C secondary windings during the time the A secondary is trying to deliver current to the welding

(a)

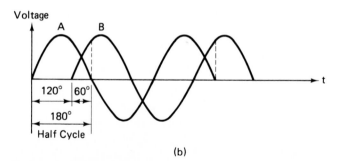

(b)

Figure 7-11. (a) The three-phase welding transformer with the primary winding connected in delta. During one welding current pulsation the X ignitrons fire, and during the next current pulsation the Y ignitrons fire. (b) The firing delay angle for the B phase must be no less than 60°; this ensures that the B winding is not energized before the A winding is deenergized.

electrodes. The A secondary must be able to get a clear "shot" at the electrodes. This is why the B and C primary windings must not be energized while the A primary winding is energized. Of course, this argument works the same way when the B phase is driving the electrodes, or when the C phase is driving the electrodes.

The requirement that only one phase be energized at any instant can be met by designing the SCR gate control circuits so that the ignitrons have a firing delay angle *no less than* 60° (conduction angle no greater than 120°). Figure 7-11(b) shows why this is so. When one transformer phase energizes, it is bound to deener-

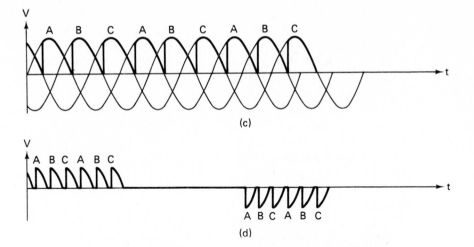

Figure 7-11. (c) Welding current waveform for a 60° firing delay angle. Only one pulsation is shown. (d) Welding current waveform for a 90° firing delay angle. Two pulsations are shown, illustrating the current reversal from one pulsation to the next.

gize when the phase voltage which is driving it crosses into the negative region. This always occurs 60° after the *next* phase voltage crosses into the positive region. This is shown clearly by Fig. 7-11(b). Therefore, if the firing of the ignitrons is delayed by at least 60°, it is impossible for any given transformer phase to begin conducting until the preceding phase has stopped conducting.

This complete idea is illustrated in Fig. 7-11(c), which shows all three phase voltages. In that waveform, the phase *A* voltage is shown passing into the negative region 60° after the phase *B* voltage has passed into the positive region. By properly sizing the SCR gate control components, it is possible to prevent ignitron *B* from firing during the first 60° of the cycle of the phase *B* voltage. This ensures that the phase *A* voltage has gone negative by the time the phase *B* ignitron fires, guaranteeing that the *A* phase of the transformer is deenergized before the *B* phase is energized.

The argument given here for the *A-B* phase relationship also holds true for the *B-C* phase relationship and for the *C-A* phase relationship.

In Fig. 7-11(c), the firing delay angle is exactly 60°. It would not have to be *exactly* 60° of course. The only requirement is that it be *not less than* 60°.

Reversal of welding current flow direction from one current pulsation to the next is accomplished by alternating between ignitron-SCR pairs. During one welding current pulsation, the ignitron-SCR pairs labeled *X* in Fig. 7-11(a) are sequentially fired. During the next welding current pulsation, the ignitron-SCR pairs labeled *Y* are sequentially fired. That is, during one welding current pulsation, the A_X pair is fired, then the B_X pair is fired, then the C_X pair is fired, and this sequence is repeated as many times as called for by the HEAT SELECTOR SWITCHES. During the next welding current pulsation, the A_Y pair is fired, then the B_Y pair, then the C_Y pair, and this sequence is repeated as many times as the HEAT SELECTOR SWITCHES are set for. The resulting waveform is illustrated in Fig. 7-11(d), this time with a 90° firing delay angle.

Since the actual Weld Power Circuit contains six ignitron-SCR pairs, the control circuit drawn in Fig. 7-10(a) is actually repeated six times. Also, the alternating between the *X* ignitron-SCR pairs and the *Y* ignitron-SCR pairs, which causes welding current reversal, is controlled by the Heat-Cool Stepping and Gating Circuit discussed in Sec. 7-7. Additions would be necessary to that circuit to enable it to alternate between the *X* and *Y* pairs. Those additions will not be shown here. This is not because they are difficult to understand in themselves, but only because they would add further complexity to an already confusing circuit arrangement. A schematic diagram of the entire welding sequence control circuit is shown in Figure 7-12. In order to keep the size of the figure manageable, the Weld Power Circuit is not included in the figure. Instead, the "enable the SCR gate control circuits (HI)" line is shown leaving Fig. 7-12 from the extreme right-hand side. This line signals the Weld Power Circuit, telling it when to start and stop the actual welding operation, as explained in Sec. 7-9.

Figure 7-12(a). The welding sequence control circuit in its entirety. Each of the nine subcircuits is enclosed by a dashed line and labeled.

Figure 7-12(b).

Figure 7-12(c).

256

Figure 7-12(d).

QUESTIONS AND PROBLEMS

1. Is it necessary to relieve the DISENGAGE ELECTRODES hydraulic line when pressuring up the ENGAGE ELECTRODES hydraulic line of Fig. 7-1(b)? Explain.

2. Name the five main intervals of an automatic welding sequence, in order. Explain what happens during each interval.

3. Why is the **Hold** interval necessary?

4. Name the two subintervals of the **Weld** interval. Explain what happens during each one.

5. Give an approximate figure for how much time each of the five intervals takes.

6. In this system, what is the longest possible time the **Hold** interval could last?

7. Repeat Question 6 for the **Release** interval.

8. Repeat Question 6 for the **Squeeze** interval.

9. Repeat Question 6 for the **Heat** subinterval.

10. Repeat Question 6 for the **Cool** subinterval.

11. Repeat Question 6 for the **Weld** interval.

Questions 12–15 can be answered by referring solely to Fig. 7-3.

12. When the Interval Time Counter has counted out, which subcircuit does it pass this information to?

13. To which subcircuits does the Interval Stepping Circuit and Decoder send the information about the interval the system is currently in?

14. Which line is used to send count pulses to the Heat-Cool Counter?

15. Which line is used to tell the Interval Stepping Circuit to step into a new interval?

16. What conditions are necessary to drive the output of OR1 LO? Express your answer in basic system terms, not in terms of other logic gates. That is, do not say simply that the I1 output must go LO; tell what must happen physically in the system to *make* the I1 output go LO.

17. What conditions are necessary to energize RLW, the LIFT WHEEL relay? Same instructions as for Question 16.

18. What conditions are necessary to energize REWE, the ENGAGE ELECTRODES relay. Same instructions as for Question 16.

19. During which intervals are count pulses delivered to the Interval Time Counter by way of NOR3?

20. What is the purpose of NOR1 and the **Release** terminal connection to NOR1 in Fig. 7-4? Why couldn't we just run the SC2 output into OR1 and eliminate NOR1?

21. Why is the **Release** terminal wired to the I3 input instead of the **Standby** terminal? What would happen during the automatic cycle, if the **Standby** terminal were wired there by mistake?

22. Why is it not necessary to disable OR4 during the firing time of the INTERVAL STEPPING ONE-SHOT? (At first consideration, it appears that it *would* be necessary in order to prevent count pulses from entering the counter during the presetting operation.)

23. When is the INTERVAL STEPPING ONE-SHOT triggered from its T_1 terminal? When is it triggered from its T_2 terminal?

24. During which interval(s) does the output of AND3 go HI?

25. In the Interval Stepping Circuit, FFA receives a negative edge at its CLK terminal every time the system is about to step into a new interval. When exactly does FFB receive a negative edge at its CLK terminal? Repeat the question for FFC.

26. Explain why only one diode is necessary for the decoding of the **Release** interval state in the Interval Stepping Circuit Decoder. Why not two or three diodes like all the other states?

27. Figure 7-5 shows a diode decoding matrix specially built for this decoding job. Is this special matrix *really* necessary, or could you make a standard BCD to 1-of-10 decoder work? Explain carefully.

28. Suppose the system has just entered **Hold** and the settings on the HOLD SELECTOR SWITCHES are being preset into the Interval Time Counter. The selector switches are set to provide a hold time of 47 cycles. Identify the level of each one of the units preset lines, D, C, B, and A, and also each one of the tens preset lines, D, C, B, and A.

29. Under what conditions does the output of OR5 go LO?

30. Under what conditions does the output of OR6 go LO?

31. Under what conditions does the output of OR7 go LO?

32. The preset number entered into the Interval Time Counter at the beginning of the **Weld** interval does not represent how many ac line cycles are necessary to time the counter out. What *does* that number represent?

33. When does the output of AND4 go HI. When does it return to the LO level? Same instructions as for Question 16.

34. When is the HEAT-COOL ONE-SHOT triggered from its T_1 terminal? When is it triggered from its T_2 terminal?

35. Does the WELDING CURRENT PULSATION ONE-SHOT fire as the system *enters* the **Heat** subinterval or as the system *leaves* the **Heat** subinterval?

36. When is the HEAT-COOL FLIP-FLOP held cleared by a LO signal on its CL terminal?

37. What conditions are necessary to drive the output of AND6 HI? Same instructions as for Question 16.

38. At what instant does the counting actually take place when welding current pulsations are counted by the Interval Time Counter? Does it take place on the positive-going edge of the WELDING CURRENT PULSATION ONE-SHOT output pulse or on the negative-going edge?

39. Why is it necessary to wire the $\bar{Q}$ output of the HEAT-COOL ONE-SHOT to the input of AND6? That is, why can't we just leave AND6 as a two-input gate and forget about the top input?

40. If it were desired to adjust the system controls to deliver 24 current pulsations during the **Weld** interval, with each pulsation consisting of 15 cycles of current flow followed by 36 cycles of no current flow, explain how the operator should adjust the following six selector switches: WELD UNITS, WELD TENS, HEAT UNITS, HEAT TENS, COOL UNITS, and COOL TENS.

41. What conditions are necessary to drive the NAND5 output LO? Same instructions as for Question 16.

42. Explain the function of inverters 15 and 17.

43. Referring to Fig. 7-10, explain why it is impossible for the UJT to fire when the incoming power line polarity is wrong for firing the ignitron.

44. To increase the average welding current during a current pulsation, should the resistance of R_5 be increased or decreased? Explain.

45. Why is it impossible to fire the UJT when the "enable the SCR gate control circuit" terminal is LO?

46. Exactly how much time elapses between the T_1 secondary voltage crossing through zero and ZD1 clipping the waveform at $+15$ V?

47. Explain why C_1 is in a discharged state at the beginning of every positive half cycle of the T_1 secondary. That is, why doesn't the capacitor start out with a residual charge left over from the previous positive half cycle?

48. What is the peak current through the collector of Q_2 when it shorts out ZD1?

49. Find the minimum and maximum charging time constants for C_1.

50. Would the circuit of Fig. 7-10 work properly if the T_2 secondary winding were reversed? Explain.

51. Figure 7-10(a) is drawn for the simplified situation in which the welding current direction is *not* reversed. Assume this means that the X ignitron-SCR pairs are the ones being used. Would anything be different for the Y ignitron-SCR pairs? That is, should anything in the schematic be changed for the Y pairs? Exactly what?

52. Is there anything in the circuitry of Chapter 7 that tells you the phase relationship between the 60-Hz count pulses and the ac power line?

8

Operational Amplifiers

The term *operational amplifier* refers to a high-gain dc amplifier with a *differential input* (two input leads, neither of which is grounded). Although discrete operational amplifiers are built, designers of industrial electronic circuitry have started to utilize integrated circuit operational amplifiers extensively, almost to the exclusion of discrete operational amplifiers. We will therefore confine our efforts to IC operational amplifiers, hereafter called *op amps*.

An IC op amp is a complete prepackaged amplifier whose operating characteristics and behavior depend almost entirely on the few *external* components connected to its terminals. That is, the voltage gain, input impedance, output impedance, and frequency bandwidth depend almost solely on stable external resistors and capacitors. This means that the various amplifier characteristics can be tailored to fit a particular application merely by changing a few components, without redesigning the entire amplifier. It is this versatility and ease of adjustment that makes op amps popular in industrial control.

OBJECTIVES

After completing this chapter, you will be able to:

1. Explain why and how an IC op amp can be tailored to suit a particular application
2. Give the approximate range of values of each important amplifier characteristic of a typical op amp, such as output power rating, differential voltage gain, input impedance, common mode rejection ratio, etc.
3. Describe the offset problem of op amps and show how it can be corrected
4. Explain the operation of the following frequently encountered op amp circuits: the voltage comparer, the inverting amplifier, the summing circuit (weighted and unweighted), the noninverting amplifier, the differential amplifier, and the voltage-to-current converter
5. Calculate feedback resistor sizes to produce any desired voltage gain (or weighting factor) of the amplifying circuits above
6. Explain the operation of op amp integrators and differentiators and calculate feedback resistor and capacitor sizes to produce any desired time constant

8-1 OP AMP CHARACTERISTICS

All op amps have at least five terminals. These terminals are identified on the op amp symbol given in Fig. 8-1(a).

An op amp has two power supply terminals, one connected to positive supply voltage and the other connected to negative supply voltage. It also has two input terminals; One is called the *inverting input*, marked −, and the other is called the

Figure 8-1. (a) Schematic symbol and terminal names of an op amp. (b) Op amp circuit showing how the positive and negative supplies and the load are connected. The common terminal of the supplies is considered to be circuit ground. Therefore the output terminal can deliver either a positive or negative voltage across the load.

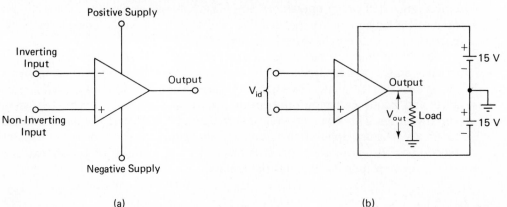

(a)   (b)

noninverting input, marked $+$. There is a single output terminal to which the load is connected. The other side of the load must be connected to ground. There is no direct ground connection to the op amp itself, but there is an implied ground reference due to the two supply voltages being referenced to ground. This becomes clearer in Fig. 8-1(b), which shows positive and negative supply voltages applied to the op amp, and a load connected up. The ± 15-V supply voltage shown in Fig. 8-1(b) is a common supply voltage value for op amps.

Most op amps have other terminals besides these basic five. These other terminals are used for the frequency compensation components, offset adjustment components, etc. We will ignore these other terminals for the time being and concentrate only on the five shown in Fig. 8-1.

8-1-1 Basic capabilities

Let us first get an idea of the overall capabilities and limitations of a typical op amp. An op amp cannot deliver much output power, since it is only a small integrated circuit. A typical device might have a maximum output voltage of about 15 V and a maximum output current of 20 mA. This would yield an output power given by

$$P_{\text{out}} = (V_{\text{out}})(I_{\text{out}}) = (15 \text{ V})(20 \text{ mA}) = 300 \text{ mW}.$$

There are some op amps available which can deliver several watts of output power, but these are the exception rather than the rule.

Op amps can handle frequencies down to 0 Hz. That is, they are dc amplifiers. As far as their high-frequency response is concerned, op amps typically have an *upper cutoff frequency* of about 1 MHz. (Upper cutoff frequency is the frequency at which the voltage gain equals 0.707 times its midrange value.) Special-purpose op amps are available having upper cutoff frequencies of several hundred megahertz.

The fairly wide bandwidth (the range of frequencies the amplifier can handle) is another useful feature of op amps. If a particular circuit application calls for a narrow bandwidth, this can be provided simply by connecting the proper external components to the frequency compensation terminals (not shown in Fig. 8-1). This ease of adjustment of upper cutoff frequency is another useful aspect of op amps.

8-1-2 Open-loop voltage gain

The open-loop voltage gain of an op amp is the differential gain of the amplifier with no external components connected to it. It is the ratio of V_{out} to V_{id} in Fig. 8-1(b), or

$$A_{\text{VOL}} = \frac{V_{\text{out}}}{V_{id}},$$

where A_{VOL} stands for the open-loop voltage gain and V_{id} is the voltage between the differential input terminals. The open-loop voltage gain of an op amp is quite large, usually falling between 10,000 and 200,000. This means that it doesn't take much input voltage across the differential input terminals (V_{id}) in order to drive the amplifier into saturation. To get an idea of just how small V_{id} tends to be, consider this example.

Example 8-1

A certain op amp has an open-loop voltage gain of $A_{VOL} = 50,000$. If the supply voltages are ± 15 V, what value of V_{id} will saturate the amplifier?

Solution. For most op amps, the saturation voltages are around 80% of the supply voltages. That is, the output cannot attain a positive or negative voltage level greater than 80% of the supply voltage. For this example,

$$V_{sat} = (0.80)(\pm 15 \text{ V}) = \pm 12 \text{ V},$$

$$V_{id} = \frac{\pm 12 \text{ V}}{50,000} = \pm 0.24 \text{ mV} = \pm 240 \text{ } \mu\text{V}.$$

Thus only 240 μV across the differential input terminals is sufficient to drive the amplifier into saturation.

The fact that V_{id} is so small gives us a useful tool for quickly analyzing many op amp circuits. The tool is to assume that since V_{id} is so small anyway, we might as well assume that it is zero. Thus we call the differential input voltage a *virtual zero*. This idea will become very handy to us later.

In Fig. 8-1(b), V_{out} is positive if V_{id} is positive on the bottom and negative on the top. That is, if the more positive side of V_{id} is applied to the noninverting input (the + input), then V_{out} will be positive on the top of the load resistor and negative on the bottom of the load resistor.

V_{out} will be negative if V_{id} is positive on top and negative on bottom in Fig. 8-1(b). That is, if the more positive side of V_{id} is applied to the inverting input (the − input), then V_{out} will be negative on the top of the load resistor.

8-1-3 Input resistance

Input resistance is the equivalent resistance seen by an external source between the differential input terminals, as Fig. 8-2 suggests.

The input resistance of an op amp is usually rather high. Popularly used op amps have an input resistance, R_i, between 10 K Ω and 1 M, Ω with most op amps

Figure 8-2. Illustration of the input resistance (R_i) of an op amp. Because R_i is quite large, the current between the input terminals (I_{id}) is very small.

tending toward the higher end of that range. An R_i of 250 K is usually a good estimate of an op amp's input resistance if no specification sheet is available.

The fact that R_i is so high means that the amount of current that flows between the differential input terminals is quite small. Considering the op amp of Example 8-1, if R_i equals 250 K, then even when V_{id} is large enough to saturate the amplifier, the differential input current is given by

$$I_{id} = \frac{V_{id}}{R_i} = \frac{240 \ \mu V}{250 \ K} = 0.98 \text{ nA}.$$

That is, the current flowing from one input terminal to the other is less than one-*billionth* of an amp. The fact that I_{id} is so small gives us another helpful tool for understanding the workings of op amp circuits. The tool is to assume that since I_{id} is so small anyway, we might as well assume that it is zero. This assumption will be handy later on.

Notice in Fig. 8-2 that the power supply voltage connections have not been drawn. This is a common practice since it is assumed that anyone who looks at an electronic drawing containing op amps knows that the supply connections are present in the real circuit. To keep the schematic drawing uncluttered, the supply connections are often left out.

8-1-4 Other considerations concerning op amps

Output Resistance. The output resistance of op amps ranges from about 50 Ω to about 200 Ω depending on type. An op amp's effective output resistance is reduced when external feedback components are connected, so in most applications using feedback the overall output resistance is considered to be 0 Ω.

Common Mode Inputs. The op amp is basically a *differential* amplifier, so it should ideally amplify only the *difference* between the voltages appearing at the two inputs. This means that any signal component which appears simultaneously at both inputs should be completely eliminated and should not affect the output signal in any way. This situation is illustrated in Fig. 8-3(a).

Figure 8-3. A 3-V dc signal appearing simultaneously at both input terminals of an op amp. Such common mode signals are rejected by the op amp, while the differential signal is amplified and passed to the output.

The noninverting input has an input signal of 3 V dc, while the inverting input has an input signal of 3 V dc *plus* 0.2 mV p-p ac. An ideal op amp would amplify the difference between the two signals, namely 0.2 mV p-p and would completely ignore the 3-V dc *common mode signal*, the signal which appears at both input terminals. In real life, however, a little bit of the common mode signal does appear at the output terminal. For most op amps the portion that appears at the output is only about one-thousandth (or less) of the common mode signal appearing at the inputs. The exact factor by which the common mode signal is attenuated is called the *common mode gain*, symbolized A_{VCM}. In equation form,

$$A_{\text{VCM}} = \frac{\text{common mode output voltage}}{\text{common mode input voltage}}.$$

The lower the value of A_{VCM}, the better the op amp.

A more often used specification regarding an op amp's ability to ignore a common mode signal is its *common mode rejection ratio*, abbreviated CMRR. It is defined as the ratio of open-loop differential gain to open-loop common mode gain, or

$$\text{CMRR} = \frac{A_{\text{VOL}}}{A_{\text{VCM}}}.$$

For an op amp with an open-loop differential gain of 50,000, as suggested earlier, and a common mode gain of 0.001, the common mode rejection ratio is given by

$$\text{CMRR} = \frac{A_{\text{VOL}}}{A_{\text{VCM}}} = \frac{50,000}{0.001} = 50,000,000.$$

In plain terms, this means that the op amp amplifies a differential signal 50 million times as much as it amplifies a common mode signal. The higher the CMRR, the better the op amp.

The Offset Problem.   One other aspect of op amps deserves mention before we proceed to op amp circuit applications. This is that op amps are subject to *offset*. Offset is the problem which causes the output voltage to be nonzero when the input voltage is zero. In graphical terms, the transfer curve does not pass through the origin. Offset is illustrated in two different ways in Fig. 8-4.

In Fig. 8-4(a), both input terminals are tied together and connected to ground. Therefore $V_{id} = 0$ V. Yet the output voltage, which should be 0 V, is actually 1.6 V. The figure of 1.6 V was picked at random; it could be any voltage between $-V_{\text{supply}}$ and $+V_{\text{supply}}$. The output offset voltage varies from one individual op amp to another of the same type number.

The offset problem is illustrated graphically in Fig. 8-4(b). The ideal op amp would exhibit an output-to-input relationship as shown by the dashed line. The ideal curve passes through the origin, meaning that 0 volts of input yields 0 volts of output. The real transfer curve is shown by the solid line, which demonstrates that $V_{\text{out}} = +1.6$ V when $V_{id} = 0$ V. Of course an actual curve might show up anywhere on the graph, above or below the ideal curve.

The offset problem may sound serious, but it is actually not hard to deal with

Figure 8-4. (a) Schematic diagram which shows offset in an op amp. (b) Transfer curve showing op amp offset. In an ideal op amp, the transfer curve should pass through the origin, but this usually doesn't happen in reality.

in most situations. It is solved by applying to the op amp a small fixed input voltage of such a size and polarity that the offset is cancelled. The methods of doing this will be illustrated as we go along.

8-2 THE VOLTAGE COMPARER

As mentioned earlier, the IC op amp is normally used with external components connected to it. These external components provide *negative feedback* which serves to reduce the overall voltage gain to a reasonable value. The voltage gain when external feedback components are used is called *closed-loop voltage gain*, symbolized A_{VCL}.

An op amp is seldom used in the open-loop configuration because its open-loop voltage gain is so high that it is difficult to balance the output voltage between the positive and negative saturation points. One application in which it *is* used open loop is the *voltage comparer*. A voltage comparer (also called a *comparator*) does what its name implies. It compares one voltage to another voltage and signals which one is greater. For an example of a voltage comparer, refer to Fig. 8-5(a). In this figure, the inverting input is connected to ground and the external V_{in} source is connected to the noninverting input. The circuit makes a comparison between V_{in}

(a)   (b)

(c)

Figure 8-5. Op amp comparers. (a) Comparing V_{in} to 0 V. The output signal switches abruptly from about -12 to $+12$ V when V_{in} becomes slightly more positive than 0 V. (b) Again comparing V_{in} to 0 V, but the input-output relationship is reversed. The output signal switches from $+12$ to -12 V as V_{in} passes through zero, going positive. (c) Comparing V_{in} to some other reference voltage. The reference voltage can be varied by adjusting the pot.

and ground (0 V). If V_{in} is greater than 0 V, meaning that it is positive, the output will go to positive saturation, about $+12$ V. If V_{in} is less than 0 V, meaning that it is negative, the output will go to negative saturation, about -12 V. As long as V_{in} exceeds a fraction of a millivolt, the output will go to one or the other of these extremes because of the very large A_{VOL}. In this way the output voltage of the voltage comparer indicates the result of the comparison.

A voltage comparer with reversed output signals is shown in Fig. 8-5(b). If V_{in} is greater than 0 V, the output goes to -12 V. If V_{in} is less than 0 V, the output goes to $+12$ V.

A voltage comparer which compares V_{in} to some voltage other than 0 V is shown in Fig. 8-5(c). By adjusting the potentiometer, any voltage from -15 to

+15 V can be applied to the inverting input. If V_{in} is greater than the pot setting, V_{out} goes to +12 V. If V_{in} is less than the pot setting, V_{out} goes to −12 V.

The op amp voltage comparer finds application in many industrial circuits which must constantly monitor some varying voltage signal and signify whether it is greater than or less than some other reference voltage.

8-3 THE OP AMP INVERTING AMPLIFIER

An op amp inverting amplifier is an amplifier whose output polarity (or phase) is the opposite of the input polarity (or phase). That is, if the input is a dc signal, the output is an amplified dc signal of the opposite polarity. If the input is an ac signal, the output is an amplified ac signal 180° out of phase with the input. A diagram of an inverting amplifier is shown in Fig. 8-6(a). In this application, we

Figure 8-6. (a) An op amp inverting amplifier showing how the external resistors are hooked up. (b) Current directions and the voltage polarities in an op amp inverting amplifier, assuming that V_{in} is positive relative to ground.

have external components (resistors) connected to the op amp terminals to provide negative feedback. This arrangement is called *closed-loop* operation, as mentioned before. In closed-loop operation, the voltage gain and input impedance of the overall amplifier are almost completely dependent on the size of the external resistors, but independent of the op amp itself. Let us see why this is the case.

Figure 8-6(b) shows the voltages and currents that exist in an inverting amplifier. Assume for a moment that V_{in} is positive relative to ground. A positive voltage will tend to deliver current into the amplifier through resistor R_{IN}. This current is labeled I_{in} in Fig. 8-6(b). As I_{in} flows through R_{IN}, it creates a voltage drop across R_{IN} which is positive on the left and negative on the right. This voltage drop tends to be just slightly less than V_{in} itself, causing V_{id} to assume some tiny value, positive on the top and negative on the bottom.

Now recall two tools of understanding that we established in Sec. 8-1:

a. The differential input voltage across an op amp is virtually zero.
b. The current that flows into the differential input terminals of an op amp is virtually zero.

Since the $+$ input is grounded, tool **a** tells us that the $-$ input is virtually at ground potential also. It is customary to refer to the inverting input in this circuit as *virtual ground*. In actuality, the $-$ input may be a few millionths of a volt away from ground, but this is a negligible amount.

If the $-$ input is assumed to be at ground potential, then the voltage drop across R_{IN} must equal V_{in}. In equation form, using Kirchhoff's voltage law,

$$V_{R_{IN}} = V_{in} - V_{id} = V_{in} - 0\,\text{V},$$

$$V_{R_{IN}} = V_{in}.$$

The input current I_{in} is given by Ohm's law,

$$I_{in} = \frac{V_{R_{IN}}}{R_{IN}} = \frac{V_{in}}{R_{in}}. \tag{8-1}$$

For example, if $R_{IN} = 20$ K and $V_{in} = 4$ V dc, then

$$I_{in} = \frac{4\,\text{V}}{20\,\text{K}} = 0.2\,\text{mA dc}.$$

Now consider the output of the amplifier. Remember that V_{out} is a negative voltage since V_{in} was positive and the amplifier inverts its input. Since the $-$ input is at virtual ground potential, the voltage drop across feedback resistor R_F must equal the output voltage. This can be seen by applying Kirchhoff's voltage law around the loop. In equation form,

$$V_{R_F} = -V_{out}.$$

The current flow through R_F is given by Ohm's law,

$$I_F = \frac{V_{R_F}}{R_F} = \frac{-V_{out}}{R_F}. \tag{8-2}$$

Now use tool **b**, which says that I_{id} in Fig. 8-6(b) is virtually zero. Kirchhoff's current law tells us that

$$I_{in} = I_{id} + I_F = 0 + I_F,$$

$$I_{in} = I_F. \tag{8-3}$$

Combining Eqs. (8-1), (8-2), and (8-3), we get

$$I_{in} = \frac{V_{in}}{R_{IN}} = I_F = \frac{-V_{out}}{R_F},$$

which yields

$$\frac{V_{out}}{V_{in}} = A_{VCL} = \frac{-R_F}{R_{in}}. \tag{8-4}$$

This means that the closed-loop voltage gain of an op amp inverting amplifier depends solely on the external resistors R_F and R_{IN}. From now on, when we speak of voltage gain in regard to an op amp circuit, it will be understood that we mean *closed-loop* voltage gain.

Example 8-2

The inverting amplifier in Fig. 8-6(a) has $R_F = 50$ K, $R_{IN} = 8$ K, and $R_{load} = 10$ K. $A_{VOL} = 35,000$.

a. What is the output voltage if $V_{in} = -200$ mV?

b. What input voltage would be required to drive the output to positive saturation? Assume the supply voltages are ± 15 V.

Solution

a. The voltage gain of the amplifier depends solely on R_F and R_{IN}, not on A_{VOL} or R_{load}:

$$A_{VCL} = \frac{V_{out}}{V_{in}} = \frac{-R_F}{R_{IN}} = -\frac{50 \text{ K}}{8 \text{ K}} = -6.25,$$

$$V_{out} = (-6.25)(-200 \text{ mV}) = +1.25 \text{ V}.$$

b. If the amplifier is into positive saturation, the output voltage would have to go to approximately $+12$ V. It should be understood that this is only an approximation and that the actual saturation voltage would vary with temperature and with load resistance:

$$V_{in} = \frac{V_{out}}{-6.25} = \frac{12 \text{ V}}{-6.25} = -1.92 \text{ V}.$$

To get an intuitive feel for what an inverting amplifier does, consider its behavior this way:

a. When a positive input voltage V_{in} is applied, it tends to drive the inverting input positive.

b. When the inverting input goes positive, V_{out} goes negative. The negative potential at V_{out} causes current to flow through R_{IN} and R_F (they are equal currents), which tends to *lower* the voltage at the inverting input.

c. V_{out} will take on whatever value it has to in order to *lower the voltage at the inverting input to virtual ground*. V_{out} *must* do this, because if it *doesn't* bring V_{id} down to virtual ground, V_{id} will tend to drive V_{out} even more negative.

A numerical example will make the above idea clear. Consider the inverting amplifier in Fig. 8-7(a). It has a closed-loop voltage gain given by

$$A_V = \frac{-4 \text{ K}}{1 \text{ K}} = -4,$$

so if $V_{in} = +1$ V, $V_{out} = -4$ V. Let us suppose for a minute that V_{out} didn't want to go to -4V but wanted to be -3 V instead. Would this be possible?

The answer is no, because if V_{out} went to -3 V, the situation would be as shown in Fig. 8-7(b). The 1-K and 4-K resistors are effectively in series since the op amp itself draws no current. From Ohm's law, the current through these resistors would

(a)   (b)

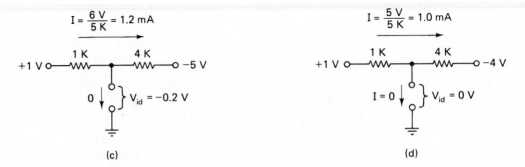

(c)   (d)

Figure 8-7. (a) An inverting amplifier with a closed-loop voltage gain of 4. For a V_{in} of +1 V, V_{out} must equal −4 V. (b) Showing why a V_{out} of −3 V is impossible. The output voltage is not negative enough to balance the system. (c) Showing why a V_{out} of −5 V is impossible. The output voltage is too negative to balance the system. (d) Showing why a V_{out} of −4 V is the only result possible. The output voltage is just right for balancing the system.

equal the total voltage drop, 4 V, divided by the series resistance, 5 K. That is,

$$I = \frac{4\text{ V}}{5\text{ K}} = 0.8\text{ mA.}$$

The drop across the 1-K resistor would then be

$$V_{1\text{ K}} = (0.8\text{ mA})(1\text{ K}) = 0.8\text{ V.}$$

This would result in a V_{id} of

$$V_{id} = V_{in} - V_{1\text{ K}} = 1.0\text{ V} - 0.8\text{ V} = 0.2\text{ V.}$$

A V_{id} of +0.2 V is a far cry from virtual ground, and it would drive V_{out} further negative. Therefore the circuit will not be content to settle at this point but will try to drive the output more negative.

Now let us suppose that the output wanted to go past −4 V, all the way to −5 V. Would this be possible?

The answer again is no, because then the situation would be as shown in Fig. 8-7(c). The current would equal the total voltage drop, 6 V, divided by the total series resistance, 5 K:

$$I = \frac{6 \text{ V}}{5 \text{ K}} = 1.2 \text{ mA},$$

$$V_{1\text{ K}} = (1.2 \text{ mA})(1 \text{ K}) = 1.2 \text{ V}.$$

This would result in a V_{id} of

$$V_{id} = V_{\text{in}} - V_{1\text{ K}} = 1.0 \text{ V} - 1.2 \text{ V} = -0.2 \text{ V}.$$

Once again, a V_{id} of -0.2 V is nowhere near virtual ground, so it would drive V_{out} more positive (less negative). Therefore the circuit will not be content to settle at this point either, but would drive the output more positive than -5 V (less negative).

The only acceptable output voltage is -4 V, as shown in Fig. 8-7(d):

$$I = \frac{5 \text{ V}}{5 \text{ K}} = 1.0 \text{ mA},$$

$$V_{1\text{ K}} = (1.0 \text{ mA})(1 \text{ K}) = 1.0 \text{V}.$$

Then V_{id} is given by

$$V_{id} = V_{\text{in}} - V_{1\text{ K}} = 1.0 \text{ V} - 1.0 \text{ V} = 0 \text{ V}.$$

Therefore the differential input voltage is at virtual ground, and it will not try to drive V_{out} in either direction. The circuit will settle at this point.

Keep in mind that V_{id} is not *absolutely* 0 V, just very close to 0 V. If V_{id} were *absolutely* 0 V, there could be no output voltage.

8-4 THE PHASE INVERTER

A special case of an op amp inverting amplifier is the *phase inverter*. In a phase inverter, R_{IN} equals R_F, so the voltage gain equals 1. V_{out}, however, is inverted from V_{in}. If V_{out} is an ac waveform, V_{out} is a duplicate waveform except that it is of opposite polarity, or shifted by 180°.

8-5 OP AMP SUMMING CIRCUITS

An op amp can be connected to perform the mathematical operation of addition. That is, it can "add" together two or more dc voltages or ac waveforms. When it is connected to perform this function, the circuit is called a *summing circuit*, or a *summer*.

A true summing circuit is shown in Fig. 8-8(a). With all the resistors equal, the output equals the sum of $V_1 + V_2$, but is inverted. This relationship is written in equation form beneath the circuit schematic.

$-V_{out} = V_1 + V_2$

If all Resistors are Equal.

(a)

$-V_{out} = \dfrac{40\ K}{10\ K}(V_1 + V_2)$

$-V_{out} = 4(V_1 + V_2)$

(b)

$-V_{out} = \dfrac{40\ K}{10\ K}(V_1) + \dfrac{40\ K}{20\ K}(V_2)$

$-V_{out} = 4(V_1) + 2(V_2)$

(c)

$-V_{out} = \dfrac{R_F}{R_1}V_1 + \dfrac{R_F}{R_2}V_2 + \dfrac{R_F}{R_3}V_3$

(d)

Figure 8-8. Op amp summing circuits. (a) A true summing circuit, having no gain. (b) A summing circuit which also provides gain. (c) A summing circuit in which different inputs have different weights. (d) A three-input summing circuit and its generalized input-output equation.

The output-input equation for a summing circuit can be derived by using the two tools of understanding that were used in Sec. 8-3. The voltage drop across each input resistor equals the input voltage applied to it, since $V_{id} = 0$ V. The current through R_F equals the sum of the input currents through the input resistors, since $I_{id} = 0$. In equation form,

$$I_{R_1} = \frac{V_1}{R_1} \quad \text{and} \quad I_{R_2} = \frac{V_2}{R_2},$$

$$I_F = I_{R_1} + I_{R_2} = \frac{-V_{out}}{R_F}.$$

Combining the above equations yields

$$\frac{-V_{out}}{R_F} = \frac{V_1}{R_1} + \frac{V_2}{R_2}$$

which can be written

$$-V_{out} = \frac{R_F}{R_1}V_1 + \frac{R_F}{R_2}V_2. \tag{8-5}$$

If all resistors are equal, as in Fig. 8-8(a), this equation reduces to

$$-V_{out} = V_1 + V_2.$$

If a positive output voltage is desired rather than a negative one, it is a simple matter to put this output into a phase inverter.

An op amp summing circuit which also provides voltage gain is shown in Fig. 8-8(b). Equation (8-5) always expresses the output voltage in terms of the input voltages. For the resistor values of Fig. 8-8(b), Eq. (8-5) reduces to the equation given beneath that circuit schematic.

An op amp summing circuit which provides different *weighting factors* for the input voltages is shown in Fig. 8-8(c). Input V_1 is more heavily weighted than input V_2. That is, V_1 is "more important" than V_2 in determining V_{out}, since it is amplified by a factor of 4, whereas V_2 is amplified by a factor of only 2. The equation is given beneath the circuit schematic.

Op amp summing circuits can handle any number of inputs. A three-input summing circuit and its output-input formula are given in Fig. 8-8(d). The main idea to keep in mind is that the input signals feeding through *smaller* input resistors are more heavily weighted since they experience a larger voltage gain.

Example 8-3

For the circuit of Fig. 8-8(d), $R_1 = 5$ K, $R_2 = 8$ K, $R_3 = 6$ K, and $R_F = 24$ K. The input voltages are $V_1 = -0.5$ V, $V_2 = +0.7$ V, and $V_3 = -1.4$ V. Find V_{out}.

Solution

$$-V_{out} = \frac{R_F}{R_1}(V_1) + \frac{R_F}{R_2}(V_2) + \frac{R_F}{R_3}(V_3),$$

$$-V_{out} = \frac{24 \text{ K}}{5 \text{ K}}(-0.5 \text{ V}) + \frac{24 \text{ K}}{8 \text{ K}}(+0.7 \text{ V}) + \frac{24 \text{ K}}{6 \text{ K}}(-1.4 \text{ V}),$$

$$-V_{out} = -2.4 \text{ V} + 2.1 \text{ V} - 5.6 \text{ V} = -\mathbf{5.9} \text{ V},$$

$$V_{out} = +5.9 \text{ V}.$$

8-6 THE NONINVERTING AMPLIFIER

An op amp wired as a noninverting amplifier is shown in Fig. 8-9(a). As its name implies, the output voltage of a noninverting amplifier has the same polarity as the input voltage.

Once again, let us use our two tools seen earlier. The fact that V_{id} is virtually 0 V means that the voltage appearing at the inverting input is the same as V_{in}.

Figure 8-9. (a) An op amp noninverting amplifier, showing how the external resistors are connected. (b) Current directions and voltage polarities in a noninverting amplifier, assuming that V_{in} is positive relative to ground.

Therefore the voltage drop across R_{IN} equals V_{in}, since the left side of R_{IN} is grounded. Refer to Fig. 8-9(b). The current through R_{IN} is given by

$$I_{in} = \frac{V_{in}}{R_{IN}}.$$

Now consider the output of the circuit. Since the right-hand side of R_F is connected to the output and the left-hand side of R_F is virtually at V_{in} potential, the voltage drop across R_F is given by

$$V_{R_F} = V_{out} - V_{in}.$$

From Ohm's law,

$$I_F = \frac{V_{R_F}}{R_F} = \frac{V_{out} - V_{in}}{R_F}.$$

Now making use of the fact that $I_F = I_{in}$ because $I_{id} = 0$, we can say that

$$\frac{V_{out} - V_{in}}{R_F} = \frac{V_{in}}{R_{IN}}$$

$$\frac{V_{out}}{R_F} = \frac{V_{in}}{R_F} + \frac{V_{in}}{R_{IN}} = \frac{V_{in}(R_F + R_{IN})}{(R_F)(R_{IN})}$$

$$\frac{V_{out}}{V_{in}} = A_{VCL} = \frac{R_F + R_{in}}{R_{IN}} = \frac{R_F}{R_{IN}} + 1. \tag{8-6}$$

Again, the closed-loop voltage gain depends solely on the values of external resistors, not on the op amp itself.

Example 8-4

For the noninverting amplifier shown in Fig. 8-9(a), $R_{IN} = 1.2\ \text{K}$. Find R_F so that the gain of the amplifier is 50.

Solution

$$A_V = \frac{R_F}{R_{IN}} + 1,$$

$$50 = \frac{R_F}{1.2\text{ K}} + 1,$$

$$R_F = (50 - 1)(1.2\text{ K}) = \textbf{58.8 K}.$$

The noninverting amplifier has an advantage over the inverting amplifier in that it has a much higher input impedance. The closed-loop input impedance of a noninverting amplifier is much greater than the op amp's open-loop input resistance, which is fairly high to begin with.

The closed-loop input impedance of an *inverting* amplifier, on the other hand, is equal to R_{IN}, since the right-hand side of R_{IN} is virtually grounded [Fig. 8-6(a)].

8-7 CORRECTING THE OFFSET PROBLEM

It was pointed out in Sec. 8-1-4 that real-life op amps do not necessarily give 0 V of output for 0 V of input. This was termed *offset*, and it is a problem which must be dealt with in op amp circuitry. The way offset is corrected is by applying a small fixed voltage to the differential input of the amplifier. The small fixed voltage is called *input offset voltage*. It is adjusted to the proper magnitude and polarity to exactly cancel the offset tendency of the op amp. That is, the input offset voltage is adjusted until $V_{out} = 0$ when $V_{in} = 0$. The most popular methods of applying the input offset voltage are shown in Fig. 8-10.

In Fig. 8-10(a), the + input, instead of being grounded, receives a small fixed dc voltage determined by the pot setting. The pot wiper provides the input offset voltage. Figure 8-10(b) shows a noninverting amplifier. Normally R_{IN} would be grounded on the left side, but here it receives an input offset voltage from the 10-K pot wiper.

Sometimes an op amp has special *null adjustment* terminals, as illustrated in Fig. 8-10(c). In this case, it is not necessary to tamper with the external input circuitry to cancel the offset. You simply connect a pot to the two null terminals, connect the pot wiper to the negative supply voltage, and adjust until the offset is cancelled out. To understand why this cancels the offset, it would be necessary to look at the internal construction of the op amp's circuitry. If you are curious about this, refer to any good book devoted solely to op amps.

8-8 THE OP AMP DIFFERENTIAL AMPLIFIER

Sometimes it is necessary to amplify the voltage difference between two input lines, neither of which is grounded. In this case, the amplifier is called a *differential amplifier*. An op amp differential amplifier is shown in Fig. 8-11.

(a)

(b)

Null Terminals

10 K Pot

$-V_s$

(c)

Figure 8-10. Correcting the offset problem. (a) In an inverting amplifier, the pot wiper applies a small input offset voltage to the + input terminal. This cancels the inherent offset of the op amp itself. (b) The same idea applied to a noninverting amplifier. (c) Op amp with special nulling terminals. This makes it unnecessary to introduce a pot into the input circuitry.

In most cases, $R_1 = R_2$ and $R_F = R_D$. If this is so, the output-input equation is

$$V_{out} = \frac{R_F}{R_1}(V_2 - V_1). \tag{8-7}$$

Equation (8-7) tells us that the op amp differential amplifier amplifies the *difference* between the two input lines and has a voltage gain dependent solely on the external resistors, as usual.

When an op amp differential amplifier is used, there is a limit to the amount of *common mode* voltage that can be applied to the two inputs. Exceeding this maximum common mode input voltage can ruin an op amp. Therefore it is not sufficient to be concerned only with the difference between V_2 and V_1. One must also be concerned with the voltage that V_2 and V_1 have in common. An op amp's data sheet will always specify its maximum common mode input voltage.

Figure 8-11. Op amp differential amplifier. In a differential amplifier, neither input is tied to circuit ground. This can reduce the amount of noise injected into the amplifier, because any noise appears simultaneously on both input terminals. Because the noise is a common mode signal, the amplifying circuitry rejects it.

8-9 AN OP AMP VOLTAGE-TO-CURRENT CONVERTER

Occasionally in industrial electronics, it is necessary to deliver a current which is proportional to a certain voltage, even though the load resistance may vary. If the load resistance stayed constant, there would be no problem. The load current would be proportional to the applied voltage just naturally, by Ohm's law. However, if the load resistance varies from unit to unit, or if it varies with temperature or with age, then delivering a current exactly proportional to a certain voltage is not so easy. A circuit which can perform this job is shown in Fig. 8-12. It is called a *voltage-to-current converter*. This circuit is able to convert voltage to current because of the virtual zero across the differential inputs. That is, if V_{in} appears at the + input, then a voltage virtually equal to V_{in} must appear at the − input. The

Figure 8-12. Op amp voltage-to-current converter. The important idea about the voltage-to-current converter is that the load current is fixed by V_{in} even if there are variations in the load itself.

current through R_1 is determined by Ohm's law,

$$I_{R_1} = \frac{V_{in}}{R_1},$$

so I_{R_1} will never change as long as R_1 itself doesn't change.

Because of the fact that virtually no current flows between the inverting and noninverting inputs, we can say that

$$I_{R_1} = I_{load}.$$

Therefore,

$$I_{load} = \frac{V_{in}}{R_1}. \tag{8-8}$$

Equation (8-8) holds true *irrespective of the load's resistance*. The load current is guaranteed to be proportional to the input voltage under *any* conditions of load resistance (within limits).

Another good thing about the op amp voltage-to-current converter is that it can be driven by a voltage source which is not itself capable of supplying the load current called for by Eq. (8-8). This is because the voltage source only has to drive a noninverting op amp, whose input impedance is very high (many megohms). The load current itself is supplied by the op amp.

8-10 OP AMP INTEGRATORS AND DIFFERENTIATORS

Besides being able to perform the mathematical operations of addition (summing circuits) and multiplication (amplifiers), op amps can also perform the advanced mathematical operations of integration and differentiation. While these functions are not so commonplace as the more basic functions, they are still an important part of the industrial use of op amps.

In simplest terms, a differentiator is a circuit whose output is proportional to *how quickly the input is changing*. An integrator is a circuit whose output is proportional to *how long the input has been present*.*

Figure 8-13(a) shows an op amp differentiator. The differentiator can be thought of intuitively this way:

a. If V_{in} is a steady dc voltage, capacitor C will charge to V_{in}, there will be no current through C or R_F, so V_{out} will be 0 V.
b. If V_{in} is a slowly changing voltage, the voltage across capacitor C will always be slightly less than V_{in} since it has probably had ample opportunity to charge up. This means that only a small current will be flowing in the capacitor leads and through R_F. Therefore V_{out} will be small.
c. If V_{in} is a rapidly changing voltage, then the capacitor voltage will be considerably less than V_{in} since it probably hasn't had time to charge. This will result in a large current flow through C and R_F and a large V_{out}.

*This definition is quite an oversimplification of the action of an integrator. In fact, the definition is really only accurate if the input voltage is unvarying dc.

$$-V_{out} = R_F C \frac{dV_{in}}{dt}$$

(a)

$$-V_{out} = \frac{1}{R_{in}C} \int V_{in}\ dt$$

$$-V_{out} = \frac{1}{R_{in}C}\ V_{in}\ (t)$$
if V_{in} is dc.

(b)

Figure 8-13. (a) Op amp differentiator. Ideally, the output voltage is proportional to the rate of change of the input voltage. (b) Op amp integrator. Ideally, if V_{in} is a dc signal, the output voltage is proportional to the amount of time the input has been present.

The precise output-input relationship of an op amp differentiator is expressed by the formula beneath the differentiator schematic.

Figure 8-13(b) shows an op amp integrator. An integrator can be thought of intuitively this way:

a. If V_{in} has just very recently appeared at the input terminal, current has not been flowing through R_{IN} for very long. Therefore current has not been flowing through C for very long either, and C has not charged very far. The voltage to which C

has charged equals the output voltage V_{out}, since one side of C is connected to the output and the other side of C is connected to virtual ground. Since capacitor voltage is small if V_{in} has just recently appeared, V_{out} is also small.

b. If V_{in} has been present for some time, current has been flowing through R_{IN} and C for a while. This means that C has had time to charge considerably and therefore has considerable voltage across its plates. Since V_{out} equals capacitor voltage, V_{out} is a considerable voltage (not small).

c. The longer V_{in} persists, the more capacitor C will charge, and the greater will be the output voltage. Thus V_{out} is proportional to *how long* V_{in} has been present.

The general output-input relationship for an op amp integrator is expressed in the first formula beneath the integrator schematic in Fig. 8-13(b). The output-input relationship for the special case of an unvarying dc input voltage is expressed in the second formula.

QUESTIONS AND PROBLEMS

1. Over what approximate range is the output voltage of the op amp in Fig. 8-1(b) able to swing?

2. Explain the difference between *open-loop voltage gain* and *closed-loop voltage gain*.

3. What does it mean to say that an op amp has a high CMRR?

4. Give a typical value of the input resistance of an op amp.

5. Give a typical value of the A_{VOL} of an op amp.

6. About how much output power can a typical op amp deliver?

7. Generally speaking, what technique is used to adjust an op amp's frequency response?

8. Is it possible to take an op amp having an upper cutoff frequency of 2 MHz and raise its cutoff frequency to 3 MHz by using frequency-compensating capacitors and resistors

9. Explain why the $-$ input in the inverting amplifier of Fig. 8-6 is "virtually grounded" although it is not actually grounded.

10. What is the approximate input impedance (resistance) seen by the signal source in the inverting amplifier of Fig. 8-6?

11. Which amplifier has the higher input resistance, the inverting amplifier or the noninverting amplifier? Explain why.

12. When would offset be a more serious problem, when amplifying a dc signal or when amplifying an ac signal? Explain your answer.

13. If a 2-V p-p, 100-Hz sine wave were applied to *both* inputs of the op amp in Fig. 8-1(a), what would the output voltage waveform look like?

14. If the differential amplifier of Fig. 8-11 has a voltage gain of 20 and a differential input of $V_2 - V_1 = 3V$ p-p, describe and sketch the output waveform.

15. Suggest a simple way in which the offset adjustment could be made "finer" in Fig. 8-10(a) or (b). "Finer" means that a given amount of rotation of the pot shaft produces a smaller change in offset.

16. Explain intuitively why the voltage-to-current converter of Fig. 8-12 is able to deliver an unvarying current to a variable resistance load.

17. Which is more likely to occur with an uncorrected op amp, a positive output offset voltage or a negative output offset voltage?

18. What is better, a large A_{VCM} or a small A_{VCM}? Explain why.

19. If the op amp differentiator of Fig. 8-13(a) had a sawtooth input voltage waveform, what would the output voltage waveform look like?

20. Repeat Question 19 for a square-wave input.

21. If the op amp integrator of Fig. 8-13(b) had a square-wave input voltage, what would the output voltage waveform look like?

22. Using schematic diagrams, show how you would experimentally measure an op amp inverting amplifier's A_{VCL}. State what laboratory instruments you would need to take your measurements.

23. Repeat Question 22 for each of the following parameters: A_{VCM}, input impedance, $V_{out(sat)}$, $V_{out(offset)}$, and upper cutoff frequency.

In Questions 24–33, assume that the output offset has been corrected to zero.

24. The voltage comparer of Fig. 8-5(c) has its potentiometer wiper adjusted two thirds of the way from the left terminal. If V_{in} is a 12-V p-p sine wave, graph V_{out}.

25. The op amp inverting amplifier of Fig. 8-6(a) has $R_{IN} = 5$ K and $R_F = 35$ K.
    a. What is V_{out} if $V_{in} = +1.2$ V?
    b. What is V_{out} if $V_{in} = -0.65$ V?
    c. For part b, what voltage would a VOM-type voltmeter indicate at the $-$ input relative to ground?
    d. If $V_{in} = -3.9$ V, what is V_{out}?

26. The inverting amplifier of Fig. 8-6(a) has $R_{IN} = 1.5$ K.
    a. Choose a variable resistance R_F such that A_{VCL} can be varied from 0 to 20.
    b. How would you make A_{VCL} variable from 10 to 60?
    c. What is the input impedance?

27. In the summer of Fig. 8-8, $R_1 = 10$ K, $R_2 = 10$ K, and $R_F = 50$ K. $V_1 = +2.5$ V and $V_2 = -0.9$ V. Find V_{out}.

28. Design a three-input weighted summing circuit which gives the following output-input relationship: $-V_{out} = 4V_1 + 2V_2 + V_3$.

29. In the noninverting amplifier of Fig. 8-9(a), $R_{in} = 6$ K, $R_F = 48$ K, $R_{load} = 500\ \Omega$, and $V_{in} = +750$ mV dc.
    a. Find A_{VCL}, V_{out}, I_F, and I_{load}.
    b. What voltage would be indicated at the $-$ input relative to ground using a VOM-type meter?
    c. How much output power is the op amp delivering?

30. Design a noninverting amplifier having the following characteristics: $A_{VCL} = 25$, maximum undistorted output power $= 250$ mW into a 200-Ω load (undistorted means that the output voltage waveform is not clipped due to saturation), and input impedance greater than 1 M.

31. It is desired to build a voltage-to-current converter to drive a 1200-Ω load (nominal resistance). The proportionality factor is to be 5 mA/V.

    a. Find R_1 to provide this proportionality factor.

    b. If $V_{in} = 7.2$ V, find I_{load} if $R_{load} = 1200 \Omega$.

    c. If $V_{in} = 7.2$ V, find I_{load} if R_{load} changes to 1150 Ω.

    d. Repeat parts b and c for a V_{in} of -10.5 V.

32. For the differentiator of Fig. 8-13(a), $R_F = 20$ K and $C = 0.075$ μF. V_{in} is a triangle wave centered on ground, with a magnitude of 8 V p-p, at a frequency of 400 Hz. Graph the output voltage waveform, with the voltage and time axes properly scaled and marked.

33. For the integrator of Fig. 8-13(b), $R_{in} = 1$ M and $C = 2$ μF. $V_{supply} = \pm 15$ V. V_{in} is a step function, 5 V tall. (Imagine a switch closing and causing $+5$ V to be suddenly applied to the input.) Graph the output voltage waveform, with the voltage and time axes properly scaled and marked.

SUGGESTED LABORATORY PROJECTS

PROJECT 8-1: AN OP AMP INVERTING AMPLIFIER

Construct the inverting amplifier of Fig. 8-6. Use $R_{IN} = 1$ K and $R_F = 12$ K. $R_{load} = 1.5$ K. Use supply voltages of ± 15 V. To construct the circuit, refer to the pin diagram of the op amp you use. Install all compensating resistors and capacitors as suggested by the manufacturer.

1. Observe the output offset when $V_{in} = 0$ V. Substitute several different op amps and observe the random nature of the offset.

2. Vary the offset by attaching a potentiometer as shown in Fig. 8-10(c) if the op amp has provisions for such a pot. If not, use the method of Fig. 8-10(a). Observe how the offset can be corrected to zero.

3. Applying a 1-V p-p, 1-kHz sine wave to the input, observe the output. Determine the actual closed-loop voltage gain and compare to formula prediction. If a dual-trace scope is available, compare the phase of the output and input. Is this what you expected?

4. Slowly increase the magnitude of V_{in} until the output starts saturating. Is this saturation voltage about what you expected?

5. With the scope on its most sensitive voltage scale, observe the voltage at the $-$ input. Is this about what you expected? Find the open-loop voltage gain, A_{VOL}. Compare to manufacturer's specifications.

6. Making sure the magnitude of V_{in} stays constant, raise the frequency of V_{in} until V_{out} falls to 0.707 of the value it had at 1 KHz. This frequency is the upper cutoff frequency. Compare your result to the manufacturer's specification for the compensating components you used. (The specification sheet may use the term *3 db down frequency* instead of the term *upper cutoff* frequency.)

PROJECT 8-2: AN OP AMP NONINVERTING AMPLIFIER

Note: For instructions on how to measure the input impedance and output impedance of an amplifier, refer to D. L. Metzger, *Electronic Circuit Behavior*, Prentice-Hall, 1975, pp. 165–166.

Construct the noninverting amplifier of Fig. 8-9(a). Use ± 15-V supplies. $R_{in} = 4.7$ K, $R_F = 5.6$ K, and $R_{load} = 1.5$ K. Zero the offset as in Project 8-1.

1. Apply a 3-V p-p, 500 Hz input. Measure the magnitude and phase of V_{out} and find A_{VCL}. Does this agree with your formula prediction?

2. Try to measure the input impedance of the amplifier. Since the input impedance is extremely high, it will probably be impossible to measure it accurately, but at least you can show that it is higher than some large value (say 10 MΩ). Do not use a pot to perform this test since a pot will pick up and inject too much noise into the op amp. Use a fixed resistor in place of the pot.

3. Reduce the input signal until V_{out} is less than 100 mV$_{pp}$. Measure the output impedance of the amplifier. Is this about what you expected?

4. Connect the scope directly across the $+$ and $-$ differential inputs and measure V_{id}. Is this what you expected?

5. Remove the 1.5-K load and consider the 5.6-K as the load. This circuit is now a voltage-to-current converter. With a 1-mA ammeter inserted in series with the 5.6-K resistor, apply a V_{in} of 4.5 V dc. Does the load current agree with what you predict? Vary the load resistance above and below 5.6 K. Observe the effect on load current. Do you see what this circuit does? Do you know any other circuits which can deliver a constant current proportional to the input voltage in the face of a varying load resistance?

PROJECT 8-3: AN OP AMP INTEGRATOR

Construct the op amp integrator of Fig. 8-13(b) and carefully cancel the offset. Put a closed switch in parallel with the capacitor to prevent it from charging. At the instant the switch is opened, C will start charging, and the integration will begin. Use $C = 1\ \mu$F, nonpolarized, and $R_{in} = 1.5$ M.

1. Predict how fast the integrator will integrate. Predict how long it should take to integrate up to 10 V with $V_{in} = -1$ V.

2. Apply $V_{in} = -1$ V, open the switch, and measure the time required to integrate to 10 V. Compare to your prediction.

3. Change C to 0.05 μF and R_{in} to 10 K. Apply a 6-V p-p square wave at 300 Hz to the input. Predict what the output waveform should look like. Then open the switch and observe the output waveform, and compare to your prediction.

4. Change the input voltage waveform to a train of positive pulses, 1 V peak, at 1 kHz. Make sure there are no negative pulses applied to the input. This can be done by using a true pulse generator or by using a square-wave generator with its negative half cycle blocked by a diode.

Open the shorting switch and let the integration begin. You should see a staircase waveform at the output, rising to the saturation voltage of the op amp. It will only occur once per switch opening. Try to explain why this waveform occurs.

9
Feedback Systems
and Servomechanisms

In all the industrial systems we have discussed so far there has been one thing in common: They have not been *self-correcting*. Self-correcting as used here refers to the ability of a system to monitor or "check" a certain variable in the industrial process and *automatically*, without the intervention of a human, correct it if it is not acceptable. Systems which can perform such self-correcting action are called *feedback systems* or *closed-loop systems*.

When the variable which is being monitored and corrected is an object's physical position the feedback system is assigned a special name; it is called a *servo system*. The basic characteristics and components of closed-loop systems are the subjects of this chapter.

OBJECTIVES

After completing this chapter, you will be able to:

1. Explain the generalized closed-loop block diagram and state the purpose of each one of the blocks.

2. State the characteristics which are used to judge a closed-loop control system; that is, describe what makes the difference between a good system and a bad system.

3. List the five general closed-loop control modes and explain how each one acts to correct the system error.

4. Cite the reasons why the On-Off control mode is the most popular.

5. Define the term *proportional band*, and solve problems involving proportional band, full scale range of the controller, and the control limits.

6. Discuss the problem of *offset* in proportional control, and show why it cannot be eliminated in a strict proportional controller.

7. Explain why the proportional plus integral control mode overcomes the offset problem.

8. Describe the effects of changing the integral time constant (reset rate) in a proportional plus integral controller.

9. Explain the advantage of the proportional plus integral plus derivative control mode over simpler control modes. State the process conditions which require the use of this mode.

10. Describe the effects of changing the derivative time constant (rate time) in a proportional plus integral plus derivative controller.

11. Define and give examples of the three different types of delay exhibited by industrial processes, namely, time constant delay, transfer lag, and transportation lag.

12. Interpret a table which relates the characteristics of an industrial process to the proper control mode for use with that process.

13. Interpret a table which relates the characteristics of an industrial process to the correct settings of proportional band and reset rate.

9-1 OPEN-LOOP VERSUS CLOSED-LOOP SYSTEMS

Let us start by considering the essential difference between an open-loop system (not self-correcting) and a closed-loop system (self-correcting). Suppose that it is desired to maintain a given constant liquid level in the tank in Fig. 9-1(a). Liquid enters the tank at the top and flows out via the exit pipe at the bottom.

One way to attempt to maintain the proper level is for a human being to adjust the manual valve so that the rate of liquid flow into the tank exactly balances the rate of liquid flow out of the tank when the liquid is at the proper level. It might require a bit of hunting for the correct valve opening, but eventually the human could find the proper position. If he then stands and watches the system for a while and sees that the liquid level stays constant, he may conclude that all is well, that the proper valve opening has been set to maintain the correct level. Actually, as long as the operating conditions remain precisely the same, he is right.

The problem is that in real life, operating conditions don't remain the same. There are numerous subtle changes that could occur that would upset the balance he has worked to achieve. For example, the supply pressure on the upstream side

(a)

(b)

Figure 9-1. System for maintaining the proper liquid level in a tank. (a) An open-loop system; it has no feedback and is not self-correcting. (b) A closed-loop system; it has feedback and is self-correcting.

of the manual valve might increase for some reason. This would increase the input flow rate with no corresponding increase in output flow rate. The liquid level would start to rise and the tank would soon overflow. (Actually, there would be some increase in output flow rate because of the increased pressure at the bottom of the tank when the level rises, but it would be a chance in a million that this would exactly balance the new input flow rate.)

An increase in supply pressure is just one example of a change that would upset the manual adjustment. Any temperature change would change the liquid viscosity and thereby change the flow rates; a change in a system restriction downstream of the exit pipe would change the output flow rate, etc.

Now consider the setup in Fig. 9-1(b). If the liquid level falls a little too low, the float moves down, thereby opening the tapered valve to admit more inflow of liquid. If the liquid level rises a little too high, the float moves up, and the tapered valve closes a little to reduce the inflow of liquid. By proper construction and sizing of the valve and the mechanical linkage between float and valve, it would be possible to control the liquid level very close to the desired point. (There would have to be some slight deviation from desired liquid level to cause the valve opening to change.) With this system the operating conditions can change all they want. No matter which direction the liquid level tries to stray from the desired point and no matter what the reason for the straying, the system will tend to restore it to that desired point.

Our discussion to this point has dealt with the specific problem of controlling the liquid level in a tank. However, in the general view, many different industrial control systems have certain things in common. No matter what the exact system, there are certain relationships between the controlling mechanisms and the controlled variable that never differ. We try to illustrate these cause-effect relationships by drawing block diagrams of our industrial systems. Because of the general "sameness" among different systems, we are able to devise *generalized* block diagrams that apply to *all* systems. Such a generalized block diagram of an open-loop system is shown in Fig. 9-2(a).

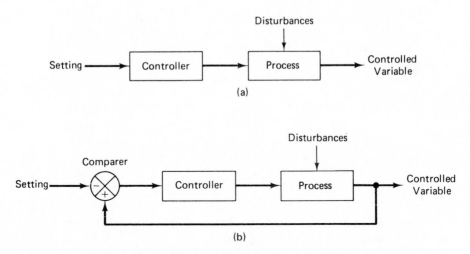

Figure 9-2. Block diagrams which show the cause-effect relationships between the different parts of the system. (a) For an open-loop system. (b) For a closed-loop system.

Now let us try to relate the blocks in Fig. 9-2(a) to the physical parts in the manual control valve system of Fig. 9-1(a). Figure 9-2(a) shows that a *controller* (in our example the manual valve) affects the overall *process* (in our example the pipes that carry the liquid and the tank containing the liquid). The arrow leading from the controller box to the process box just means that the controller somehow "sends signals to" or "influences or affects" the process. The controller box has an arrow pointing into it called the *setting*. This means that the human operator must somehow supply some information to the controller (at least once) that indicates what the controller is supposed to do. In our example the setting would be the position of the valve handle. The process box has an arrow pointing into it called *disturbances*. This means that external conditions can upset the process and affect its outcome. In our example the disturbances are the changing conditions mentioned earlier such as pressure changes, viscosity changes, etc. The *controlled variable* arrow stands for that variable in the process that the system is supposed to monitor and correct when it needs correcting. In our example the controlled variable is the liquid level in the tank.

The block diagram is basically only a cause-effect indicator, but it shows rather clearly that for a given setting the value of the controlled variable cannot be reliably known. Disturbances that happen to the process make their effects felt in the output of the process, namely in the value of the controlled variable. Because the block diagram of Fig. 9-2(a) does not show any lines *coming back around* to make a circle, or to "close the loop," such a system is called an *open-loop system*. All open-loop systems are characterized by the inability to compare the actual value of the controlled variable to the desired value and to take action based on that comparison.

On the other hand, the system containing the float and tapered valve of Fig. 9-1(b) is represented in block diagram form in Fig. 9-2(b). In this diagram the setting and the value of the controlled variable are *compared to each other* in a *comparer*. The output of the comparer represents the *difference* between the two values. The difference signal then feeds into the controller, allowing the controller to affect the process. The fact that the controlled variable comes back around to be compared with the setting makes the block diagram look like a "closed loop." A system that has this feature is called a *closed-loop system*. All closed-loop systems are characterized by the ability to compare the actual value of the controlled variable to its desired value and automatically take action based on that comparison.

For our example of the float level control in Fig. 9-1(b), the *setting* represents the location of the float in the tank. That is, the human operator selects the level that he desires by locating the float at a certain height above the bottom of the tank. This setting could be altered by changing the length of rod *A* which connects the float to horizontal member *B* of the linkage in Fig. 9-1(b).

The comparer in the block diagram is the float itself in our example. The float is constantly aware of the actual liquid level, because it moves up or down according to that level. It is also aware of the setting, which is the desired liquid level, as explained above. If these two are not in agreement, the float sends out a signal which depends on the magnitude and the polarity of the difference between them. That is, if the level is too low, the float causes horizontal member *B* in Fig. 9-1(b) to be displaced (rotated) counterclockwise; the amount of counterclockwise displacement of *B* depends on how low the liquid is.

If the liquid level is too high, the float causes member *B* to be displaced clockwise. Again, the amount of displacement depends on the difference between the setting and the controlled variable; in this case the difference means how much higher the liquid is than the desired level.

Thus the float in the mechanical drawing corresponds to the comparer block in the block diagram of Fig. 9-2(b).

The controller in the block diagram is the tapered valve in the actual mechanical drawing. The valve opens and closes to raise or lower the liquid level, in the same way that the controller in Fig. 9-2(b) sends an output signal to the process to affect the value of the controlled variable.

In our particular example, there is a fairly clear correspondence between the physical parts of the actual system and the blocks in the block diagram. In some systems, the correspondence is not nearly this clear-cut. It may be difficult or impossible to say exactly which physical parts comprise which blocks. One physical part may perform the function of two different blocks, or it may perform the function of one block and a portion of the function of another block. Because of the difficulty in stating an exact correspondence between the two system representations, we will not always attempt it for every system we study.

The main point to be realized here is that when the block diagram shows the value of the controlled variable being fed back and compared to the setting, the system is called a closed-loop system. As stated before, such systems have the ability to automatically take action to correct any difference between actual value and desired value, no matter why the difference occurred.

9-2 CLOSED-LOOP SYSTEM DIAGRAMS AND NOMENCLATURE

9-2-1 The general closed-loop block diagram

A more detailed general block diagram which adequately describes most closed-loop systems is shown in Fig. 9-3.

The ideas embodied in this general system block diagram are as follows: A certain process variable which is being controlled (temperature, pressure, fluid flow rate, chemical concentration, humidity, viscosity, mechanical position, mechanical speed, etc.) is measured and fed into a comparer. The comparer, which may be mechanical, electrical, or pneumatic, makes a comparison between the measured value of the variable and the set point, which represents the desired value of the variable. The comparer then generates an error signal, which represents the difference between measured value and desired value. The error signal is considered to be equal to the measured value minus the desired value, so if measured value is too large, the error signal is positive, and if measured value is too small, the error signal is of negative polarity. This is expressed in the equation

$$\text{error} = \text{measured value} - \text{set point}.$$

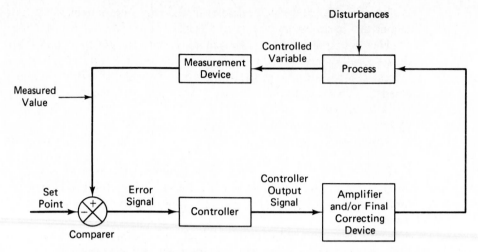

Figure 9-3. The classic block diagram of a closed-loop system. This generalized drawing gives the names of the main parts (blocks) of a closed-loop system. It also gives the names of the signals which are sent between the various blocks.

The controller, which also may be either electrical, mechanical, or pneumatic, receives the error signal and generates an output signal. The relationship between the controller output signal and the error signal depends on the design and adjustment of the controller and is a detailed subject in its own right. All closed-loop controllers can be classified into five classes or *modes* of control. Within the modes, there are certain variations, but these variations do not constitute an essential control difference. Mode of control has nothing to do with whether the controller is electrical, mechanical, or pneumatic; it depends only on *how drastically* and in *what manner* the controller reacts to an error signal. More precisely, it depends on the mathematical relationship between the controller's output and its input (its input is the error signal). We will take a close look at the five control modes in Secs. 9-4 through 9-8.

Figure 9-3 shows that the controller output is fed to a final correcting device; amplification may be necessary if the controller output signal is not powerful enough to operate the final correcting device. The final correcting device is very often an electric motor, which may be used to open or close a valve, move some mechanical object in one direction or the other, or any similar function. The final correcting device might also be a solenoid valve, a pneumatically actuated valve or damper, or an SCR or triac to control the load power in an all-electric system.

An example of the need for amplification would be a situation in which the controller output is a low-voltage, low-current signal (as would be obtained from a Wheatstone bridge) and the final correcting device is a 2-hp motor. Obviously, the low power signal from a Wheatstone bridge cannot drive a 2-hp motor directly, so an electronic power amplifier would be inserted between them. We will explore and study final correcting devices in Chapter 10.

The measurement device in Fig. 9-3 might be a thermocouple, a strain gage, a tachometer, or any of the numerous devices which can make a measurement of a variable. Very often the measurement device gives an *electrical* output signal (usually a voltage) even though it is measuring a nonelectrical quantity. Measurement devices which do this are called *transducers*. We will take a detailed look at several transducers in Chapter 11.

9-2-2 Nomenclature used in closed-loop systems

Unfortunately, the terms used to describe what's happening in a closed-loop control system are not universally agreed upon. We will now look at the various words and phrases that are used by different people, and we will decide on which ones we will use.

Starting from the left in Fig. 9-3, the term *set point* is also called the "set value," "setting," "desired value," "ideal value," "command," and "reference signal." We will use the term *set point* in this book.

The *comparer* is also called a "comparator," "error detector," "difference detector," etc. We will use the term *comparer*.

The *error signal* also goes by the names of "difference signal," "deviation," and "system deviation." We will use *error signal*.

The *final correcting device* is also called the "correcting element" and the "motor element." We will use *final correcting device*.

The *controlled variable* is sometimes called the "controlled condition," "output variable," "output condition," "process variable," etc. We will use *controlled variable*.

The *measurement device* is also referred to as the "detecting device," "detector," or "transducer." We will use the term *measurement device* in most cases, but when we wish to emphasize the ability of the measurement device to convert a nonelectrical signal into an electrical signal, we will use *transducer*.

The *measured value* is sometimes called the "actual value." We will use *measured value*.

9-2-3 Characteristics of a good closed-loop system

It may seem obvious that the measure of a "good" closed-loop control system is its ability to bring the measured value in close agreement with the set point. In other words, a good system reduces the error signal to zero, or almost to zero. The final difference between measured value and set point that the system allows (that it cannot correct) is usually called *offset*. Therefore a good system has low offset. We are now using the word offset to have a different meaning than it had when applied to op amps in Chapter 8.

There are other features of a closed-loop system that are also important, in some cases even more important than low offset. One of them is speed of response. If conditions occur that drive the measured value out of agreement with the set point, a good system will restore the agreement quickly. The quicker the restoration, the better the system.

It is possible to design systems which have low offset and fast speed of response, but they sometmes tend to be *unstable*. Unstable means that the system causes violent large variations in the value of the controlled variable as it frantically "hunts" for the proper controller output. It occurs because the system overreacts to an error, thereby causing an even larger error in the opposite direction. It then tries to correct the opposite error and overreacts again going in the other direction. When this occurs, the system is said to be oscillating. The oscillations usually eventually die out, and the system settles down at the correct value of controlled variable. In the meantime, though, the process has been effectively out of control, and bad consequences may result.

In certain cases the oscillations may not die out at all. They may continue getting larger and larger until the process is permanently out of control. If the control system is a mechanical positioning system (a servo system), these oscillations may cause the mechanism to shake itself apart and destroy itself.

As can be seen, then, a good system is one which is stable. The less violent the oscillations in controlled variable, the more stable it is, and the better the system.

9-3 EXAMPLES OF CLOSED-LOOP CONTROL SYSTEMS

It is normally easiest to see the correspondence between actual physical components and the generalized block diagram of Fig. 9-3 when the system is a servo system. To learn to recognize the block diagram functions of the system components, we will now consider some examples of servo systems.

9-3-1 A simple rack and pinion servomechanism

Figure 9-4 shows a linear positioning system. The pointer is attached to a thin cord which runs over a fixed pulley, around the movable pulley, and over another fixed pulley and attaches to the object which is to be positioned. The object sits on a rack whose pinion is driven by the motor. If the pointer is moved to the left on the scale, the movable pulley is drawn up by the cord, causing the potentiometer wiper to move up by the same amount. When the potentiometer contact is no longer in the center, the unbalanced bridge circuit delivers an input voltage to the amplifier. The output of the amplifier runs the motor, which drives the object to the left. When the object has moved the same distance as the pointer, the movable pulley returns to its rest position, and the potentiometer contact is centered once again. The bridge goes back into balance, causing zero input voltage to the amplifier, which then stops the motor.

It can be seen that whenever the bridge becomes unbalanced, it will send a low

Figure 9-4. A mechanical positioning system using a rack and pinion. This is a simple example of a servomechanism.

power signal to the amplifier, which will amplify it to drive the motor. The motor moves the controlled object into such a position that the bridge is put back into balance. Since the bridge is only balanced when the movable pulley is in its rest position, the controlled object always moves exactly the same distance as the pointer, because only by doing so can it return the movable pulley to the rest position.

In this system, the position of the pointer represents the set point. The position of the object represents the controlled variable. The cord and pulley arrangement represents the comparer, with the instantaneous movable pulley position being the error signal. The bridge circuit is the controller, and the controller output signal is the voltage applied to the input of the amplifier. The motor with rack and pinion arrangement represents the final correcting device.

9-3-2 A profile cutting machine

The same idea applied to a more elaborate mechanism is illustrated in Fig. 9-5. This system is a profile cutting machine. A pattern piece, or model, is fastened to the mounting support, as is the uncut workpiece. The mounting support is then slowly moved to the left. As it moves, the motor-driven cutting tool cuts an identical profile in the workpiece.

The system works like this. The rigid feeler is held tight against the pattern by the action of the tension spring on the right-hand side of the pivot. As the feeler moves up and down, its motion is transmitted to the movable pulley through the cord attached to the right-hand side of the feeler arm. This movable pulley is attached to the wiper of a potentiometer, so as the pulley moves away from its

Figure 9-5. A profile cutting system. This is an example of a more complex servomechanism.

centered position, the bridge becomes unbalanced. The unbalanced bridge drives an amplifier, which drives the servo motor. The servo motor causes the movable frame to move up or down the proper amount to return the movable pulley to its centered position. As the frame moves, it causes the rotating cutting tool to cut into the workpiece. As the cutting tool duplicates the position of the feeler, the profile it cuts duplicates that of the pattern.

In this feedback system, the set point is the pattern depth, or feeler position. The controlled variable is the position of the cutting tool, or, equivalently, the position of the movable frame. All other system parts serve the same block diagram function that they did in Fig. 9-4.

9-3-3 A bimetal temperature control system

Figure 9-6 shows a popular method for temperature control which is used for home heating systems and in some industrial systems. The spiral bimetal strip is immersed in the medium whose temperature is being controlled. Because the two component metals have different coefficients of expansion, the spiral either unwinds or winds up tighter as the metal temperature changes. Assume in this example that the spiral strip is constructed with the metal with the greater coefficient of expansion on the inside, so the spiral *unwinds* as the temperature *rises*. Attached to the end of the spiral is a mercury switch, a sealed glass bulb containing liquid mercury and two electrodes. Mercury, although it is a liquid under standard conditions, is a metal and is an excellent electrical conductor. When the mercury switch tips to the right (rotates clockwise) the mercury drains to the right-hand side of the bulb and breaks the electrical connection between the electrodes. When the switch tips to the left (rotates counter clockwise) the mercury flows to the left-hand side of the bulb and makes an electrical connection between the electrodes.

Figure 9-6. A closed-loop system for controlling temperature.

When the mercury switch is open, the gas solenoid is deenergized, and the gas control valve closes, stopping the flow of natural gas through the pipe to the burner. When the mercury switch closes, it energizes the gas solenoid, opening the gas control valve and allowing natural gas to flow to the burner.

The shaft to which the center of the bimetal spiral strip is attached is itself rotatable. The position of this shaft sets the initial position of the spiral strip. The initial position of the spiral strip determines the desired temperature.

Here is how the system works. If the temperature is below the desired control temperature, the bimetal spiral strip will tend to be wound up. This causes the mercury switch to close, energizing the gas solenoid and turning on the burner. As the temperature rises due to the heat released by burning the natural gas, the bimetal strip unwinds. At a certain temperature, the strip will have unwound enough to open the mercury switch. This truns off the burner. With the burner off, the temperature slowly drops until it causes the strip to wind up enough to close the mercury switch. The burner then comes back on to drive the temperature back up. In this manner the system will continue to maintain the actual temperature near the desired temperature.

The rotatable shaft attached to the setting pointer comprises the set point in the generalized block diagram. To raise the set point, the setting pointer is moved to the right. The measured value in the block diagram is the amount that the bimetal spiral strip is wound up. The comparer is the mercury bulb; the position of the mercury could be considered the error signal. The mercury in combination with the electrodes is the controller. The final correcting device is the gas solenoid valve-burner combination.

You may take the point of view that the solenoid valve is part of the controller and that the burner alone represents the final correcting device. This viewpoint is also reasonable and could easily be adopted. This just goes to point out that there may be no definite one-to-one correspondence between the actual system components and the blocks in the generalized block diagram of Fig. 9-3. The correspondence may be quite blurry.

9-3-4 A pressure control system using a motor-driven positioner

Consider the control system shown in Fig. 9-7. The pressure at a certain point in a process chamber is to be maintained at a desired value. The method of adjustment is the variable position damper which is controlled by a slow-moving positioning motor. If the damper in the inlet duct is opened somewhat, the pressure in the chamber tends to rise. If the damper closes to restrict the inflow, the pressure in the chamber tends to drop. As is commonly done, the positioning motor is driven by an amplifier whose input voltage comes from a Wheatstone bridge.

The process pressure is detected by a *bellows*. As the pressure rises, the bellows expands, causing its left face to push against the compression spring. The bellows is linked to the wiper arm of the pressure error potentiometer, so that as pressure increases, the wiper arm moves toward the top in Fig. 9-7. Therefore if the system experiences a disturbance which causes the pressure to rise too far above the desired value, the bellows will move the wiper arm of the pressure error potentiometer up. This will cause a temporary imbalance of the bridge circuit, so a voltage will be applied to the amplifier input.

The amplifier output will then drive the motor in the correct direction to move the wiper arm of the left-side potentiometer up. As this is taking place the motor linkage is closing off the control damper. When the movement of the left-side potentiometer wiper equals the movement of the right-side potentiometer wiper, the bridge is back in balance, and all movement stops. The damper ends up in a more closed position, thereby limiting the pressure increase to a small amount.

In this system the set point is the adjustment screw on the compression spring,

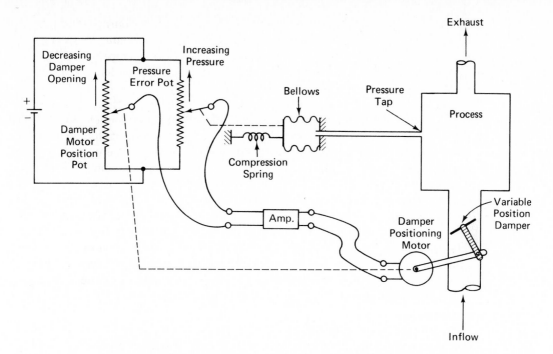

Figure 9-7. A closed-loop system for controlling pressure in a process chamber.

which can alter the force that the spring exerts on the face of the bellows. The measurement device is the bellows itself. Low pressure causes the bellows to collapse, moving to the right; high pressure causes the bellows to expand, moving to the left. The comparer is the combination of compression spring, bellows, and potentiometer arm. The position of the potentiometer arm represents the error signal. Assuming that the exact middle position means zero error, then positions toward the top indicate positive errors (measured value greater than set point) and positions toward the bottom indicate negative errors (measured value less than set point).

The combination of Wheatstone bridge, amplifier, and positioner motor can be considered to comprise the controller. The variable position damper is the final correcting element.

9-4 MODES OF CONTROL IN INDUSTRIAL CLOSED-LOOP SYSTEMS

As mentioned in Sec. 9-2, the *manner* in which the controller reacts to an error signal is an indication of the mode of control. It is somewhat difficult to make hard-and-fast classifications of control modes, but it is generally agreed that there are five basic modes: They are

a. On-Off
b. Proportional
c. Proportional plus integral
d. Proportional plus derivative
e. Proportional plus integral plus derivative

The above list of modes is arranged in order of complexity of the mechanisms and circuitry involved. That is, the mode at the top, On-Off, is the simplest to implement; as you move down in the list, the physical construction of the controllers gets more complex. Naturally the more complex modes of control are also more difficult to understand.

In general, the more difficult the control problem, the farther down the list you must go to find the appropriate control mode. However, in many industrial processes the control need not be very precise; or the nature of the process may be such that precise control is easy to accomplish. In these situations, the simpler control modes are completely adequate. In fact, the simplest method, On-Off, is by far the most widely used. It is inexpensive, reliable, and easy to adjust and maintain.

In this book we are concentrating on electrical and electronic industrial control, so specific examples of the various control modes will be electric controllers. The principles involved are the same when discussing pneumatic, hydraulic, or mechanical controllers, although the methods of implementation are altogether different, naturally.

In the succeeding sections of this chapter, Secs. 9-5 through 9-9, we will study each of the five control modes. We shall start with the simplest and work up to the most complex. Each of the five control modes is explained in terms of *temperature* being the controlled variable. Temperature control is easier to visualize than most other variables. However, keep in mind that the principles discussed in this chapter apply just as well to the control of other process variables besides temperature.

9-5 ON-OFF CONTROL

In the *On-Off control mode*, the final correcting device has only two positions or operating states. For this reason, On-Off control is also known as *two-position control* and also as *bang-bang control*. If the error signal is positive, the controller sends the final correcting device to one of its two positions. If the error signal is negative, the controller sends the final correcting device to the other position. On-Off control can be conveniently visualized by considering the final correcting device to be a solenoid-actuated valve, as seen in Sec. 9-3-3. When a valve is actuated by a solenoid, it is either fully open or completely closed; there is no middle ground. Therefore a solenoid actuated valve fits perfectly into an On-Off control system. A graph of final correcting device position (percent valve opening) for ideal On-Off control is given in Fig. 9-8(a). In this figure, the controlled variable is considered to be temperature, with the set point equal to 120°F. As can be seen, if the measured

Figure 9-8. Graphs pertaining to the On-Off control mode. (a) Valve position vs. measured temperature, with a set point of 120° F. (b) Actual measured temperature vs. time and valve opening vs. time. In the graph of valve opening versus time, the solid line is for a snap-action valve and the dashed line for a slow-acting valve.

value of the temperature is less than 120°F by even a tiny amount, the valve is positioned 100% open. If the measured value of the temperature is greater than 120°F by even a tiny amount, the valve is 0% open, or completely shut off.

Figure 9-8(b) shows a typical graph of measured value of temperature versus time, with valve position plotted against the same time axis. Notice that the actual temperature tends to oscillate around set point. This is a universal characteristic of On-Off control. This particular graph shows a 4°F overshoot in the positive direction and a 4°F undershoot in the negative direction. These particular values are picked at random. The actual overshoot depends on the complete nature of the system and may be different in the positive and negative directions (overshoot may be different from undershoot).

Overshoot occurs because the process cannot respond instantly to the change in valve position. When temperature is climbing, it's because the rate of heat input is greater than the rate of heat loss in the process. Quickly shutting off the control valve cannot reverse that trend instantly, because there will be residual heat energy built up in and around the heating device which must diffuse through the process chamber. As this residual heat is distributed, it temporarily continues to raise the temperature.

In the same way, a temperature downtrend cannot be reversed instantly because it takes a while to distribute new heat energy throughout the process. Until the distribution can occur, the downtrend will continue, causing undershoot.

To be sure, the system can be designed to keep the magnitude of the oscillations small, but this tends to cause more frequent cycling. This aggravates the other disadvantage of On-Off control, namely the wear on the correcting device caused by the frequent operation. In this specific example, the solenoid valve will wear out sooner if its frequency of opening and closing is higher.

The graph of valve position in Fig. 9-8(b) reflects the fact that the valve is wide open when the temperature is below set point and completely closed when the temperature is above set point. The dashed lines are for the case in which the valve is not a snap-action valve. This is often encountered when the valve is physically large. Large heavy valves cannot be successfully operated by snap action but must be operated somewhat slowly. A geared-down motor and linkage is the most effective method of actuating such valves.

9-5-1 Differential gap

No On-Off controller can exhibit the ideal behavior sketched in Figs. 9-8(a) and (b). All On-Off controllers have a small *differential gap*, which is illustrated graphically in Fig. 9-9(a).

Differential gap of an On-Off controller is defined as the smallest range of values the measured value must pass through to cause the correcting device to go from one position to the other. Differential gap is defined specifically for On-Off control; there is no such thing as differential gap in other control modes. It is often expressed as a percent of full scale.

Differential gap is an expression of the fact that the measured value must rise above set point by a certain small amount (the error signal must reach a certain positive value) in order to close the valve. Likewise, the measured value must drop below set point by a certain small amount (the error signal must reach a certain negative value) in order to open the valve. In the example given in Fig. 9-9, the actual measured temperature must rise above set point by 3°F to close the valve, and it must drop below set point by 3°F to open the valve. Therefore the smallest possible temperature change which can drive the valve from open to closed is 6°F. The differential gap is thus 6°F.

Differential gap can also be expressed as a percent of full controller range. If the controller had a range of, say, 60°F to 300°F, then the size of its range would be 240°F (300°F − 60°F). A temperature of 6°F would represent 2.5% of full control range, because

$$\frac{6°F}{240°F} = 0.025 = 2.5\%$$

Therefore in this case the differential gap could be expressed as 2.5% instead of 6°F.

The practical effect of differential gap is shown in the time graphs of Fig. 9-9(b). As can be seen, the magnitude of oscillation is larger, but the frequency of oscilla-

Figure 9-9. Graphs illustrating differential gap in On-Off control. (a) Valve opening vs. temperature. Set point is 120°F, and the differential gap is 6°F. (b) Actual measured temperature vs. time and valve position vs. time, with a 6°F differential gap.

tion is smaller. Differential gap is thus a curse and a blessing. It is a curse because the instantaneous measured value can stray farther from set point, but it is a blessing because wear on the correcting device is reduced.

In many On-Off controllers, the differential gap is fixed. If so, it is usually less than 2% of full scale. Some On-Off controllers have adjustable differential gap so the user can select the amount that suits his application.

If you are familiar with magnetic materials and circuits, you will recognize that differential gap in an On-Off controller has the same effect as hysteresis in a magnetic core. In general, when the switching point of the dependent variable depends not only on the *value* of the independent variable but also on the *direction of approach*, we say that hysteresis exists. Recall that we also saw hysteresis in triac power control in Chapter 5. In magnetism, the dependent variable is flux density (B), and the independent variable is magnetomotive force (H). In the On-Off control mode the dependent variable is the final correcting device position (valve either open or closed), and the independent variable is the error signal.

A good example of an On-Off controller is the temperature control system utilizing the spiral strip and mercury switch in Sec. 9-3-3. Other On-Off control systems will be shown in Chapter 12.

9-6 PROPORTIONAL CONTROL

In the proportional control mode, the final correcting device is not forced to take an all-or-nothing position. Instead, it has a *continuous* range of possible positions. The exact position that it takes is *proportional* to the error signal. In other words, the output from the controller block is *proportional* to its input.

9-6-1 Proportional band

Assuming that the final correcting device is a variable-position valve controlled by a slow gear motor and linkage, we can illustrate the effects of proportional control by drawing a graph of percent valve opening versus temperature. This is done in Fig. 9-10(a). To visualize what is happening, imagine that the valve is controlling the flow of fuel to a burner. This setup is illustrated schematically in Fig. 9-10(b). When the valve opening gets larger, more fuel is delivered, and more heat is released into the process. Therefore the process temperature tends to rise.

Figure 9-10. The proportional control mode. (a) Graph of valve position vs. measured temperature. Valve position is *proportional* to the error signal. (b) Layout of the control system.

(a)

(b)

When the valve opening gets smaller, less fuel is delivered to the burner, and the process temperature tends to fall.

Figure 9-10(a) shows the proportional relationship between percent valve opening and error signal. Study this graph carefully. To get started, imagine that the set-point temperature is now 180°F. Furthermore, let us assume that the process temperature is being maintained right on 180°F with the valve 40% open. There would be no way to know that this would be the case, since the percent of valve opening which is needed in order to maintain 180°F will depend on many unpredictable process conditions. Such things as ambient temperature, rate of heat consumption by the load, fuel supply pressure, heating capacity of the fuel, etc., will all have an effect on what valve opening will be necessary. Therefore, let us just *assume* that a 40% valve opening is correct.

Now if something happens to cause the measured temperature to change, the valve will assume a new position according to the graph of Fig. 9-10(a). If the temperature should somehow drop to 175°F, the valve will open to the 60% point. This will cause the temperature to subsequently rise back toward 180°F. If the original drop in temperature had been more drastic, say down to 170°F, the valve would have opened 80%. Therefore the controller is responding not only to the fact that the measured temperature was too low; it is responding to the *amount* of the error. The more serious the error, the more drastic the corrective action. This is the essential difference between proportional control and On-Off control.

The word proportional is correctly applied in this situation because the amount of correction introduced is in proportion to the amount of the error. When the error is 5°F (measured value equals 175°F), the valve travels from 40% open to 60% open; this means it travels 20% of its full range. If the error is twice as large, namely 10°F (measured value equals 170°F), the valve travels from 40% open to 80% open, or 40% of its full range. Thus the corrective action is also twice as large when the error is twice as large. In general, a given percent change in error brings about a proportional percent change in valve position.

In the example shown in Fig. 9-10(a), a measured temperature of 165°F or less causes the valve to go 100% open, and a measured temperature of 190°F or more causes the valve to go 0% open. The difference between these two points is called the *proportional band of control*. The proportional band in this case is 25°F. Within that band, the valve response is proportional to temperature change; outside that band the valve response ceases because it has reached its limit.

Usually, proportional band is expressed as a percent of the full-scale range of the controller. If the controller set point can be adjusted anywhere between 60°F and 300°F, as assumed earlier, it has an adjustment range of 240°F. The proportional band expressed as a percent would be given by

$$\frac{25°F}{240°F} = 0.104 = 10.4\%$$

The formal definition of proportional band is this: Proportional band is the percent of full controller range by which the measured value must change in order to cause the correcting device to change by 100%. Most proportional controllers have an adjustable proportional band, usually variable from a few percent up to a few hundred percent.

Figures 9-11(a)–(f) give graphical representation of different proportional band adjustment settings, both on a degrees Fahrenheit basis and on a percent of full controller range basis. Any two graphs drawn alongside each other are drawn for the same percents proportional band, but the one on the left is plotted versus temperature and the one on the right is plotted versus percent of full controller range. That is, the graph in Fig. 9-11(a) shows percent of valve opening versus *temperature* for a 20% proportional band, while the graph in Fig. 9-11(b) shows percent of valve opening versus *percent of full range of the controller*, also for a 20% proportional band.

The graphs in Figs. 9-11(c) and (d) show the same things but for a 50% proportional band. The graphs in Figs. 9-11(e) and (f) show the same things again but for a 100% proportional band.

In all cases, the controller range is assumed to be from 60°F to 300°F, giving a full range of 240°F. In all these graphs, the vertical axis indicates percent valve opening, as stated above.

In all the graphs we have tacitly assumed that the measured temperature is being maintained at the set point of 180°F with a 40% valve opening until a disturbance occurs which upsets the measured temperature.

These graphs deserve careful study. By studying and thoroughly understanding the specific numbers indicated on these graphs, you can get a firm grasp of the meaning of proportional band.

Here is a step-by-step interpretation of the graph in Fig. 9-11(a):

a. If the measured temperature is 180°F, the valve will take a position 40% open.

b. If the temperature drops below 180°F, the valve will start opening farther. At 172°F, for example, the valve will be about 57% open. At 164°F, the valve will be about 73% open. These points can be read from the graph. When the measured temperature drops to 151.2°F, the valve will be 100% open. Any further drop in temperature below that point will cause no further corrective action, since the valve has reached its limit. (Hopefully, if the system is designed properly, the temperature should be able to recover from this level and start rising back up toward 180°F with the valve 100% open.) The temperature which causes a 100% valve opening (151.2°F) is specifically marked on the horizontal temperature axis in Fig. 9-11(a).

c. If the measured temperature should rise above 180°F for any reason, the valve will start closing down to less than 40% open. For example, if the temperature should reach 188°F, the valve would throttle back to about 23% open, in an effort to drive the temperature back down toward 180°F. If the temperature should somehow hit 199.2°F, the valve will go fully closed, or 0% open. The exact temperature which causes 0% valve opening is specifically marked on the horizontal temperature axis in Fig. 9-11(a). Beyond this temperature, control is lost because the valve has

Figure 9-11. Graphs of percent valve opening vs. temperature and also vs. percent of full control range. (a) Proportional band = 20%. Valve opening plotted vs. temperature. (b) Proportional band = 20%. Valve opening plotted vs. percent of full control range. (c) Proportional band = 50%. Valve opening plotted vs. temperature. (d) Proportional band = 50%. Valve opening plotted vs. percent of full control range. (e) Proportional band = 100%. Valve opening plotted vs. temperature. (f) Proportional band = 100%. Valve opening plotted vs. percent of full control range.

reached its limit. However, with the valve fully closed and no fuel entering the burner in Fig. 9-10, the temperature is bound to start dropping back down toward 180°F.

To demonstrate that the performance shown by Fig. 9-11(a) constitutes a 20% proportional band, look at the calculations done next to that graph. The range of temperatures which causes the valve to drive from fully open to fully closed is 151.2°F to 199.2°F, which is a span of 48°F. A span of 48°F represents 20% of the total range of the controller, since

$$\frac{48°F}{240°F} = 0.2 = 20\%.$$

Therefore Fig. 9-11(a) represents a 20% proportional band.

To illustrate how the percent calculations are made in the graphs on the right-hand side, namely Figs. 9-11(b), (d), and (f), here is the derivation of the 58% figure and the 38% figure in Fig. 9-11(b).

The temperature of 199.2°F is 139.2°F higher than the lowest temperature in the range of the controller (60°F). That is,

$$199.2°F - 60°F = 139.2°F.$$

To calculate what percent this figure is of the full controller range, we say

$$\frac{139.2°F}{240°F} = 0.58 = 58\%.$$

Therefore the percent of full controller range which causes the valve to go fully closed is 58%, and this is specifically marked on the horizontal axis of Fig. 9-11(b).

The above calculations are then repeated for the fully *open* valve position:

$$151.2°F - 60°F = 91.2°F,$$

$$\frac{91.2°F}{240°F} = 0.38 = 38\%.$$

The exact percent of full controller range that causes the valve to go fully open is therefore 38%, and this is specifically marked on the horizontal axis of Fig. 9-11(b).

To be sure you understand the meaning of proportional band, you should check the calculations done alongside Fig. 9-11(d) for a 50% proportional band. You should then do the appropriate calculations for Fig. 9-11(f) and verify the numbers marked on the horizontal axis of Fig. 9-11(f).

In Figs. 9-11(e) and (f), which are for a 100% proportional band, the lines have been *extrapolated* below 60°F. The extrapolations are drawn dashed because such temperature measurements are impossible; the controller cannot detect temperatures below 60°F. However, it is convenient to imagine these temperatures anyway, because it makes the calculations for proportional band easier to demonstrate. In a real-life situation, this would mean that the valve could never open up to 100%. The error necessary to open the valve that far is beyond the range of the controller. The maximum valve position in this situation would be 90% open. You should check and verify the calculations presented alongside Figs. 9-11(e) and (f).

Variations in Process Conditions. In all the graphs of Fig. 9-11 it is assumed that a set-point temperature of 180°F can be accomplished by a 40% valve opening. Remember that this could change drastically as process conditions change. For example, it might require a 65% valve opening to maintain the temperature at 180°F under heavier load conditions; it might even require a 90% valve opening to maintain a temperature of 180°F under very heavy load conditions. If these different load conditions actually existed, the *slopes* of the lines in Fig. 9-11 would remain the same, but their *horizontal locations* on the graphs would change. This idea is illustrated for a 20% proportional band in Fig. 9-12.

Figure 9-12. Graphs of valve opening vs. temperature with a 20% proportional band for three different process conditions.

Here is the interpretation of the graphs in Fig. 9-12. The left graph is for a 40% valve opening to produce a measured temperature of 180°F, and that graph is just a repeat of Fig. 9-11(a). The center graph in Fig. 9-12 is for the situation in which process conditions have changed so that a 65% valve opening is necessary to produce a temperature of 180°F. Notice that the 180°F temperature line intersects the center graph at 65% valve opening.

The center graph indicates that the valve will be fully open at 163.2°F and will be fully closed at 211.2°F. The proportional band of temperatures is *still* 48°F, which is 20% of full controller range. The only thing different between the left graph and the center graph is the *horizontal location*.

The right-hand graph in Fig. 9-12 is for the situation in which process conditions have changed more drastically, so that a 90% valve opening is necessary to produce an actual temperature of 180°F. Notice that the 180°F temperature line intersects the right-hand graph at 90% valve opening.

In the right-hand graph, the fully open temperature is 175.2°F, and the fully closed temperature is 223.2°F. The temperature band is still 48°F, and the proportional band is therefore still 20%.

9-6-2 The effects of proportional control

Let us now discuss the control effects of using the proportional control mode. As might be expected, it eliminates the permanent oscillation that always accompanies On-Off control. There may be some temporary oscillation as the controller homes in on the final control temperature, but eventually the oscillations die out if the proportional band is adjusted properly. However, if the proportional band is set too small, oscillations may occur anyway, because a very small proportional band makes proportional control behave almost the same as On-Off control. You should think carefully about the last sentence. If you understand what proportional band means, you will understand why that statement is true.

Thus we can see that the proportional control mode has one important advantage over On-Off control. It eliminates the constant oscillation around the set point. It thereby provides more precise control of the temperature, and reduces wear and tear on the valve. A variable position valve moves only when some sort of process disturbance happens, and even then it moves in a less violent manner than a snap-action valve. Its life expectancy is therefore much longer than that of a solenoid snap-action valve.

Figure 9-13 shows some typical responses of a proportional temperature controller to a load disturbance. In each case in Fig. 9-13, a load disturbance has occurred which tends to drive the temperature down. Figure 9-13(a) shows the response for a narrow proportional band (10%). The approach to the control position is fast, but once it is there, the temperature oscillates a while before settling down.

In Fig. 9-13(b) a medium proportional band (50%) causes a slower approach to the control point, but it almost eliminates oscillation.

The behavior of a large proportional band (200%) is shown in Fig. 9-13(c). The time for the system to reach control point is longer, but once it is there, the temperature does not oscillate at all.

If you pay careful attention to the meaning of the graphs in Fig. 9-13, you may

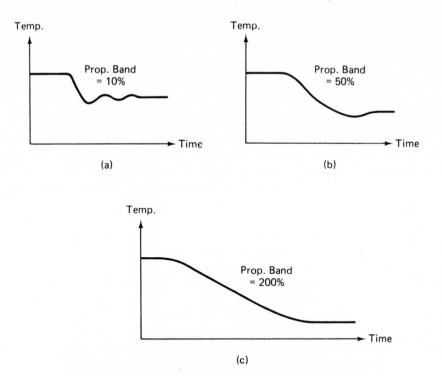

Figure 9-13. Graphs of temperature vs. time after a load disturbance. (a) Narrow proportional band (10%). (b) Medium proportional band (50%). (c) Wide proportional band (200%).

become concerned. The graphs of temperature versus time in Fig. 9-13 are showing
that after a load disturbance the actual measured temperature *does not return to
the original control value.*

Now one reasonable expectation from a temperature controller is that it return
the actual temperature to the set point after a load change. Figure 9-13 shows that
a proportional controller does not do this. Furthermore, the wider the proportional
band, the greater the difference between the two control values before and after
the disturbance.

To understand why this is so, imagine again that a proportional controller is
maintaining a control temperature of 180°F with the control valve 40% open. If a
disturbance occurs which causes the temperature to drop (an increase in rate of
heat loss through the walls of the chamber, for example), the valve will drive open.
The increased fuel flow will tend to drive the temperature back up toward 180°F,
but it can never recover fully, because the increased fuel flow is now a permanent
requirement. The control valve must *permanently remain farther open* to meet the
increased demand for heat input. Since the percent opening of the valve is proportional to the error signal, a permanently increased valve opening can only happen
if there is a permanently increased error.

Looking at it another way, we cannot expect the temperature to recover fully to 180°F; because if it did, the valve would return to 40% open just like it was before the increased heat demand. If the valve is in the same position as before, how will it satisfy the process demand for more heat input?

Either way we look at it, the temperature cannot recover fully to its predisturbance level. With a narrow proportional band, the recovery is closer, because it takes only a small increase in error to create a large change in valve position. Therefore the increased heat demand can be met with very little permanent error introduced. On the other hand, if the proportional band is large, it takes a relatively large error to produce a given amount of change in valve position. Therefore the increased heat demand can only be met by introducing a larger permanent error. These ideas are illustrated in Fig. 9-13; the larger proportional bands create larger permanent errors.

This serious shortcoming of the proportional mode of control means that proportional control is not very useful except in a certain few types of processes. As a general statement, it may be said that proportional control works well only in systems where the process changes are quite small and slow. It helps if the disturbances occur slowly, because then the proportional band can be adjusted quite narrow, since there is not much oscillation produced by a *slow* process change. The only objection to a narrow proportional band is that it may cause oscillation of the controlled temperature. If it is possible to adjust the proportional band nice and tight, the permanent error can be kept small.

9-6-3 Offset in proportional control

We have been considering the problem of failure of the actual temperature to return to the original control value after a system disturbance. However, we have purposely avoided the problem of controlled temperature error before a disturbance. In other words, we have not asked the question "Does the actual measured temperature agree with the set point *before* a system disturbance happens?" The answer to this question is "probably not." There is only one unique set of circumstances under which a proportional controller can ever cause exact agreement between measured temperature and set point. The chance of hitting that set of circumstances is remote. Here is why.

The design of real-life proportional temperature controllers is such that absolutely zero error signal causes a 50% opening in the control valve (this can be altered by the user, but let us consider it to be exactly 50%). The 50% figure is desirable so that the controller has available to it equal maximum corrections in both directions. That is, it has just as much correcting ability for both positive errors and negative errors. Now, under a given set of process conditions, a 50% valve opening will cause a given fixed temperature to actually occur in the process. If the set point happens to be *that particular temperature*, then the controller will hold the valve 50% open when the error reaches zero, and the 50% opening will cause the measured value to exactly agree with the set point.

For example, imagine a set of process conditions that causes the temperature to stabilize at exactly 700°F when the control valve is locked at 50% open. If it so happens that we *want* a set point of exactly 700°F (this would be a fantastic coincidence), then here is what will happen. The temperature will rise toward 700°F from below. With a measured temperature below 700°F, the error will be negative, and the valve will be more than 50% open. The closer the actual measured temperature gets to 700°F, the smaller the error gets, and the closer the valve gets to 50% open. At the point where the measured temperature hits exactly 700°F, the error will be zero, and the valve will be positioned precisely at 50% open. Since a 50% opening happens to be just what's needed to maintain a 700°F temperature, there is no further temperature change, and the system controls at that value.

Keep in mind that this is the *only* possible set point at which the controller could bring about exact agreement. Even if the set point were 705°F, the valve would have to be open more than 50% (say 50.2%) to attain that temperature. For the valve to be open 50.2%, the error signal must be nonzero. The error would be a very small negative value in this case. Thus the actual measured temperature could never climb to exactly 705°F but would have to stop at about 704.9°F in order to maintain the error necessary to keep the valve open more than 50%.

Of course, with normal luck, the set point we want will be quite a bit different from the stable temperature at 50% valve opening, so the permanent error will be larger than the 0.1°F suggested above. Just as a typical example, if the set point were 950°F, the valve might end up 75% open, with the control point at 944°F. The 6°F permanent difference between set point and control value is called *offset*. The farther the set point is from the 50%-open temperature, the worse the offset will be.

The idea of offset is shown graphically in Fig. 9-14. In Fig. 9-14(a), at the first set point a certain offset exists. When the set point is changed in the same direction as the first offset, the new set point results in a worse offset.

In Fig. 9-14(b), when the set point is changed in the opposite direction to the first offset, the new set point causes a better (smaller) offset.

9-6-4 An electric proportional temperature controller

An example of a proportional temperature controller is illustrated in Fig. 9-15. Two equal potentiometers are arranged in a bridge configuration, with the centers of both pots grounded. The pot on the right is called the *error pot*, and the pot on the left of the bridge is called the *valve position pot*. Assume for a moment that the proportional band adjustment is dialed completely out (shorted out). Then whatever position the wiper of the error pot assumes, the wiper of the valve position pot must assume that same position. If the error pot wiper moves up 200 Ω, for example, the unbalanced bridge will deliver a signal to the electronic amplifier. This will drive the motor in such a direction that the wiper of the valve position pot will also move up. When the valve position pot has moved up the same 200 Ω, the bridge is

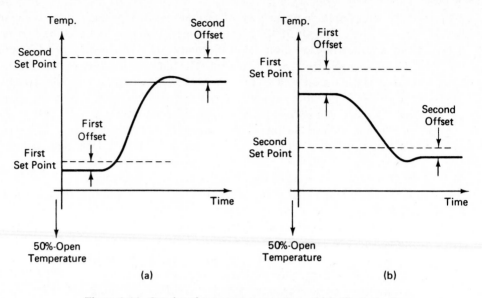

Figure 9-14. Graphs of temperature vs. time which illustrate the offset problem in proportional control. In both graphs the valve is assumed to be more than 50% open. (a) Offset gets worse (larger) when the set point is moved farther above the 50%-open temperature. (b) Offset gets better (smaller) when the set point is moved closer to the 50%-open temperature.

back in balance, the input is removed from the amplifier, and the motor stops. Thus the electronic amplifier and gear motor will force the valve position pot to follow the error pot.

The wiper of the error pot is positioned by a pressure bellows expanding against a set-point spring. As the process temperature changes, the pressure of the fluid in the filled sensing bulb will change. This pressure change is imparted to the flexible bellows through capillary tubing. Higher temperature causes the bellows to expand to the left, *against* the set-point spring. Lower temperature causes it to retract to the right, *aided* by the compressed set-point spring. The motion of the bellows is imparted to the wiper of the error pot. The set point is adjusted by adjusting the compression of the spring. Higher set points require tighter compression, and lower set points mean relaxed compression. When the actual temperature is above set point (positive error), the error pot wiper moves above midpoint. When the actual measured temperature is below set point (negative error), the error pot wiper moves below midpoint. The distance the error pot wiper moves from its midpoint is proportional to the magnitude of the error.

It can be seen that for any given amount of error there is a given position of the control valve which will balance the bridge, and that valve position is proportional to the error.

To make the controller more sensitive (decrease its proportional band), we

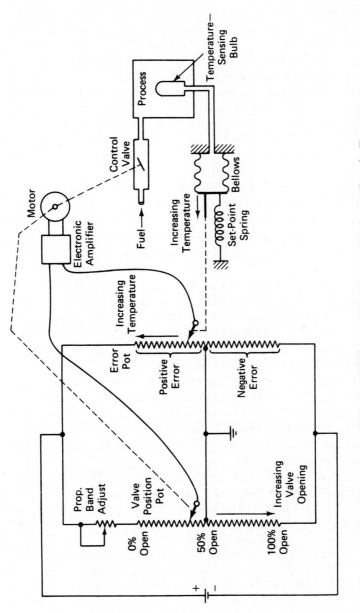

Figure 9-15. An electric method of implementing the proportional control mode. Temperature is the controlled variable.

decrease the voltage across the valve position pot. This is accomplished by adjusting the *proportional band adjust pot* in the upper left of Fig. 9-15. As this resistance is raised, the voltage across the valve position pot decreases. When that happens, the wiper of the valve position pot must move *farther* to balance a given movement in the error pot wiper. Looking at it another way, it takes less movement of the error pot to produce a given movement in the valve position pot. This means that the proportional band has been reduced.

As an example, suppose the error pot has 20 V applied to it from end to end but that the valve position pot has only 10 V from end to end due to the setting of the proportional band adjust pot. If the error pot wiper now moves 100 Ω, the valve position pot wiper must move 200 Ω to create an equal voltage at its wiper terminal and null the amplifier input voltage. Since a 200-Ω change in the valve position pot represents a given valve opening, it can be seen that a given change in valve percent opening has been accomplished with a smaller temperature error than would otherwise be necessary. Thus the proportional band has been reduced.

A careful study of Fig. 9-15 reveals why a permanent offset occurs with this type of electric proportional controller. Suppose we are controlling right on set point with the valve 50% open. If a load disturbance occurs which drives the temperature up, the error pot will move up a certain distance. The valve position pot must follow it up the same distance, because of amplifier-motor action. This causes a reduction in fuel flow, which tends to reduce the temperature. As temperature starts to fall back toward the set point, the error pot travels back toward its center position, and the valve position pot follows it, thus reopening the valve a little bit. This will continue until a point is found where any further reduction in temperature will cause the valve to be open enough to drive the temperature up once again. When that point is found the system stabilizes, and all pot movement stops. Unfortunately, this stabilization point must necessarily be a little bit above the set-point temperature. In other words, the error pot never makes it back to its center point. It cannot make it back to center point because if it did the valve would be 50% open again, and we already know that with the valve 50% open, the temperature rises. That was the beginning premise of the discussion.

Therefore the system stabilizes at a new control point which is a little higher in temperature than the original set point. At stabilization, the error pot is a little above center, indicating a positive error; the valve position pot is also a little above center, indicating an opening of a little less than 50%. Permanent offset has set in.

9-7 PROPORTIONAL PLUS INTEGRAL
CONTROL

In Section 9-6 we showed that proportional control eliminates oscillation in the measured variable and reduces wear on the control valve, but introduces permanent offset in the measured variable. For this reason, it is not too useful in most systems. Strict proportional control can be used only when load changes are small and slow and the variation in set point is small. This point was made in Sec. 9-6. For the more common process situation, in which load changes are large and rapid and set point may be varied considerably, the *proportional plus integral mode* of control is better suited. Proportional plus integral control is also called proportional plus reset control.

In proportional plus integral control, the position of the control valve is determined by two things:

a. The magnitude of the error signal. This is the proportional part.
b. The *time integral of the error signal*; in other words, the magnitude of the error multiplied by the *time* that it has persisted. This is the integral part.

Since the valve can respond to the time integral of the error, any permanent offset error that results from proportional control alone is eventually corrected as time goes by. It can be thought of this way: The proportional control part positions the valve in proportion to the error that exists. Then the integral control part senses that a small error (offset) is still persisting. As time passes, the integral part moves the valve *farther* in the same direction, thus helping to reduce the offset. The longer the error persists, the more additional distance the valve moves. Eventually, the error will be reduced to zero, and the valve will stop moving. It stops moving because as more time passes, the time integral of the error does not increase anymore, due to the fact that the error is now zero.

To understand the action of the integral part of such a controller, it is helpful to study a schematic diagram which shows how one could be built. Refer to Fig. 9-16. It shows the same controller as Fig. 9-15 except that an integral part has been added to make it into a proportional plus integral controller.

The best way to picture the action of this proportional plus integral controller is to focus on the *RC* circuit attached to the wiper of the valve position pot. Recall that a capacitor never charges instantly and sometimes takes rather a long time to build up any appreciable amount of voltage. This is the case in this circuit, because the *RC* time constant is rather large. When the valve position pot wiper moves off its center point and applies a voltage to the *RC* circuit, at first the entire wiper voltage appears across *R* because the capacitor *C* has no charge at all. As time passes *C* charges up, thus *reducing* the voltage across *R*. The voltage across *R* is equal to the wiper voltage (potential difference between wiper and ground) minus the capacitor voltage. As capacitor voltage builds up, resistor voltage decreases.

Now imagine that the controller is controlling right on set point (zero error) with the valve 50% open. As in Sec. 9-6, assume that the proportional band adjustment pot is shorted out. If a process disturbance occurs which drives the temperature up, the error pot will move up a certain distance. The valve position pot must follow it up the same distance, because of the amplifier-motor action. Thus the percent valve opening is reduced, the temperature is *paritally* corrected, and an offset error sets in. The offset error is due to the fact that the error pot must *remain* off center in order to hold the valve slightly closed, as explained in detail in Sec. 9-6-4.

Figure 9-16. An electric method of implementing the proportional plus integral control mode.

To give concreteness to our discussion of integral control, assume a specific situation. Let us assume that the voltage on the wiper of the error pot is $+1$ V relative to ground and that the voltage on the valve position pot is also $+1$ V relative to ground. Thus the voltage applied to the amplifier is the difference between these, which is 0 V. The motor therefore stops.

As time passes, C will start charging, $+$ on the top and $-$ on the bottom. This reduces the voltage across R, say to 0.75 V. The voltage into the amplifier is now the difference between 1.0 and 0.75 V, which is 0.25 V. This 0.25 V is amplified and causes the motor to run further in the *same direction* (closing the valve). The valve position pot will move up until the wiper voltage is 1.25 V, which will once again null the amplifier. Therefore the fuel flow is further reduced, and the temperature moves closer to set point. The error pot wiper voltage is now reduced as the temperature error approaches zero.

As time continues to pass, C continues to charge, thereby constantly reducing the voltage across R, which is the signal on one of the amplifier input leads. As long as the error is not zero, the voltage across R can be reduced to *less than* the error pot wiper voltage as time passes; this will continue to drive the valve position pot in the up direction, closing the valve farther and farther. Eventually, the temperature will be reduced back to set point, causing the error pot to return to the center. This applies 0 V on the amplifier input lead attached to the error pot wiper. At that time, the capacitor will just have reached full charge, and the voltage across R will be zero, causing 0 V on the other amplifier input lead. The valve therefore stops in the correct position to maintain the temperature right at the set point.

The final position of the pots is now quite different from what it would be for strict proportional control. The error pot wiper is centered, and the valve position pot wiper is moved up far enough to establish the proper fuel flow to the process. There is no way to know beforehand at what percent opening it will settle.

It can be seen that the position of the control valve is *initially* determined by the proportional part of the control, but *finally* settles in a position partially determined by the integral part of the control. The relative importance of the proportional and integral parts can be varied by adjusting the resistor R. In most controllers, R is a potentiometer, so the RC time constant can be adjusted. When the time constant is large (large R), the integral part is less effective (slower to make its effects felt). When the time constant is small (small R), the integral part is more effective. Figure 9-17 shows the control effect of changing the integral time constant.

In most industrial controllers, the integral time constant is not used as a reference. Instead, the *reciprocal* of the integral time constant is the variable which is talked about. This variable is termed *reset rate*. The term reset rate can be rather confusing if you are used to thinking in terms of time constants. However, just remember that when reset rate is low (large time constant) the integral part is slow to make its effects felt by the process. When reset rate is high (small time constant), the integral part of the control is quick to make its effects felt by the process.

Figure 9-17(a) shows a graph of valve position (percent opening) versus time

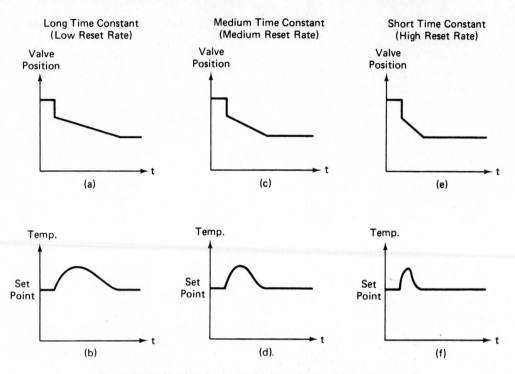

Figure 9-17. Graphs of valve position vs. time and actual temperature vs. time after a disturbance. The control mode is proportional plus integral. (a) and (b) Long integral time constant. (c) and (d) Medium integral time constant. (e) and (f) Short integral time constant.

for a large integral time constant (low reset rate). Farther up on the vertical axis means increased valve opening, and farther down on the vertical axis means reduced valve opening.

Figure 9-17(b) shows actual measured temperature versus time for the large integral time constant. The graphs in Figs. 9-17(a) and (b) should be considered as a pair. The graphs in Figs. 9-17(c) and (d) comprise another pair, for a medium time constant, and those in Figs. 9-17(e) and (f) comprise a third pair, for a small time constant.

Look first at Figs. 9-17(a) and (b) for the large time constant (low reset rate). As can be seen, the valve makes an initial change in position due to the sudden appearance of an error when a load disturbance takes place. The proportional part of the controller provides this initial change. Thereafter, the valve slowly closes farther in an effort to correct the resulting offset. Because of the slow reaction of the integral part, the actual temperature is slow to come back to set point, as shown in Fig. 9-17(b).

In Fig. 9-17(c), the valve reacts more quickly to the offset error because of the medium integral time constant. The temperature therefore is quicker to return to set point in Fig. 9-17(d).

In Fig. 9-17(e), the valve reacts very quickly to offset error because of the small integral time constant (high reset rate). The temperature quickly returns to the set point in Fig. 9-17(f).

In Fig. 9-17, the valve is shown going farther *closed.* This corresponds to an initial disturbance which drove the measured temperature process *higher.* If the initial process disturbance had been in the other direction, tending to drive the actual temperature lower, the valve would have gone farther *open,* but the same general behavior would have been shown.

All the graphs in Fig. 9-17 are somewhat idealized. In real life, the temperature would not recover to set point so smoothly. Instead, it would make some oscillations on the way back to the set point, and it would probably make at least one oscillation around the set point once it had recovered. The graphs in Fig. 9-17 are drawn idealized to more clearly illustrate the effect of varying the reset rate.

There is a limit to how high the reset rate can be adjusted. If it is made too large, the temperature can break into prolonged oscillations after a disturbance.

The proportional plus integral mode of control will handle most process control situations. Large load changes and large variations in set point can be controlled quite well, with no prolonged oscillations, no permanent offset, and quick recovery after a disturbance.

9-8 PROPORTIONAL PLUS INTEGRAL PLUS DERIVATIVE CONTROL

Although proportional plus integral control is adequate for most control situations, it is not adequate for *all* situations. There are some processes which present very difficult control problems that cannot be handled by proportional plus integral control. Specifically, here are two process characteristics which present such difficult control problems that proportional plus integral control may not be sufficient:

a. Very rapid load changes
b. Long time lag between applying the corrective action and the appearance of the results of that corrective action in the measurement

In cases where either of these two problems prevails, the solution may be *proportional plus integral plus derivative control.* The term *derivative control* is also called *rate control.** In proportional plus integral plus derivative control the corrective action (the position of the valve) is determined by three things:

*Do not confuse this "rate control" with the phrase "reset rate." *Rate control* refers to control having a derivative part. *Reset rate* refers to the adjustment of the integral time constant in integral control. It is unfortunate that the pioneers in process control technology used the same word to convey such different ideas, because now we are stuck with it.

a. The magnitude of the error. This is the proportional part.

b. The time integral of the error or the magnitude of the error multiplied by the time that it has persisted. This is the integral part.

c. The time rate of change of the error; a rapidly changing error causes a greater corrective action than a slowly changing error. This is the derivative part.

In an intuitive sense, the derivative part of the controller attempts to "look ahead" and foresee that the process is in for a bigger change than might be expected based on present measurements. That is, if the measured variable is changing very rapidly, it is a pretty sure bet that it is going to try to change by a large amount. This being the case, the controller tries to "outguess" the process by applying more corrective action than would be called for by proportional plus integral control alone.

As before, to understand what derivative control does, it is helpful to study a schematic diagram of how a derivative controller could be built. To avoid getting mixed up between integral and derivative parts, we will first show a proportional plus derivative controller schematic in Fig. 9-18. The full proportional plus integral plus derivative controller is shown in Fig. 9-19.

9-8-1 A proportional plus derivative
electric controller

Proportional plus derivative control, as illustrated in Fig. 9-18, is very seldom used in industrial temperature control. It is presented here only to explain the derivative part of a proportional plus integral plus derivative controller. Proportional plus derivative control *is* popular in industrial servo control systems, however.

Focus on the *RC* circuit connected between the valve position pot wiper and ground. Notice that the positions of resistor and capacitor have been reversed from that of the integral controller in Fig. 9-16. Again, remember the fact that it always takes a certain amount of time to charge a capacitor through a resistor.

If a disturbance occurs that drives the process temperature up, the error pot will move up a certain distance. The valve position pot will attempt to follow it because of the amplifier-motor action. However, to null the amplifier input voltage, the voltage across the capacitor must equal the error pot wiper voltage. Since the voltage across *C* lags behind the valve position pot wiper voltage due to the *RC* time constant delay, the valve position pot must *overcorrect*. That is, it must move up farther than it normally would in order to null the amplifier.

Furthermore, the amount by which it overcorrects depends on how rapidly the error is changing. If the error is changing slowly, the position pot wiper will follow slowly, and the capacitor will have time to keep almost fully caught up with the voltage on the position pot wiper. Therefore not much overcorrection is necessary.

On the other hand, if the error is changing rapidly, the position pot wiper will

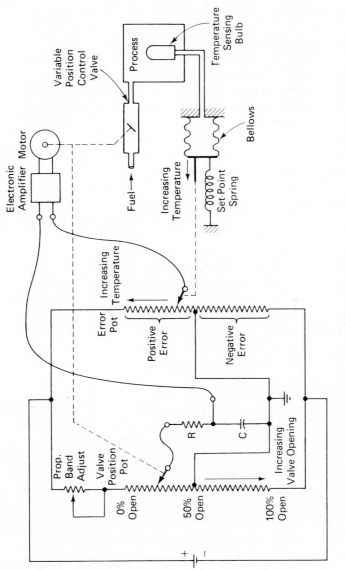

Figure 9-18. An electric method of implementing the proportional plus derivative control mode.

follow rapidly, and the capacitor will lag far behind the position pot wiper voltage. Therefore a large overcorrection is necessary to keep the amplifier nulled (to keep the capacitor voltage equal to the error pot wiper voltage).

In this way, the derivative part of the controller responds to the *rate of change* of the error; it introduces an additional adjustment in valve opening beyond what the proportional control alone would produce. The amount of additional movement depends on the rapidity of the change in error.

In a real industrial controller, the resistor R is a potentiometer, so the derivative time constant can be varied. When the derivative time constant is small (low R), the derivative part of the control is less effective. It introduces only a small overcorrection due to rapid change in error. When the derivative time constant is large (high R), the derivative part becomes more effective. It introduces a large overcorrection when a rapid change in error occurs.

The reference variable which is commonly used when dealing with derivative control is *rate time*. Rate time is a rather complicated variable from the mathematical point of view. Nevertheless, here is its formal definition: Rate time is the amount of time allowed for the measured variable to change through the full controller range, if it is to drive the final correcting device through its full range of adjustment, assuming a 100% proportional band.

Intuitively, rate time is the amount of time by which the controller is "looking ahead" or "seeing into the future." Obviously, this is a very nonrigorous description of rate time, since nothing can really see into the future. It is best to think of rate time as being equal to the derivative time constant multiplied by a numerical constant. The larger the rate time, the greater the amount of overcorrection the controller makes for a rapid change in error.

9-8-2 A proportional plus integral plus derivative electric controller

Figure 9-19 shows a schematic diagram of a complete proportional plus integral plus derivative controller. Notice that the derivative part is attached to the integral part. The output of the integral RC circuit is the input to the derivative RC circuit.

The direction of adjustment of the integral pot in order to increase the reset rate (to increase the amount of integral part contribution) is shown in the figure. Also, the direction of adjustment of the derivative pot in order to increase the rate time (to increase the amount of derivative part contribution) is shown in Fig. 9-19.

The operation of the controller in Fig. 9-19 can be understood by combining the explanations of the proportional plus integral controller with the proportional plus derivative controller.

The graphs of Fig. 9-20 show the control effect of changing the derivative time constant (which changes the rate time).

Figures 9-20(a) and (b) show the valve position and measured temperature for a large and rapid load change with derivative control removed. As can be seen, the

Figure 9-19. An electric method of implementing the proportional
plus integral plus derivative control mode.

initial error is quite large and consequently takes a long time to be corrected.

In Figs. 9-20(c) and (d), the derivative time constant (rate time) is small, and the initial error is not so large because the initial valve correction is greater. The controller has introduced overcorrection because it recognized that the initial rapid rate of change in measured temperature portended a large total temperature change unless special corrective steps were taken. Because the initial error is smaller, recovery to set point is earlier.

In Figs. 9-20(e) and (f) the derivative contribution has been increased by raising the derivative time constant. Therefore the initial error is even less than before because more initial valve overcorrection is provided. With the reduced initial error, the time to recover and stabilize at set point is reduced even more than before.

Just as there was a limit with reset rate, there is a limit to how far the rate time can be increased. Prolonged oscillations around the set point can occur if too much derivative control is introduced, that is, if rate time is set too high.

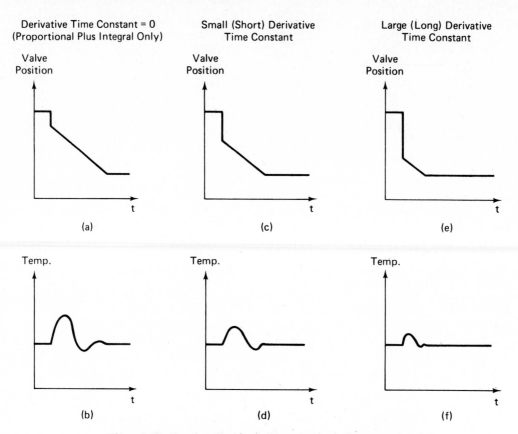

Figure 9-20. Graphs of valve position vs. time and measured temperature vs. time following a disturbance. The control mode is proportional plus integral plus derivative. (a) and (b) Zero derivative time constant. (c) and (d) Short derivative time constant. (e) and (f) Long derivative time constant.

9-9 PROCESS RESPONSE

In Secs. 9-4 through 9-8 we have concentrated on the action of the controller block in the generalized block diagram of Fig. 9-3. No matter what particular mode of control is being used, it can in fairness be said that the controller is doing the "thinking" for the entire system. The controller is the component that sends out orders to the final correcting device, based on its assessment of the direction and size of the error. We have seen that sophisticated controllers may also consider the length of time that the error has persisted as they decide how to adjust the final correcting device. Some controllers may also consider how rapidly the error is changing as they decide what orders to send to the final correcting device. The controller does all this according to a predetermined plan which existed in the mind of the system designer and also in the mind of the person who made the final adjustments (proportional band, reset rate, etc.).

It should be apparent, however, that the action of the controller does not describe the whole picture. The reaction of the *process itself* to the final correcting device is just as important as the action of the controller in determining the overall behavior of the system. In this section we will discuss the response characteristics of typical industrial processes and show how these characteristics affect overall system response.

9-9-1 Time constant delay (process reaction delay) in industrial processes

The most obvious characteristic of industrial processes is that they require a certain amount of time to fully respond to a change in input. For example, in the process illustrated in Fig. 9-21(a), a liquid is being heated by a steam heating coil while it is being agitated. The liquid is admitted at the inflow pipe on the lower left of the tank, and it exits at the outflow pipe at the upper right. Assume that the controlled variable is the liquid temperature, and try to imagine what will happen

Figure 9-21. (a) Temperature control system for understanding process reaction delay. (b) Graph of temperature vs. time following a disturbance, illustrating the effects of thermal capacity and thermal resistance.

(a)   (b)

if there is a sudden increase in steam flow through the heating coils (with consequent increase in average steam temperature).

The liquid temperature will not instantly increase to a new value but will rise more or less according to the curve of Fig. 9-21(b). The reason for the delay is that the tank of liquid has what is called *thermal capacity* and the heat transfer apparatus has what is called *thermal resistance*. The thermal capacity is an expression of the idea that a certain quantity of heat energy (BTUs) must be added to the tank before the temperature can rise a given amount. Thermal resistance is an expression of the idea that all mediums have a natural reluctance to carry heat energy from one point to another; in this case the transfer of heat energy is to take place from the hot steam, through the metal walls of the heating coils, and into the surrounding liquid.

Thermal capacity is analogous to electrical capacitance. Both concepts are expressions of the fact that the relevant *quantity* (coulombs of charge in the electrical case, BTUs of heat in the thermal case) must be transferred before the relevant *potential* (voltage change in the electrical case, degrees of temperature change in the thermal case) can be built up.

Thermal resistance is analogous to electrical resistance. Both concepts are expressions of the fact that a certain *potential difference* (voltage drop in the electrical case, temperature difference in the thermal case) is necessary to cause a certain *rate of flow* (amperes of current in the electrical case, BTUs per second of heat flow in the thermal case) to be established. We are well acquainted with the fact that a certain amount of time is necessary for the voltage across a capacitor to build up to a steady value if a resistor-capacitor circuit is subjected to a sudden change in driving voltage. The exact same situation prevails in the thermal case. A certain amount of time is necessary for the thermal capacity (the tank of liquid) to rise to a steady temperature when a thermal resistance-capacity system is subjected to a sudden change in temperature difference. In the same way that the capacitor will experience 63% of its total voltage change in one electrical time constant, the thermal capacity will experience 63% of *its* temperature change in one thermal time constant. The larger the thermal resistance, the larger the thermal time constant, and the more time is necessary to reach a final steady temperature value. The same holds true for the thermal capacity; the larger the capacity, the more time is required to bring the temperature up to a steady value. In the example of Fig. 9-21(b), the thermal time constant equals about 20 sec; about five time constants, or 100 sec, are required for the temperature to reach the new value.

The thermal time constant depends on the thermal resistance and the thermal capacity, as stated in the above paragraph. Thermal resistance depends on the thermal conductivity of the metal in the heating coils, the thickness of the coil walls, and the surface area of the coils. The thermal capacity depends on the size of the tank (amount of liquid present) and the specific heat of the liquid.

The point of this whole discussion is that in a temperature control process there is a time delay between the application of corrective action and the appearance of the final result of that corrective action.

This delay is called *time constant delay* or *process reaction delay*. We will normally use the term process reaction delay, unless we wish to specially emphasize its equivalence to *RC* time constant behavior.

Virtually all industrial processes, not only thermal ones, show this type of delay. In many cases the delays are measured in seconds. Some processes have process reaction delays of a few minutes, and some have process reaction delays in the range of 15–30 min. Occasionally you may come across industrial processes that have process reaction delays of an hour or more.

9-9-2 Transfer lag

In some thermal processes there is more than one resistance-capacity combination. An example of such a process is shown in Fig. 9-22(a). Natural gas is burned inside the radiant tube heaters on each side of the furnace. The heat is carried through the tube walls and is transferred to recirculating air passing over the tubes. The fan forces the heated air through distribution nozzles and on to the metal billets which are being heated. In this arrangement the response of the billet temperature to a change in fuel input is even more drastically retarded, as shown in the curve of Fig. 9-22(b). In fact, the response is not even the same shape as the time constant curve of Fig. 9-21(b). The reason for the more retarded temperature response is that there are now two thermal resistance-capacity combinations in series with each other. The first involves the thermal resistance of the radiant tube walls and the capacity of the recirculating air. The second thermal time constant involves the thermal resistance and thermal capacity of the metal itself. The circuit of Fig. 9-22(c) is the electrical equivalent of the thermal process in Fig. 9-22(a). R_1 represents the thermal resistance of the radiant tube walls, and C_1 represents the thermal capacity of the recirculating air; R_2 represents the thermal resistance of the metal comprising the billet, while C_2 is the thermal capacity of the billet. It is pretty evident by looking at the electrical circuit that C_1 must charge first before C_2 can start to charge. Therefore the charging of C_2 is considerably delayed after the application of the input signal when the switch goes closed. The same problems occur in the thermal process. The billet temperature cannot start to rise until the recirculating air temperature has risen, and of course the recirculating air temperature cannot rise instantly after an increase in the heat input to the radiant tubes. Whenever there are two thermal time constants, the process is referred to as a *two-capacity* process, and the delay is referred to as *transfer lag*.

As a general rule, transfer lag is a much more serious problem than the simple single time constant delay (process reaction delay) seen in the system of Fig. 9-21. This is because transfer lag causes the measured process temperature to initially respond very slowly to a corrective action. This slow initial response is shown clearly in Fig. 9-22(b), in which the temperature has gone through only about 10% of its total change in the first minute following the correction. Single time constant delay, by contrast, allows the measured temperature to respond quite quickly immediately following a correction. In fact, the response is quickest immediately after the correc-

Figure 9-22. (a) Temperature control system for understanding transfer lag. (b) Graph of temperature vs. time following a disturbance, illustrating the effects of *two* thermal capacities and *two* thermal resistances. Transfer lag is present. (c) The electrical analogy.

tive action occurs, as shown in Fig. 9-21(b). This is the same behavior as seen in the universal time constant curve of Fig. 3-27.

Long process reaction delays cannot necessarily be considered a problem at all, since they help prevent overshoot of the measured temperature. Long transfer lags, on the other hand, always constitute a difficult control problem.

Figure 9-23 shows the effects of various process constructions. Figure 9-23(a)

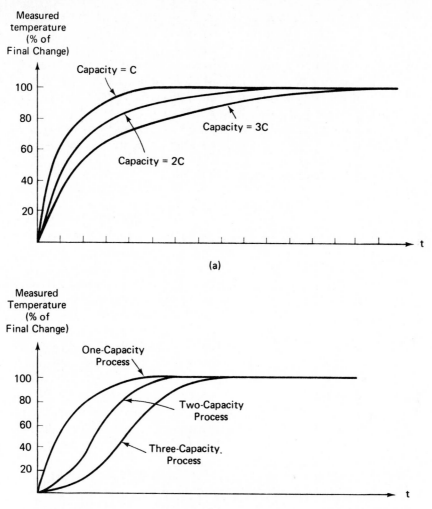

Figure 9-23. Graphs of actual temperature vs. time following a disturbance for different system constructions, illustrating the severe effect of transfer lag. (a) Effect of increasing the amount of thermal capacity in a one-capacity system. (b) Effect of increasing the number of thermal capacities in the system. The transfer lag problem is worsened when more capacities are present.

shows the effect of increasing the capacity in a one-capacity process, assuming the process resistance is held constant. Figure 9-23(b) shows the effect of increasing the number of resistance-capacity combinations in the process.

Note especially the response of the process temperature immediately following the corrective action (near the zero point on the time axis). In this time area, the effect of transfer lag is very severe compared to the effect of simply increasing the time constant in a single-capacity process.

The same principles which apply to thermal processes also apply to other types of industrial processes, as we saw in Sec. 9-9-1. Pressure control processes, liquid level control processes, and all other industrial processes have their associated resistances and capacitances, and they often suffer from two or more resistance-capacity combinations. Because of this, they are subject to the same transfer lag problems which affect thermal processes. The graphs in Fig. 9-23 could apply to *any* industrial process, no matter what the controlled variable happens to be.

9-9-3 Transportation lag and dead time

When transfer lag is present the controlled variable takes some time to reach its new steady value after the controller sends an order to the correcting device, but at least *some* partial response is felt immediately. This is clearly shown in Figs. 9-22 and 9-23. A more difficult control problem occurs when *absolutely no response whatsoever* is felt in the controlled variable for a certain time period after the controller signals the correcting device. This situation usually occurs when the physical location of the correcting device is far away from the physical location of the measuring device. The system shown in Fig. 9-24 is an example of such an arrangement.

Assume that the mixing/heating tank must be located 30 ft away from the point where the hot mixture is to be used. There might be some practical reason for this. For example, it may be that the mixing/heating tank must be located indoors and the discharge opening must be some distance away outdoors. Since there may be some cooling during the long pipe run, the temperature is measured and controlled at the discharge point rather than inside the tank. This allows the controller to eliminate the effect of cooling in the pipe, which may vary widely with outdoor temperature changes.

Since the hot mixture leaves the constant-diameter carrying pipe at a velocity of 10 ft/sec, it takes 3 sec to travel down the 30-ft-long pipe. This being the case, if the temperature in the mixing tank is changed, it will take the temperature measuring device 3 sec to realize it. This delay is completely independent of and in addition to any transfer lag that may be present in the tank. A delay of this nature is termed *transportation lag*. Other terms used are transport lag and distance/velocity lag; we shall use the term transportation lag. The actual amount of time that the correcting device change remains undetected is called the *dead time*. Transportation lags are always associated with the controlled medium moving from one physical location to another in the process. The slower the speed of movement, the worse the transportation lag. The farther the distance between the two points, the worse the transportation lag. The effect of transportation lag is shown in Fig. 9-24(b); the dead time is 3 sec. Transportation lag presents a difficult control problem, even worse than transfer lag.

In industrial control systems, dead time can arise for reasons other than trans-

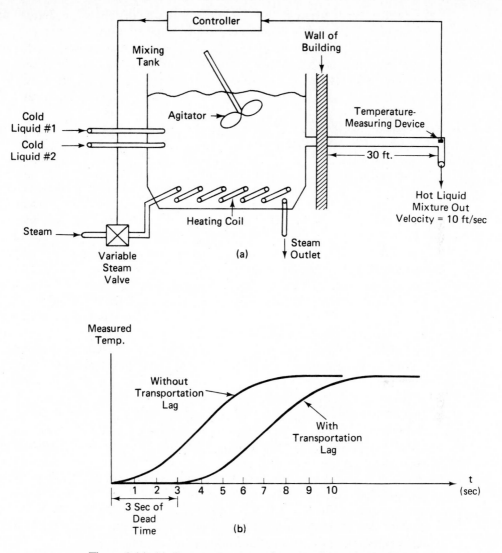

Figure 9-24. (a) Temperature control system for understanding transportation lag. (b) Graphs of measured temperature vs. time following a disturbance, with and without transportation lag. A dead time of 3 sec occurs when transportation lag is present. Transfer lag is present in both cases.

portation lag. For example, in a position control system, there is always some sort of gearing present. As you may know, all gears are subject to *gear backlash* to some extent. Gear backlash is the problem of gear teeth meshing imperfectly, so that the driving gear must turn through a small initial angle before its teeth make contact with the driven gear. Because of this, the controller in a servo system may cause the servo motor to start turning, but the resultant motion of the load is delayed until the gear teeth make contact. The result is a period of dead time. If the gear train is complex, with several gear combinations between the servo motor and the controlled object, the gear backlash problem is magnified. Dead time can be a serious problem in servo systems of this kind.

9-10 RELATIONSHIP BETWEEN PROCESS CHARACTERISTICS AND PROPER MODE OF CONTROL

Generally speaking, the characteristics of the process being controlled determine which mode of control is best suited to that process. In Sec. 9-9 we discussed three important process characteristics, time constant delay, transfer lag, and transportation lag (dead time). We saw in Secs. 9-6, 9-7, and 9-8 that the size and speed of load disturbances were also rather important characteristics of the process. These five process characteristics determine the nature and difficulty of the control job and therefore which mode of control is required.

Of course the desired accuracy of control is also a prime determining factor in the choice of control mode; if the measured variable can be allowed to deviate from set point by wide margins without harm to the product, there is no sense installing a controller capable of keeping the deviation small. In such a case, it doesn't matter very much how nasty the process characteristics are; a simple On-Off controller will suffice.

Table 9-1 summarizes the relationship between process characteristics and control mode. Naturally this table is somewhat rough.

The various control modes are listed in the left column. The other columns describe the conditions which will allow that control mode to be successful. The column entries in the four far-right columns of the table describe the *maximum limit* for that particular characteristic. The entry in the "time constant delay" column describes the *minimum limit*.

For example, if the entry in the dead time column is "moderate," it means that control mode will work successfully if dead time is either moderate or short or nonexistent (short or nonexistent dead time is easier to handle than moderate dead time).

However, if the entry in the process reaction delay column is "long," it means that the process reaction delay must be long, and not short or moderate. (In many cases, a fast response causes serious overshoot, as mentioned in Sec. 9-9-2.)

As Table 9-1 shows, On-Off control is generally acceptable only under simple process conditions. It works only when the process reaction delay is fairly long (slow response). A short process reaction delay causes excessive overshoot and undershoot with On-Off control.

Proportional control can tolerate long or moderate process reaction delays because it continually repositions the final correcting device as the controlled vari-

TABLE 9-1. The types of processes which can be successfully controlled by each of the five basic control modes.

Control Mode	Process Reaction Delay (Mininum)	Transfer Lag (Maximum)	Dead Time (Maximum)	Size of Load Disturbance (Maximum)	Speed of Load Disturbance (Maximum)
On-Off	Long only (cannot be short)	Very short	Very short	Small	Slow
Proportional only	Long or moderate (cannot be too short)	Moderate	Moderate	Small	Slow
Proportional plus integral	Any	Moderate	Moderate	Any	Slow
Proportional plus derivative	Long or moderate (cannot be too short)	Moderate	Moderate	Small	Any
Proportional plus integral plus derivative	Any	Any	Any	Any	Any

able approaches the set point after a disturbance. Therefore it is not so likely to produce excessive overshoot as is On-Off control. Moderate transfer lag and dead time can be handled by a proportional-only controller. However, long transfer lag and/or long dead time produce sustained cycling. This occurs because the controller does not throttle back on the final correcting element until it is too late. That is, if the lag is too great, by the time the controller realizes that the controlled variable is returning to set point, it has already permitted too much energy to enter the process. The process inertia will then carry the controlled variable beyond the set point in the opposite direction, and the controller cannot do anything about it until the excess energy has dissipated. This sets the stage for sustained cycling.

Proportional plus integral control can handle any process reaction delay and any size of load distrubance. The integral part of the control continually repositions the final correcting element until the set point is attained, no matter how large the load change. Because of this, the proportional band can be adjusted wider, because a narrow proportional band is no longer necessary to keep the offset small. With a wide proportional band, the controller can start to throttle back the correcting device sooner and harder as the controlled variable recovers to set point. This prevents excessive overshoot and possible cycling even if the process reaction rate is fast.

Within the proportional plus integral control mode, we can make finer distinctions among the different control situations. Table 9-2 shows the relative settings of proportional band and reset rate for a proportional plus integral controller. The proportional plus integral controller is assumed to operate under various conditions of process reaction rate and transfer lag/dead time. Transfer lag and dead time (transportation lag) have been lumped together under the term *lag* in Table 9-2.

TABLE 9-2. The proper setting of a proportional plus integral controller for different process conditions.

Process Characteristics		Controller Adjustments	
Process reaction rate	Total lag	Proportional band	Reset rate (reciprocal of integral time constant)
Slow	Short	Narrow	Fast
Slow	Moderate	Medium	Slow
Fast	Short	Medium	Fast
Fast	Moderate	Wide	Slow

Process reaction rate in Table 9-2 is the opposite of process reaction delay. That is, a short process reaction delay equates to a fast process reaction rate, and a long process reaction delay equates to a slow process reaction rate.

Returning to Table 9-1, proportional plus derivative control must have moderate or long process reaction delay, because in the absence of an integral part to take care of the offset the proportional band must be made narrow to keep offset small. With narrow proportional band, a short reaction delay can cause overshoot and cycling. However, with derivative control, rapid load changes are not harmful because the controller overcorrects when it detects a quickly changing error.

When long transfer lag and/or long dead time are present in a process, the only successful mode of control is proportional plus integral plus derivative. The proportional band is made quite wide so only a small portion of the corrective action after a disturbance is due to the proportional part. Most of the *immediate* corrective action is due to the derivative part. This allows a considerable inflow of energy to the process immediately when it is most needed. When the error stops growing and starts shrinking, the corrective action due to the derivative part goes away, leaving only the relatively small corrective action due to the proportional part. By this time, hopefully, enough time has elapsed that dead time is finished and the transfer lag is fairly far progressed. The controller can now sense that the controlled variable is recovering. There is no tendency to overshoot, though, because the position of the final correcting device is not drastically different from where it was before the disturbance. This is so because of the wide proportional band.

The integral part of the controller takes over at this point and slowly repositions the correcting device to drive the controlled variable back to set point. The reset rate is usually adjusted to be rather slow.

The effect of having the derivative part in the controller is shown in Fig. 9-25. Figure 9-25(a) shows the best response that could be obtained in a process having long transfer lag and long dead time (transportation lag) using proportional plus integral control. Figure 9-25(b) shows the response which is possible when a derivative part is added and the controller is properly adjusted.

Figure 9-25. Graphs of temperature vs. time following a disturbance in a system which has both transportation lag and transfer lag. (a) Without derivative control action, the overshoot and recovery time are great. (b) When derivative control is added, the overshoot and recovery time are lessened.

QUESTIONS AND PROBLEMS

1. Explain the difference between an open-loop system and a closed-loop system.

2. In Fig. 9-1(b), suppose that the length of the B member is fixed but that the pivot point can be moved to the left or to the right. Which direction would you move it to decrease the proportional band?

3. Explain in step-by-step detail how the closed-loop system of Fig. 9-1(b) would react if the downstream restriction in the exit pipe were decreased. Contrast this to what would happen in Fig. 9-1(a).

4. Explain the general function of the comparer in a closed-loop control system.

5. Explain the general function of the controller in a closed-loop control system.

6. Explain the general function of the measurement device in a closed-loop control system.

7. Explain the general function of the final correcting device in a closed-loop control system.

8. Under what conditions is an amplifier needed with the final correcting device?

9. Name some common final correcting devices used in industrial process control.

10. Define the term *error signal*.

11. When is the error signal considered positive, and when is it considered negative?

12. What does the idea of *control mode* mean?

13. Name the five basic control modes.

14. Describe the action of each of the five control modes. That is, tell what kind of orders the controller sends to the final correcting device for every possible error signal that it might receive.

15. Define the term *offset* as applied to closed-loop control systems.

16. What characteristics distinguish a good closed-loop control system from one which is not so good?

17. It was tacitly assumed that all three pulleys had the same diameter in Fig. 9-4. Suppose that the fixed pulley on the left and the movable pulley both have a 3-in. diameter and that the fixed pulley on the right has a 6-in. diameter. Suppose also that the cord *cannot slip* on any pulley. If the pointer is now moved 5 in. to the left, how far will the object move?

18. Suppose that the amplifier gain is extremely high in Fig. 9-4, so that even a few millivolts of input causes a large output voltage. Explain why the object's position will never stabilize but will continually oscillate back and forth, "hunting" for the right position.

19. In Fig. 9-5, would you get a more exact copy of the pattern piece if the mounting support were moved slowly or if it were moved quickly? Explain why.

20. In Fig. 9-5, suppose the pattern piece and workpiece are both 12 in. long and $\frac{1}{4}$ in. wide (into the page) and have a maximum cut depth of 6 in. The workpiece is wood, and the cutting tool is a circular steel saw. About how fast would you move the mounting support? Express your answer in inches per second or inches per minute. Try to give a justification for your estimate.

21. In Fig. 9-6, if the temperature setting pointer is moved to the right, thereby rotating the rotatable shaft counterclockwise, does this tend to wind the spiral tighter or unwind it? Will this tend to close the mercury switch or open it? Does this increase the temperature set point or decrease it? Is this how a residential thermostat acts? Compare to the one in your house.

22. What control mode is illustrated by Fig. 9-6?

23. In Fig. 9-7, if you desired to increase the set-point pressure, would you tighten the compression spring or loosen it? Explain.

24. In Fig. 9-7, assume that the amplifier's output voltage polarity is the same as its input voltage polarity (the output is positive on top if the input is positive on top). If the output voltage polarity is + on the top and − on the bottom, should the damper drive open or drive closed?

25. Explain the meaning of differential gap in On-Off control.

26. What is the important disadvantage of On-Off control compared to other control modes?

27. What advantage does On-Off control have over the other four control modes?

28. What is the most widely used control mode in modern American industry?

29. Do you ever see a solenoid-actuated valve as the final correcting element in the proportional control mode? Why?

30. What benefit arises from widening the differential gap in an On-Off controller?

31. What disadvantage arises from widening the differential gap in an On-Off controller?

32. In proportional control, if you want the controller to give a "stronger" reaction to a given amount of error, should you widen the proportional band or make it narrower? Explain.

33. If the full control range of a temperature controller is 1000°F and the proportional band is 15%, how much does the measured temperature have to change in order to drive the correcting device from one extreme position to the other?

34. The controller of Question 33 will stroke the final control valve from full closed to full open if the temperature changes by 280°F. What is the proportional band?

35. Suppose that the controller of Question 33 is controlling right on set point with the final control valve exactly 50% open. The set point is 670°F. Suppose that a process disturbance causes the measured temperature to drop to 630°F, which causes the final control valve to go 100% open (just). How wide is the proportional band?

36. The controller of Question 33 is controlling right on set point (670°F) with the final control valve 50% open. The proportional band is set at 40%. What measured temperature will cause the control valve to go fully open? What temperature will cause it to go fully closed?

37. The controller of Question 33 is controlling right on set point of 780°F with the final control valve 75% open. The proportional band is set at 25%. What measured temperature will cause the control valve to go fully open? To go fully closed?

38. A certain temperature controller has a control range of 1500°F to 2200°F. It is controlling right on a set point of 1690°F with the control valve 35% open. The proportional band is 28%. What temperature will cause the control valve to go fully closed? To go fully open?

39. The controller of Question 38 is controlling right on a set point of 1690°F with the control valve 35% open. The proportional band is 45%. What temperature will cause the control valve to go fully closed? To go fully open?

40. Which will produce a larger offset, a wide proportional band or a narrow proportional band?

41. Explain why permanent offset occurs with the proportional control mode.

42. A proportional temperature controller is controlling at 1415°F with the set point at 1425°F. The control valve is 80% open. If the set point is raised to 1430°F, will the offset become larger or smaller? Explain your answer.

43. In the proportional plus integral control mode, what two things determine the controller's output signal?

44. What beneficial result arises from using proportional plus integral control compared to straight proportional control?

45. When does a proportional plus integral controller tend to correct the offset faster, when the integral time constant is long or when it is short?

46. How is reset rate related to integral time constant?

47. In Fig. 9-16, how would you increase the reset rate, by increasing the resistance R or by decreasing it?

48. Explain why the proportional band is made narrower as the resistance of the proportional band adjust pot is increased.

49. In a proportional plus integral plus derivative controller, if you want the derivative response to be more vigorous, should you increase or decrease the rate time?

50. How would you increase the rate time in the controller of Fig. 9-18, by increasing R or decreasing R?

51. In Fig. 9-17, the valve ends up in the same final position no matter what the reset rate. Explain why this is reasonable and is to be expected.

52. In Fig. 9-17, the amount of *time* it takes the control valve to settle into its final position varies depending on the reset rate. Explain why this is to be expected.

53. In Fig. 9-20, the amount of initial change in valve position varies depending on the rate time. Explain why this is to be expected.

54. In Fig. 9-20, the maximum error after a disturbance depends on the rate time. Explain why this is to be expected.

55. Explain the meaning of transfer lag. Why does transfer lag exist in processes?

56. Why is transfer lag considered a serious process control problem, while process reaction delay (time constant delay) is not considered so serious?

57. In the liquid level control system of Fig. 9-1, what provides the system's capacity? What causes the system's resistance?

58. Explain the meaning of transportation lag. Why does transportation lag exist in some processes?

59. Which is the more serious control problem, transportation lag or transfer lag? Why?

60. Define dead time. What is the chief cause of dead time in a servo control system?

61. Generally speaking, when is the On-Off control mode acceptable?

62. Under what general conditions is it necessary to use the proportional mode of control rather than the On-Off mode?

63. Under what general conditions is it necessary to use the proportional plus integral mode of control rather than the proportional mode alone?

64. Under what general conditions is it necessary to use the proportional plus integral plus derivative mode of control rather than the proportional plus integral mode?

10
Final Correcting
Devices and Amplifiers

In very many cases, the final correcting device in a closed-loop system is a valve or valve-like device that varies the flow of a fluid into the process. This is usually the case in temperature control processes, where the heat input into the process is varied by adjusting a valve which controls the flow of combustion air or of liquid or gaseous fuel. Likewise, in pressure control processes, the pressure is usually corrected by changing a valve opening either on the inlet side or the outlet side of the process. For example, to raise the pressure in a process chamber, the valve regulating the inlet flow may be opened farther or the valve regulating the escape flow may be closed farther. In general, valves and valve-like devices such as dampers, louvers, sliding gates, etc., are the most common final correcting devices in industrial processes.

Sometimes the final correcting device is a continuously running motor, whose rotational speed determines the amount of process load. Many drying operations operate this way. The amount of heat energy supplied to the drying chamber is held constant, and the dryness of the final product is corrected by changing the rate at which the product moves through the drying chamber. For example, in grain being dried prior to storage, if the moisture content of the output grain is measured as too high, the control system could correct this condition by reducing the rate at which the grain is fed through the drier. In this case, control is accomplished by changing the load rather than by changing the fuel input. In such a system the final correcting device would be a motor, and the rotational speed of the motor would be the *manipulated variable*. The manipulated variable in any control system is that quantity which is varied so as to affect the value of the controlled variable. In any system having a valve as its final correcting device, the manipulated variable is fluid flow rate.

In some industrial control systems the manipulated variable is electric current. The most obvious example of this is an electric heating process. In an electric heating process the current might be controlled proportionally by continuously changing the firing angle of an SCR or triac. Or the current could be controlled in an On-Off manner by using some kind of switch or relay contact. In the former situation the thyristor and associated circuit components would comprise the final correcting device. In the latter situation the relay would be considered the final correcting device.

When the closed-loop system is a servo system, the controlled variable is the position of an object. The final correcting device is then a servo motor combined with a gearing system connecting the motor shaft to the object.

As the foregoing examples indicate, the final correcting devices in use in modern industry are quite varied. In this chapter we will look at a number of the commonly used final correcting devices and study their operation and characteristics.

OBJECTIVES

After completing this chapter, you will be able to:

1. List some of the commonly used final correcting devices in industrial control
2. Explain the operation of a solenoid valve and tell the circumstances in which solenoid valves are used
3. Explain the operation of a two-position motor-driven control valve, and interpret the limit switch timing diagram of such a valve
4. Explain the operation of a variable-position motor-driven valve, and tell the circumstances under which such valves are used
5. Describe the construction and operation of an electropneumatic valve operator using the balance beam principle, and tell the circumstances under which such operators are used
6. Describe the construction and operation of an electropneumatic converter and a pneumatic power positioner, and explain how they are joined together to accomplish variable-position control
7. Describe the construction and operation of an electrohydraulic valve positioner using the jet pipe principle, and tell the circumstances under which such positioners are used

8. Explain how electromagnetic relays and contactors can be used to control electric current in a control system, and discuss the difference between pick-up current and drop-out current for relays and contactors

9. Distinguish between a three-phase delta connection and a three-phase wye connection, and calculate power for each type of connection

10. Explain the theory of operation of a split-phase ac motor; explain the rotating magnetic field, the armature current behavior, and the creation of torque; and show how the phase shift is accomplished in a split-phase ac motor

11. Describe the general layout of a servo system, and state some of the benefits arising from servo systems

12. Distinguish between plain split-phase ac motors and an ac servo motor

13. Explain the function of a servo amplifier, and state its general characteristics and qualities

14. Explain the operation of some specific solid-state ac servo amplifiers, demonstrating the following features:
    a. Push-pull power output stage
    b. Chopper-stabilized dc input
    c. Voltage feedback stabilization
    d. Current feedback stabilization
    e. IC op amp front end followed by discrete driver and power output stages

15. Describe the operation of a dc servo motor, and tell the circumstances under which dc servo motors are used

16. Explain the operation of some specific dc servo amplifiers using SCRs as power control elements

10-1 SOLENOID VALVES

Figure 10-1 shows a cross-sectional view of an electric solenoid-operated valve, or a *solenoid valve* for short. In the absence of current through the solenoid coil, there will be no magnetic field to pull the armature up, so the compression spring will push the armature down. The valve stem is attached to the armature, so it also moves down and pushes the *valve plug* tight against the *valve seat*. This blocks the flow of fluid between the inlet and outlet ports. When the solenoid coil is energized and the coil conductors are carrying current, a magnetic field is established which pulls the armature up. The armature must overcome the spring force tending to push it down in order to move into the middle of the coil. As the armature moves up, it lifts the valve plug off the valve seat and opens the passageway from inlet to outlet. Solenoid valves are inherently two-position devices. That is, they are either all the way on or all the way off. Therefore they lend themselves to being used with the On-Off control mode.

Solenoid coils can be designed to operate on ac voltage or on dc voltage, but the ac design is much more common.

Ac solenoid coils have a serious weakness that dc solenoid valves don't have. If an ac solenoid valve sticks in the closed or partially closed position when power

Figure 10-1. Cross-sectional view of a solenoid valve.

is applied to the coil, the coil will probably burn up. This happens because the magnetic armature cannot enter the core of the coil, so the inductance of the coil remains low. (The inductance of an inductor depends strongly on the magnetic permeability of the core material.) With inductance low, the inductive reactance is also low, and a large ac current will flow through the coil indefinitely. This will eventually overheat the windings of the coil.

10-2 TWO-POSITION ELECTRIC MOTOR-DRIVEN VALVES

In situations where the valve is large or where it must operate against a high fluid pressure, it is better to actuate the valve by an electric motor than by a solenoid coil. In this case the valve body and stem would look like the valve shown in Fig. 10-1, but the stem would be attached to some sort of mechanical linkage which is stroked by an electric motor. Most two-position valves of this type are operated by a unidirectional *split-phase induction motor*. The motor is geared down to provide slow output shaft speed and high torque. As the output shaft rotates from 0° to 180°, the attached linkage opens the valve. As the shaft rotates from 180° back to 360°, the home position, the attached linkage closes the valve. Integral limit switches inside the motor housing detect when the valve has reached the 180° position and when it has reached the home position. A diagram showing the windings, limit switches, and controller connections of such a motor is presented in Fig. 10-2.

Here is how the motor-driven two-position system works. If the motor output shaft is in the home position, meaning that the valve is closed, LS1 is mechanically contacted, and LS2 is also mechanically contacted. The LS2 N.C. contact is thus open, and the LS1 N.O. contact is held closed. If the controller calls for the valve to open, it does so by closing contact *A*. This applies 115-V ac power to the N.O.

Figure 10-2. Circuit for operating a split-phase valve motor. The motor positions the valve (which is not shown) either closed or wide open. The limit switches are shown for the valve in the closed (home) position. The timing diagram shows how the cams operate the two limit switches.

terminal of LS1. Since the N.O. contact is closed at this time, power is applied to the motor windings, and the motor begins to turn. Shortly after it leaves the home position, LS1 is released by its cam, causing the N.O. contact to go open. However, power is maintained to the windings through the LS2 contact, which was also released by its cam. Refer to the timing diagram to see this.

When the motor output shaft reaches the 180° position, meaning that the valve is open, LS2 is once again contacted by its cam, as indicated in the timing diagram. This opens the LS2 contact and disconnects the power from the motor windings. The motor stops in this position, and the valve remains open.

When the controller wants the valve to close, it signals this by closing contact *B*. Power is applied to the motor windings through the N.C. contact of LS2, which is closed at this time. It is closed because LS2 is *not* mechanically actuated by a cam, as shown by the timing diagram. The motor turns in the same direction as before until it reaches the home position. In the home position, both LS1 and LS2 are contacted by their cams, so power is removed from the motor windings, and the motor stops. The valve is therefore closed.

Most electric motors used on two-position motor-driven valves have a full travel time of less than 30 sec. That is, it takes less than 30 sec for the motor to completely open or to completely close the valve.

Some motors are purposely made to operate very slowly, having a full travel time as long as 4 min. When a valve moves this slowly, there is a pretty good chance that the controlled variable will have recovered to set point before the valve completes its motion. If this situation is coupled with a *three-position controller*, the control mode is not really On-Off, and it is not really proportional but a compromise between the two. It is then termed *floating control*. A three-position controller is one which has three output signals instead of only two. These output signals are:

a. The measured value is too low, so drive the valve open.
b. The measured value is too high, so drive the valve closed.
c. The measured value is within the differential gap, so don't drive the valve at all but just let it sit where it is.

Floating control is considered by some people to be a sixth control mode in its own right. However, it is not as important as the five control modes discussed in Chapter 9, so we will not grant it that status.

10-3 PROPORTIONAL-POSITION ELECTRIC MOTOR-DRIVEN VALVES

In proportional control, as we have seen, there must be a method of positioning a control valve at any intermediate position. The usual method is to connect the valve to a reversible slow-speed induction motor. Figure 10-3(a) illustrates such an arrangement on a variable position damper.

(a)

(b)

Figure 10-3. Proportional position control damper driven by a split-phase motor. (a) The mechanical arrangement of the damper, the damper linkage, and the motor. Note that the motor is reversible, in contrast to the motor in Fig. 10-2. (b) The electrical circuit. The controller applies power to either terminal *A* or terminal *B*, depending on whether it wants the damper positioned more closed or more open. The potentiometer provides feedback information to the controller, telling it the present position of the damper.

When the proportional controller sees a *positive* error signal from the comparer, it applies 115-V ac power to terminal *A*. This connects winding 1 across the 115-V line and connects the phase-shift capacitor in series with winding 2; this series combination is also across the ac line. This causes the motor to run clockwise (let us assume) and causes the damper to begin closing off the opening through the duct. When the position potentiometer, which is integrally built into the motor housing, sends the proper position signal back to the controller, the proportional controller is satisfied, and it removes power from terminal *A*. This stops the motor and freezes the damper in that position.

When the proportional controller detects a *negative* error from the comparer, it applies 115-V ac power to terminal *B*. Now winding 2 is directly across the ac line, and winding 1 is placed in series with the phase-shift capacitor. This series combination is also across the ac line. This causes the motor to turn counterclockwise and causes the damper to start opening up. When the motor has turned far enough, the position potentiometer signal matches the error signal, and the proportional controller is satisfied. It removes power from terminal *B*, and the motor stops. The valve therefore stops in a position which is in agreement with the magnitude and polarity of the error signal.

The internal action of a reversible split-phase motor will be described in more detail in Sec. 10-9.

When the motor has traveled to either one of its extreme positions, either clockwise or counterclockwise, one of the limit switches will open up and remove power from the windings. After that, the motor can turn in the opposite direction only. It will do so when the controller orders it to begin turning the other way by applying power to the opposite terminal *A* or *B*.

10-4 ELECTROPNEUMATIC VALVES

For massive valves, an electric motor drive may not be practical. The inertia and breakaway friction of the valve assembly may preclude the use of an electric motor as the positioning device. In such situations, the valve is moved by pneumatic pressure or by hydraulic pressure.

10-4-1 An electropneumatic valve operator

Figure 10-4 is a schematic illustration of an electropneumatic valve operator. The final position of the valve is determined by the magnitude of the electrical input current. Here is how it works.

The balance beam is a small, lightweight, friction-free metal beam, a few inches long. It is pivoted about a fulcrum near its right-hand end. When an input current is fed in via the input terminals, the electromagnet coil establishes a magnetic field which interacts with the field of the permanent magnet. The force resulting from this interaction pulls the beam up, which tends to rotate it clockwise. The force tending to rotate the beam clockwise is proportional to the amount of current flowing through the electromagnet coil.

If the beam rotates slightly clockwise, the left end of the beam will move up and

Figure 10-4. Electropneumatic valve operator. The position of the valve stem is proportional to the electrical input signal in the upper right of the drawing.

restrict the escape of air from the nozzle. The closer the left end of the beam (called the baffle) gets to the nozzle, the less air can escape from the nozzle. As the air escape is cut off, air pressure increases in the variable pressure tube leading to the nozzle. This occurs because the movement of air through the fixed restriction is reduced, resulting in a smaller pressure drop across the restriction and consequently higher pressure downstream of the restriction.

The higher pressure in the variable pressure tube is applied to the diaphragm chamber above the valve diaphragm. This exerts a downward force on the valve stem, thereby opening the valve.* As the valve stem moves down, it causes the feedback spring to exert a counterforce on the balance beam, tending to rotate it counterclockwise. When the countertorque exerted by the feedback spring exactly balances the original torque exerted by the electromagnet, the beam balances in that position. Therefore, the final position of the valve stem and thus of the valve opening itself is determined by the electrical input signal (current) to the electromagnet.

If this apparatus were to be used with an electric proportional controller of the type shown in Fig. 9-15, the electrical input signal could be taken from the wiper of the valve position pot on the left. The voltage between that wiper and ground would be applied to the electromagnet in Fig. 10-4. The electromagnet would have to be designed to draw negligible current from the potentiometer so as not to disturb the voltage division along the length of the pot. The motor shown in Fig. 9-15 would not position the control valve directly but would only serve to rotate the potentiometer shaft. The apparatus shown in Fig. 10-4 would position the control valve.

10-4-2 An electropneumatic signal converter driving a pneumatic positioner

Figure 10-5 shows a somewhat different approach to controlling a large valve pneumatically. Again, the original input signal is an electric current through an electromagnet coil. In this design, though, the feedback to the balance beam does not come from the controlled valve itself but from a bellows element. The pneumatic output pressure is then further balanced against the mechanical position of the valve. Here is how it works.

The input current through the electromagnet coil creates an upward force on the balance beam, tending to rotate it clockwise. As it moves slightly clockwise, the nozzle/baffle arrangement causes the air pressure to increase in the tube ahead of the nozzle, as described in Sec. 10-4-1. This pressure signal is applied to the feedback bellows, which exerts a downward force on the beam, tending to rotate it counterclockwise. The beam stabilizes when the clockwise torque from the electromagnet is equal to the counterclockwise torque from the feedback bellows. Therefore the magnitude of the input current exactly determines the pressure signal applied to the bellows. This pressure signal is also taken out via the output tube for use in another location.

*This is correct as long as the valve is an *air-to-open valve*. Any valve which is *opened* by increasing air pressure applied to its diaphragm is called an air-to-open valve; the spring tends to close the valve. Any valve which is *closed* by increasing air pressure on its diaphragm is called an *air-to-close* valve; the spring tends to open the valve. The choice of which type of valve to use in an industrial system depends on whether the valve should fail open or fail closed in case of pneumatic pressure failure.

Figure 10-5. Electropneumatic signal converter and valve positioner. The apparatus on the top of the drawing converts an electrical signal (current) into a proportional pneumatic signal (air pressure). The apparatus on the bottom positions the valve in proportion to the pneumatic signal.

To summarize, it takes in an electrical input signal and sends out a proportional pneumatic output signal. The converter is designed so that the relationship between output pressure and input current is very linear.

The output pressure signal is fed to a pneumatic valve positioner, which may be located some distance from the electropneumatic converter. The output signal of the converter becomes the input signal to the valve positioner.*

The input pressure signal to the valve positioner tends to expand the input bellows to the right. The input bellows causes mechanical linkage *A* to rotate counterclockwise slightly. As this happens, the linkage shifts the spool in the small pilot cylinder. As the spool moves up, it uncovers the outlet ports of the pilot cylinder. When this happens, the blind-end port (top port) is opened to the interior of the pilot cylinder, applying high-pressure air to the blind end of the power cylinder. At the same time, the rod-end port is opened to the vent hole in the bottom of the pilot cylinder, venting the rod end of the power cylinder. The power cylinder thus strokes down, moving the valve stem down.

As the valve stem moves downward, it causes mechanical linkage *B* to rotate clockwise. This compresses the feedback spring, applying a force tending to compress the bellows. When the input pressure force is balanced by the feedback spring force, linkage *A* returns to its original position. This centers the spool in the pilot cylinder, blocking the outlet ports. The power cylinder stops stroking, and the valve freezes in that position. Therefore we have a condition in which the final valve opening is exactly determined by the input pressure signal. By properly designing the positioner mechanisms, the relationship between valve opening and input pressure can be made very linear.

The overall situation is that the valve opening is linearly related to the input current into the electropneumatic converter. This arrangement is very compatible with an electric or electronic proportional controller.

10-5 ELECTROHYDRAULIC VALVES

In control situations where the valve or damper is very massive, or where it is difficult to hold the valve in a steady position due to large irregular forces exerted by the moving fluid, a *hydraulic positioner* is the best actuator. Also, if a valve seldom moves, it may get stuck in a certain position. This can happen because dirt and debris may build up on moving linkages or shafts, making it very difficult to break them free when the valve is to be repositioned. A hydraulic positioner, with its terrific force capability, may be needed to handle this problem.

A popular electrohydraulic valve positioner, readily adaptable to a propor-

---

*The word *positioner* is generally used to mean an apparatus which uses a high-pressure *cylinder* to move a valve or valve-like device. This is in contrast to the example studied earlier in which variable pressure air was applied to a *diaphragm* to stroke the valve. When a diaphragm is used, the apparatus is generally called an *operator*. (Not all people adhere to this distinction between the words positioner and operator.)

Figure 10-6. Electrohydraulic valve positioner, using a jet pipe. The position of the valve is proportional to the electrical input signal.

tional controller, is shown in Fig. 10-6. Again, the input signal is a current through an electromagnet coil. As the current increases, a larger force is exerted to the left on the vertical balance beam. This tends to make the beam rotate counterclockwise. On the other side of the pivot point, toward the bottom of Fig. 10-6, is a jet pipe relay. Hydraulic oil at high pressure is forced through the jet pipe, emerging from the jet nozzle at high velocity. If the jet pipe is perfectly vertical, the oil stream impinges equally on the left and right orifices. Therefore there is no pressure imbalance between the two sides of the jet pipe relay, and the hydraulic piston is in force equilibrium. However, if the electromagnet coil moves the jet pipe slightly counterclockwise, the right orifice will feel more oil impinging than the left orifice. This will increase the hydraulic pressure on the top of the hydraulic cylinder and decrease the pressure on the bottom. The hydraulic cylinder will thus stroke downward.

As the cylinder rod moves downward, the feedback lever rotates clockwise, pulled by tension spring *A*. The linkage on the left of the feedback lever increases the tension in feedback spring *B*, tending to rotate the balance beam clockwise. Eventually the hydraulic piston will have moved far enough so that the torque exerted by the feedback spring exactly equals the original torque exerted by the electromagnet. At this point the balance beam returns to the vertical position, and the pressure is again equalized between the left and right sides of the jet pipe relay. The piston stops moving, and the valve remains in that position. The final valve position is therefore determined by the magnitude of the input current signal.

10-6 VALVE FLOW CHARACTERISTICS

The ideal flow characteristic in a controlled process is illustrated in Fig. 10-7. As the graph shows, the fluid flow is exactly linear with percent valve opening. That is, with the valve 20% open, the system flow is 20% of maximum; with the valve 40% open, the system flow is 40% of maximum; etc. The actual system flow characteristic depends not only on the valve flow characteristics but also on the flow characteristics of the rest of the piping system.

Figure 10-7. The ideal flow characteristic for a controlled process. In real situations, such linear response cannot be achieved.

The valve flow characteristic curves show the percentage of maximum flow versus percent of opening, for a *constant pressure drop across the valve*. In a real system, it is impossible to maintain a constant pressure drop across the valve as the valve position varies. This is so because as the valve position varies, the flow varies, causing the pressure losses in the rest of the piping system to vary also. Specifically, as flow increases, the pressure drop in the rest of the piping system increases, leaving less pressure drop across the valve. This situation is analogous to a fixed voltage source driving a series combination of a fixed resistor and a variable resistor. As the resistance of the variable resistor is decreased (analogous to opening a valve farther) the current flow increases, causing a larger voltage drop across the fixed resistor. Since there is only a given amount of source voltage to begin with, if a larger voltage drop occurs across the fixed resistor, there must be less voltage drop across the variable resistor.

The result of this phenomenon in a piping system is to make the real system flow characteristics quite a bit different from the *valve* flow characteristic. This is illustrated in Figs. 10-8(a) and (b). Figure 10-8(a) shows a perfectly linear valve

(a)

(b)

(c)

(d)

Figure 10-8. Flow characteristic curves of real systems. (a) Flow curve of a perfectly linear valve. (b) Overall system flow curve which would result from a perfectly linear valve. (c) Flow curve of a nonlinear valve. This valve gives lively response near the open end of its range and sluggish response near the closed end of its range. (d) Overall system flow characteristic which would result from using the valve in part (c). The valve nonlinearity tends to cancel the system nonlinearity, since the two nonlinearities are in opposite directions. The result is an overall system response which is nearly linear.

characteristic. If this perfectly linear valve were installed in a real system, the valve's pressure drop would be reduced at larger percent openings, so the flow response would also be reduced at the larger percent openings. In other words, at the larger openings we get less increase in flow for a given amount of change in valve position. This means that the slope of the system flow curve tapers off, as shown in Fig. 10-8(b).

The flow curve in Fig. 10-8(b) is pretty undesirable. It shows that 80% of the flow change takes place in the first 50% opening of the valve, and that only 20% of the flow change takes place in the second 50% opening of the valve. The disadvantages of this are quite evident.

The general solution to this problem is to design valves to have a flow characteristic like that in Fig. 10-8(c). This shows the valve having a flow characteristic which is *concave up*. When such a valve is installed in a real piping system, which always has a flow characteristic which is *concave down* as in Fig. 10-8(b), the resulting overall flow characteristic is fairly linear. The overall system flow characteristic is graphed in Fig. 10-8(d).

As a general rule of thumb, the greater the pressure losses in the rest of the piping system compared to the pressure drop across the valve, the more nonlinear the valve characteristic itself should be. Valve designers can vary the flow characteristics of a valve almost at will, by varying the shape of the valve plug.

With butterfly valves and louvers [Figs. 10-9(a) and (b)] it is of course impossible to vary the shape of the plug, because there isn't any plug. Instead, the poor flow characteristics of these devices [shown in Fig. 10-9(c)] are corrected by making adjustments to their operating shaft linkages. The manufacturers of hydraulic and pneumatic valve positioners furnish instructions on how to adjust the linkage which connects the positioner's cylinder rod to the shaft of the butterfly valve. By following these instructions, it is possible for the user to create linear flow characteristics.

10-7 RELAYS AND CONTACTORS

10-7-1 On-Off control of current to a load

When electric current is the manipulated variable in a closed-loop control system, the final correcting device is often a relay or a contactor. For example, in an electric heating process, the temperature might be controlled in the On-Off mode simply by opening and closing a contact leading to the heating element. This is illustrated in Fig. 10-10.

Figure 10-10(a) shows a single-phase heating element driven by a single-phase ac source. When the controller receives a *positive* error signal (measured temperature greater than set point), it deenergizes contactor coil *CA*. This interrupts the current flow to the heating element and allows the temperature to drop. When the

(a)

(b)

(c)

Figure 10-9. (a) Butterfly valve or damper. (b) Louver. (c) Typical flow curve of a butterfly valve or louver. Almost all the flow change takes place in the middle 30° of rotation; very little flow change takes place in the first 30° or the last 30°.

controller receives a negative error signal, it energizes coil CA. This closes the N.O. *CA* contact, applying power to the heating element and raising the temperature.

In an application requiring greater heat input, the heating element might be a three-phase element driven by a three-phase supply as in Fig. 10-10(b). In this case the contactor would need three contacts, in order to break each of the three lines.

The only difference between a relay and a contactor is in the current-carrying and interrupting capability of the contacts. Contactors are capable of handling large currents, whereas relays can handle only relatively small currents. The con-

(a)

(b)

Figure 10-10. Temperature control by closing and opening the circuit to a resistive heating element. (a) Contactor with one contact for a single-phase heating circuit. (b) Three contacts switched by one contactor for a three-phase heating circuit.

tactor in Fig. 10-10 could be replaced by a relay if the load current were small enough.

10-7-2 Relay hysteresis

An interesting feature of relays and contactors is that they naturally tend to provide a differential gap for On-Off control, because of the hysteresis effect inherent in their operation. To cause a magnetic relay to energize, the coil current must rise beyond a certain value, called the *pull-in current* or *pick-up current*, to move the armature and switch the contacts. However, once a relay has been pulled in, the coil current must drop below a certain *lower* value of current to cause the relay armature to return to normal. The lower value of current is called the *hold-in current* or the *drop-out current*. This action is illustrated in Fig. 10-11(a).

The reason for the difference between the pull-in and drop-out currents can be easily understood by referring to the stylized drawing of a relay in Fig. 10-11(b). The spring causes the armature to lift up away from the core when the coil is deenergized. This creates an air gap between the top of the core and the metal of the armature. When the coil current starts flowing, it must establish a strong enough magnetic field to pull the armature down against the spring tension. This proves somewhat difficult for two reasons:

a. There is an air gap in the magnetic loop; this causes the magnetic field to be weaker than it would be for a continuous loop of magnetically permeable material.
b. The attractive force between the core and armature (opposite magnetic poles) is weak because of the distance between the poles. When magnetic poles are farther apart, the attractive force between them is weaker, all other things being equal.

If the coil current is large enough, it will create a magnetic field which is strong enough to overcome these two handicaps, and the armature will be pulled down.

When the coil current begins to decrease, after the armature has moved, the above two handicaps are no longer in effect. Therefore it is easier to *hold* the armature down than it was to *pull it down initially*. Because of this, the coil current must drop considerably below the pull-in current in order to allow the armature to move back up.

10-7-3 A three-phase contactor switching between delta and wye

An interesting example of the use of a three-phase contactor as the final correcting device in an electric heating process is shown in Fig. 10-12(a). This is basically On-Off control, except that the Off position is not really *completely* Off. The idea is that when contactor *CA* is dropped, the three-phase heating elements are connected in a wye configuration, and when *CA* is picked, the heating elements are connected in a delta configuration. In the wye configuration, the power

Figure 10-11. (a) Graphical illustration of relay hysteresis. As current is increasing, the switching occurs when the pull-in value is reached. As current is decreasing, the switching occurs when the drop-out value is reached. (b) The essential parts of an electromagnet relay. Hysteresis is due to the gap between the armature and the magnetic core.

(a)

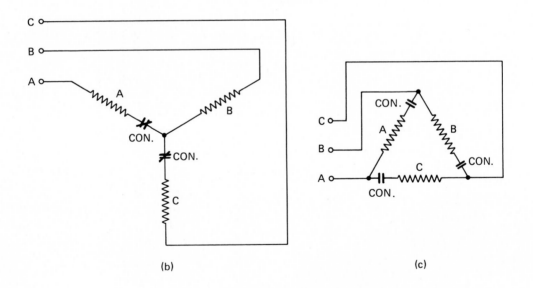

(b)

(c)

Figure 10-12. (a) A delta-wye heating circuit. (b) Simplified view of the heating circuit with the contactor deenergized. The heating elements are connected in wye, and the heat output is lower. (c) Simplified view of the heating circuit with the contactor energized. The heating elements are connected in delta, and the heat output is higher.

delivered to the three-phase heating elements is much less than the power delivered if they are connected in delta. The controller causes *CA* to drop if the measured temperature is above set point. This causes the process temperature to drop because of the reduced power input to the wye connection (with consequent reduction in heating effect). The controller causes *CA* to pick up if the measured temperature is below set point. This causes the process temperature to rise because of the increased power input to the delta connection. Of course, the system must be designed so that the heat created by a wye connection causes the temperature to decline and the heat created by a delta connection causes the temperature to rise.

To see that the elements are connected in wye with *CA* dropped, study Fig. 10-12(b). To produce Fig. 10-12(b), the normally open *CA* contacts have been removed from the schematic of Fig. 10-12(a), thus uncluttering the drawing. It is apparent that the three elements are connected in a wye configuration.

Figure 10-12(c) shows the situation with the *normally closed* contacts removed from the schematic of Fig. 10-12(a). It is apparent that the elements are connected in delta when *CA* is picked up.

We have stated that the heat created by a delta connection is greater than the heat created by a wye connection because the electrical power delivered to the elements is greater in delta than in wye. This will now be proved.

In the delta (Δ) configuration, the voltage across any one heating element is equal to the line voltage, or

$$V_\phi = V_{\text{line}} \qquad \text{for } \Delta,$$

where V_ϕ is the voltage across *one* phase of the heating load. If the resistance of one phase of the load (one single element) is symbolized R, then we can say that

$$P_\phi = \frac{V_\phi^2}{R} = \frac{(V_{\text{line}})^2}{R},$$

where P_ϕ is the average power delivered to one single phase of the load (one element). We are assuming the elements are pure resistances with unity power factor.

The total average power delivered to the entire three-phase load is simply three times the power delivered to any one phase, so total power is given by

$$P_T = \frac{3(V_{\text{line}})^2}{R} \qquad \text{for } \Delta.$$

Assuming some actual numbers, if the line voltage is 460 V ac and the resistance per phase is 25 Ω, the total power is

$$P_T = \frac{(3)(460)^2}{25} = 25,392 \text{ W} \qquad \text{for } \Delta.$$

Now for wye: The voltage across any one phase is *not* equal to the line voltage. For a balanced three-phase wye system, we can always say that

$$V_\phi = \frac{V_{\text{line}}}{\sqrt{3}} = \frac{V_{\text{line}}}{1.73} \qquad \text{for Y.}$$

Therefore, average power delivered to one phase of the load is

$$P_\phi = \frac{V_\phi^2}{R} = \frac{(V_{\text{line}}/\sqrt{3})^2}{R} = \frac{(V_{\text{line}})^2}{3R} \qquad \text{for Y.}$$

Again, the total power is just three times the power delivered to any one element, so

$$P_T = \frac{3(V_{\text{line}})^2}{3R} = \frac{(V_{\text{line}})^2}{R} \qquad \text{for Y.}$$

Assuming the same numbers as before, namely 460-V line voltage and 25-Ω heating elements, we get

$$P_T = \frac{(460)^2}{25} = 8,464 \text{ W} \qquad \text{for Y.}$$

The total heat delivered to the process by the wye connection is thus only one third as much as the heat delivered by the delta connection (8464 W/25,392 W = $\frac{1}{3}$).

10-8 THYRISTORS

When the manipulated variable is electric current and the delivery must be *continuously variable*, relays and contactors cannot do the job. In modern control systems, power thyristors, namely SCRs and triacs, are used as final correcting devices for such applications. In each of the heating systems discussed in Sec. 10-7, a thyristor could have been used instead of a magnetically operated relay or contactor.

Thyristors lend themselves quite well to proportional control of temperature. The conduction angle can be made to vary in proportion to the error between measured temperature and set point. This continuously varies the flow of current into a heating element, providing the benefits of proportional control. The integral and/or derivative control modes can be added to the system by adding the proper electronic circuitry.

Of course, not all applications of thyristors involve *closed-loop* control of current flow. There are many *open-loop* systems in which thyristors serve as the control device. In a thyristor open-loop system, current to the load is continuously variable; but of course there is no built-in comparison between measured value and set point to bring about *automatic* adjustment of the current.

As an example of an SCR in an open-loop temperature control system, simply substitute a resistive heating element for the load in Fig. 4-8 and imagine that the heating element is supplying heat energy to a process chamber. If R_V is adjusted, the firing delay angle of the SCR varies, varying the current through the heating element and thus the power delivered to the heating process. For a given set of process conditions, a given setting of R_V will provide a given process temperature. Of course, if any process condition changes (if a disturbance occurs), that particular setting of R_V will cause a different temperature to be established. In an open-loop situation, the firing delay angle of the SCR is *not* automatically altered to correct the temperature. The temperature simply lands wherever the process conditions dictate.

Thyristors have many other industrial applications besides varying current flow to heating elements. They can be used to vary the current through an electromagnet whose magnetic pulling force must be variable; they can be used to vary current to an incandescent light bulb for situations where light intensity must be varied; they can be used to vary welding current to alter the properties of a weld, as we saw in Chapter 7; they can be used as sequencing devices, as shown in Fig. 5-7. But by far their most important use is in varying the current through a motor winding to adjust the speed of the motor. This is such an important part of industrial electronics that we will devote a separate chapter to this one subject. In Chapter 13 we will treat in detail the use of power thyristors as motor speed control devices.

10-9 SPLIT-PHASE AC MOTORS

Split-phase motors were seen in Secs. 10-2 and 10-3 as motors for opening and closing flow control valves. Not only do split-phase motors handle positioning of valves, but they also perform most positioning jobs in servo systems. Most servo systems have what is called an *ac servo motor* as their final correcting device. An ac servo motor is basically a split-phase motor with some minor construction differences. In this section we will consider the operating theory and characteristics of split-phase motors. In Sec. 10-10 we will explore ac servo motors, their construction, and their operation in a mechanical positioning system.

We will treat the subjects of ac split-phase motors and ac servo motors with the assumption that you are somewhat familiar with magnetic fields and motors. If you are a complete novice in the area of motor theory, you should probably supplement these sections with some other reading and study.

In any ac induction motor, ac power is applied to the *field winding* on the stationary part of the motor (the stator). Current is then induced in the *armature winding* on the rotating part of the motor (the rotor) by transformer action. The interaction of the magnetic field created by the field winding with the current-carrying conductors of the armature winding creates mechanical forces which force the rotor to spin.

Let us concentrate first on how the field winding of a split-phase motor establishes a magnetic field, and on the behavior of that magnetic field. Then we will explore how currents are induced in the armature conductors. Finally, we will combine the magnetic field with the armature current to see how torque is produced.

The Rotating Field. Figure 10-13(a) is a schematic representation of the field winding of a split-phase motor. The view shown represents what would be seen looking down the hollow stator from one end of the motor shaft. The rotor has not been drawn in order to keep the drawing clear.

There are two windings shown in Fig. 10-13(a). Each one is usually referred to as a *winding* (singular). We have winding 1, which is driven by voltage V_1, which

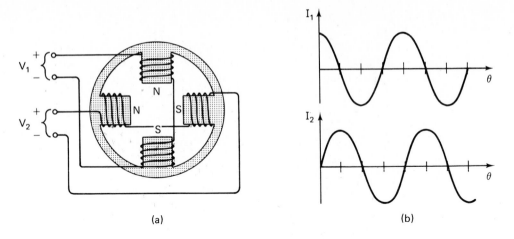

Figure 10-13. (a) Field poles and windings for a split-phase ac motor. This is a view looking down the hollow stator. (b) The phase relationship between the two winding currents. I_1 leads I_2 by 90°.

tends to establish a magnetic field in the vertical direction; and we have winding 2, driven by voltage V_2, which tends to establish a magnetic field in the horizontal direction. Naturally the ac voltages V_1 and V_2 are continually changing polarity, but it is convenient to assign them a *positive polarity*, as has been done in Fig. 10-13(a). That is, when V_1 is $+$ on the top and $-$ on the bottom, we will consider it as having a positive polarity. When V_1 is $-$ on the top and $+$ on the bottom, we will consider it as having a negative polarity. The same holds for V_2.

Now if V_1 is positive, a current will flow through winding 1 from top to bottom, thereby creating a magnetic field pointing from top to bottom. This field direction is shown in Fig. 10-14(a).

If V_2 is positive, a current will flow through winding 2 from left to right, thereby creating a magnetic field pointing from left to right. This field is shown in Fig. 10-14(c).

If V_1 and V_2 are both positive at the same time, currents will flow in *both* windings. With each winding creating an individual magnetic field, the *net* magnetic field will be a compromise between the individual fields. This means that the net magnetic field will point somewhere between the individual fields. This is shown in Fig. 10-14(b).

Somehow or another, the voltages applied to the two windings, V_1 and V_2, must be adjusted so that the winding currents are 90° out of phase, as illustrated in Fig. 10-13(b). There are several ways of doing this, which we will investigate later. For now, just remember that by some external method V_1 and V_2 must be adjusted so that I_1 and I_2 are 90° out of phase.

Now study Fig. 10-13(b) and compare it carefully with Fig. 10-14. The arrows in Fig. 10-14 show the direction of the magnetic field created by the two windings at various points along the ac cycle (various points in time). The arrow labeled F_1

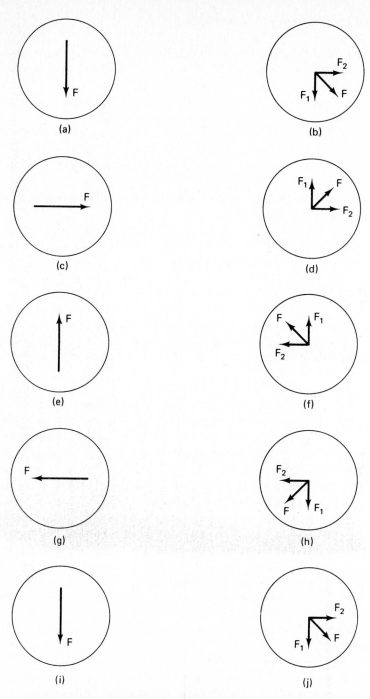

Figure 10-14. Individual fields created by the two windings (F_1 and F_2) and the net resultant field (F) at different instants in time. Each drawing represents a 45° progression in the ac cycle, or about a 2.08-millisecond progression in time for a 60-Hz ac line.

indicates the direction of the field due to winding 1; the arrow labeled F_2 indicates the direction of the field due to winding 2. The arrow labeled F indicates the overall net field direction, due to both windings.

At 0°, or at the beginning of the cycle, I_1 is maximum positive and I_2 is 0. Therefore the net field is due entirely to winding 1 and is in the direction shown in Fig. 10-14(a).

At 45° into the cycle, I_1 has decreased but is still positive, and I_2 has climbed into the positive region; both windings are contributing to the net field, which is shown in Fig. 10-14(b).

At 90°, I_1 is 0 and I_2 is maximum positive. The net field is due entirely to I_2 and is shown in Fig. 10-14(c).

At 135° into the cycle, I_1 is negative, so the field created by winding 1 is pointing *up*, as shown in Fig. 10-14(d). The field from winding 2 is still pointing to the right because I_2 is still positive. The net field is as shown.

At 180° into the cycle, I_2 has dropped to 0, and I_1 is maximum negative. The field is shown in Fig. 10-14(e).

At 225° into the cycle, I_1 is still negative, and I_2 has also moved into the negative region. Therefore the field due to I_1 points up, and the field due to winding 2 points to the left. The net field points to the upper left. This is shown in Fig. 10-14(f).

At 270° into the cycle, I_2 is maximum negative, and I_1 is 0, so the net field points to the left, as shown in Fig. 10-14(g).

At 315° into the cycle, I_1 is once again positive, and I_2 is still negative, so the net field points to the lower left, as shown in Fig. 10-14(h).

At 360° the winding currents have both returned to starting conditions, so the field is also back in the starting direction. Figure 10-14(i) shows the same net field direction as Fig. 10-14(a).

What is happening here is that the net field is rotating *around the stator*, just as if a field winding were actually physically rotating. The field is making one rotation around the stator for every cycle of the ac supply voltage. Its speed of rotation in revolutions per second is equal to the frequency of the ac voltage in cycles per second (Hz).* The strength of the magnetic field depends on the magnitude of the currents in windings 1 and 2, just as with any electromagnet.

The graphs in Fig. 10-13(b) show I_1 leading I_2 by 90°, and the diagrams in Fig. 10-14 show that the net field is rotating counterclockwise. If the relationship between the winding currents were changed so that I_2 led I_1 by 90°, the net field would rotate in the opposite direction, clockwise. You should prove this to yourself.

This completes our discussion of the action of the *field* in a split-phase motor. To sum up, we have seen that as long as both windings are carrying out-of-phase currents, the net field will rotate. The strength of the net field is determined by the amount of current flowing in the windings. The direction of rotation of the field depends on which current leads the other.

*This is true only for a two-pole motor, as shown in Fig. 10-14(a).

The Armature Conductors. When a rotor is inserted into our split-phase motor, forces will be exerted on it to make it follow the rotating field. If the field rotates clockwise, the rotor will rotate clockwise. If the field rotates counterclockwise, the rotor will rotate counterclockwise. Here is how this happens.

Figure 10-15(a) shows a three-quarter view of a *squirrel-cage rotor* for a split-phase motor. Most induction motors have rotors that look like this. (The split-phase motor is a specific example of the general class of induction motors.)

The rotor is basically a cylinder with shafts extending out both ends. (The shafts are not shown in Fig. 10-15.) The shafts are supported in either sleeves, ball bearings, or roller bearings. The rotor is thus free to spin. The rotor material is some ferromagnetic (iron-based) alloy which has good magnetic properties. The rotor has slots, or grooves, cut lengthwise in its body, in which aluminum conductors are inserted. The conductors are joined one to another at each end by an aluminum *end ring*. There is no insulation between the aluminum conductors and the iron core. However, any current which flows from one end of the rotor to the other must do so by flowing down the aluminum bars because the core is laminated. That is, it is made up of layers of core material separated by layers of insulating material. This makes it impossible for currents to flow from one end to the other through the core material.

Figure 10-15(b) shows an end-on view of the rotor with the magnetic field created by the stator windings rotating counterclockwise.* As the magnetic field does this, the magnetic flux linking each rotor loop is changing. The rotor loop consists of two side-by-side aluminum bars and the portions of the end rings which join them. Because the flux is changing, voltages are induced in the rotor loops by the same principle that makes transformers operate: namely, that a changing magnetic flux through a coil creates an induced voltage which tends to oppose the change in flux. The voltages induced in the rotor loops are such that the conductors on the one side of the cylinder are all carrying current in one direction and the conductors on the other side of the cylinder are all carrying current in the other direction. This is illustrated in Figs. 10-15(b) and (c) by the *arrowhead* and *tailfeather* technique. Arrowheads (dots) symbolize current flowing out of the page. Tailfeathers (Xs) symbolize current flowing into the page.

Because these currents are flowing, there are forces exerted on the conducting bars which cause the rotor to spin in the same direction as the rotating field. The origin of the forces can be viewed in either of two ways:

a. The currents in the rotor conductors interact with the magnetic field lines to produce mechanical forces according to the familiar *right-hand rule.* The right-hand rule says that if conventional current is flowing through a magnetic field, there will be a force exerted on the conductor; the force direction is indicated by the right thumb as the fingers of the right hand rotate the current vector (arrow) into align-

---

*This illustration is for a two-pole motor. Many ac split-phase motors have more than two poles, making the magnetic field drawing more complicated. The principle of operation, however, is exactly the same.

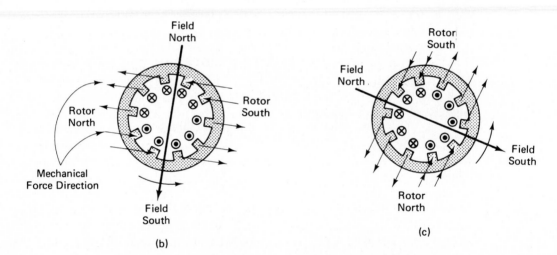

Figure 10-15. Squirrel-cage rotor. (a) Isometric view of the rotor, showing the aluminum end rings connected by aluminum conducting bars. The bars are laid into slots in the magnetic core. (b) End-on view of the rotor. The instantaneous current directions in the conducting bars are shown, along with the instantaneous force directions on the bars. The directions are correct only if the field is rotating counterclockwise. (c) The same drawing 90° later in the ac cycle. Note that the field arrow has rotated 90° from its position in part (b).

ment with the field vector. The instantaneous directions of the forces exerted on all 10 conducting bars are indicated in Figs. 10-15(b) and (c). Notice that all ten forces tend to create counterclockwise rotation. You should justify these force directions to yourself by applying the right-hand rule.

b. The current flow in the rotor conductors converts the rotor into a powerful magnet in its own right. The rotor magnetic poles then seek to align with the stator field poles according to the laws of magnetic attraction and repulsion (namely that unlike poles attract each other, and like poles repel each other). The instantaneous locations of the magnetic poles of the field winding and of the armature are also shown in Figs. 10-15(b) and (c). Notice that the rotor will attempt to follow the rotating field as the unlike poles try to come together and the like poles try to get apart.

Any way it is viewed, the rotor seeks to follow along behind the rotating field because of these forces. This is how motor torque is created.

The rotor can never catch up to the rotating field, and it can never spin at the same angular speed. If it did, there would be no *relative motion* between the magnetic field lines and the rotor conductors, and there would be no induced voltages or currents in the rotor bars. Without induced currents there would be no torque created, so the rotor would quickly drop back. For split-phase motors, the rotor spins at a speed between 85% and 95% of the speed of the rotating field, depending on the torque load it must overcome. The speed of the rotating field is called the *synchronous speed*. The difference between synchronous speed and actual rotor shaft speed is called *slip*. It is so called because the rotor is constantly "slipping behind" the rotating magnetic field. Slip is often expressed as a percent of synchronous speed. Thus if the rotor spins at 3420 rpm while the rotating field is spinning at 3600 rpm, the slip would be given by

$$\text{slip} = 3600 \text{ rpm} - 3420 \text{ rpm} = 180 \text{ rpm}.$$

The percent slip would be given by

$$\text{percent slip} = \frac{180 \text{ rpm}}{3600 \text{ rpm}} = 0.05 = 5\%.$$

Creating the Phase Difference Between the Two Winding Currents. Split-phase motors will work satisfactorily even if the phase angle between I_1 and I_2 is not exactly 90°. *Any* phase relationship which can create the effect of a rotating field will cause the motor to spin. One easy way to create a phase difference between the winding currents is to insert a capacitor in series with one of the windings and then drive both windings with the same voltage source. Figures 10-2 and 10-3 showed exactly this. The capacitor tends to cause the winding current to lead the applied voltage. It cannot cause the current to lead the voltage by exactly 90° because of the resistance and inductance of the winding, but it does establish *some* phase shift.

If the capacitor is inserted in series with winding 1, it would be indicated schematically as shown in Fig. 10-16(a). This would cause I_1 to lead I_2, as shown in Fig. 10-13(b). The rotor would then rotate counterclockwise if the windings were wrapped exactly as shown in Fig. 10-13(a).

However, if the capacitor were placed in series with winding 2, then I_2 would lead I_1, and the rotor would rotate clockwise. This situation is illustrated in Fig. 10-16(b).

(a)　　　　　　　　　　　　(b)

(c)

Figure 10-16. Creating the phase shift between winding currents. (a) With a capacitor inserted in series with winding 1, I_1 leads I_2; this causes the motor to spin in a certain direction. (b) With a capacitor inserted in series with winding 2, I_2 leads I_1; this causes the rotor to spin in the opposite direction. (c) Closing one of the switches causes one direction of rotation, while closing the other switch causes the opposite rotation.

With these relationships in mind, it is now possible to see why the motor will rotate in different directions depending on which switch is closed in Fig. 10-16(c). This is essentially the same circuit as shown in Figs. 10-2 and 10-3, and it is the most popular way to use split-phase motors in industrial control.

The use of a fixed capacitor is not the only way to create the phase difference needed to make the field rotate. Some split-phase motors *start* with a capacitor in series with one winding, and then switch that winding completely out of the circuit once the motor has accelerated up to operating speed. This can be done with a centrifugally operated switch. The winding which is switched out is called the *starting winding*, and the one which continues to drive the rotor is called the *running winding* or the *main winding*. This technique is possible because split-phase motors can often *run* on a single winding; however, they can never *start* from a standstill using a single winding.

There are many other techniques used to duplicate the action of a rotating field. Some of the methods and techniques are ingenious. Any book devoted solely to rotating machines will explain them.

10-10 AC SERVO MOTORS

As we know, when the controlled variable in the closed-loop system is a mechanical position, the system is called a servo system. Two simple servo systems were presented in Sec. 9-3 as examples of closed-loop systems. A more general arrangement of a servo system is shown in Fig. 10-17.

Figure 10-17. The general layout of a servo system.

The setting potentiometer on the left is adjusted to express the desired position of the controlled object. There would probably be some sort of scale attached to the setting pot. That scale would relate potentiometer shaft position with mechanical position of the controlled object. For example, if the servo system made it possible to position the controlled object anywhere within a 12-ft range, the setting pot dial might have markings with spacing equal to one twelfth of the total rotation of the pot shaft. Then each mark would "translate" into 1 ft of mechanical movement of the object. The operator could decide what mechanical position of the object he wanted, turn the setting pot dial to the proper number on the scale, and walk away. The servo system would handle the rest of the job of positioning the controlled object where it is supposed to be.

There are many reasons for the use of servo control systems in industry. Among them are the following:

a. The object may be very massive and/or heavy, so that a human being can not handle it directly. It can be handled conveniently only by a specially designed servomechanism dedicated to that task.

b. The object may be inaccessible, or inconvenient to get to. The remote setting and adjusting feature of the servo system is then a great benefit.

c. The object may be dangerous to be near. The remote adjustment capability allows the operator to exert control without exposing himself to danger.

For these and other reasons, servo systems are very useful in industrial control. In this section we will study the most common final correcting device of a servo system, the ac servo motor.

An ac servo motor is essentially the same as the split-phase ac motor covered in Sec. 10-9. There is one fairly important difference between an average split-phase motor and an ac servo motor. It is that the servo motor has thinner conducting bars in the squirrel-cage rotor, so the conductor resistance is higher. As we proceed with our discussion of ac servo motors, we will point out why this feature is needed.

The two windings of an ac servo motor are referred to as the *main winding* and the *control winding*. Sometimes the main winding is called the *fixed winding*. The word *phase* is frequently substituted for the word winding. Thus we hear the terms

Figure 10-18. (a) Circuit of an ac servo motor. The voltage applied to the control winding varies in proportion to the error voltage, V_e. (b) Curve of speed vs. control voltage for an ac servo motor.

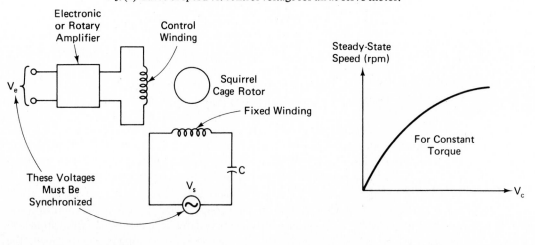

(a)   (b)

fixed phase and *control phase* to describe the two separate field windings of a servo motor.

In the great majority of servo control systems, the servo motor is not *switched* on and off as were the split-phase motors in Figs. 10-2, 10-3, and 10-16(c). Instead, a servo motor is driven as indicated in Fig. 10-18(a).

The voltage applied to the control winding, V_c, is taken from the output of an amplifier. The input to the amplifier is the error voltage, V_e, which depends on how far the object is from the desired position.

The fixed winding always has power applied to it by a fixed voltage ac source, as Fig. 10-18(a) shows. In this case there is a capacitor inserted in series with the fixed winding to shift its current by about 90°.

The method of operation of the servo motor is not hard to understand. If the difference between actual position and desired position is great, V_e will be large (see Fig. 10-17). If V_e is large, V_c will also be large, and the control winding current will be large. This will cause the servo motor to run at high speed. As the difference between the actual position and desired position decreases, the error voltage V_e also decreases, as Fig. 10-17 shows. Therefore V_c decreases, reducing the control winding current and making the motor run slower. The relationship between rotational speed and control voltage is shown in Fig. 10-18(b).*

When the position of the object is precisely correct, the error voltage V_e will shrink to zero. The amplifier will have no input signal, so V_c will also go to zero. The motor curve shows that the motor stops when $V_c = 0$, so the object will stop moving when it gets to the desired position. Of course, the servo motor *must* stop when V_c gets to zero. Recall that some split-phase motors can continue to run with only one winding powered up. This is called *single-phasing*. Servo motors must *not* be able to single-phase. This is one of the reasons for creating the high-resistance conductor bars in the rotor of an ac servo motor. High-resistance bars prevent single-phasing.

Observe the note in Fig. 10-18(a) which says that V_e and V_s must be synchronized. This means that they must be derived from the same ultimate voltage source, so their phase relationship is either in phase or 180° out of phase. This is necessary so that the final field winding currents are approximately 90° out of phase with each other. In this example, the phase-shift capacitor C causes the fixed winding current to lead the applied voltage by about 90°. The control winding current, on

---

*The curve in Fig. 10-18(b) shows the *steady* speed which would eventually be established if the control voltage stayed constant long enough for the motor to stabilize, versus the size of the control voltage. Of course, as the positioned object homes in on the desired position in a real servo system, the control voltage doesn't stand still; it is constantly shrinking toward zero. Also, this curve is for constant shaft torque, which never occurs in a real servo system; the torque delivered by the motor shaft is reduced as the object gets closer to the set position. Therefore the curve in Fig. 10-18(b) is not a valid graph of speed versus control voltage for a servo motor operating in a real servo system. It is only intended to show the characteristics of the motor under ideal and somewhat artificial conditions. Curves of this type are necessary, however, or how could we compare one motor to another?

the other hand, is roughly in phase with the control winding voltage, which is in phase with the error voltage.

The control voltage is in phase with the error voltage, assuming the amplifier processes the input signal without introducing any phase shift of its own. This last requirement, that the amplifier introduce no phase shift, is easy to accomplish at the low frequencies used for servo systems (60 Hz in most industrial systems).

The end result is one of these two situations:

a. If the error voltage is *in phase* with V_s, the control winding current will *lag* the fixed winding current by about 90°, and the motor will turn in one direction (assume clockwise).

b. If the error voltage is *180° out of phase* with V_s, the control winding current will *lead* the fixed winding current, and the motor will turn in the other direction (counterclockwise).

These two situations are illustrated graphically in Fig. 10-19. The winding currents have been assumed to be in phase with the voltages applied to the windings, which is a convenient assumption to make for purposes of understanding servo motor operation. It is not quite true in real life, however. Also, the fixed winding current is shown leading the source voltage by exactly 90°; this is another assumption which is not quite true in real life but is convenient for explanatory purposes.

In Fig. 10-19(a), on the left, V_e is in phase with V_s. Two graphs are plotted for two different error voltages, V_{e_1} and V_{e_2}, corresponding to two different distances of the controlled object from the desired position. The fact that V_e is in phase with V_s means that the controlled object is in one particular direction away from the desired position (let us say to the east).

In Fig. 10-19(b), on the right, V_e has been taken as 180° out of phase with V_s. This would mean that the controlled object is in the opposite direction away from the desired position (to the west). Again, graphs have been drawn for two different distances from the desired position. For a small distance away, the error voltage is V_{e_1}. It results in small control winding current, I_{c_1}, which causes slow motor speed. For a large distance away, the error voltage is V_{e_2}. It results in large control winding current, I_{c_2}, which causes faster motor speed. The symbol I_F stands for current through the fixed winding; note that it is always 90° ahead of V_s.

Occasionally a servo motor will have no series capacitor in series with the fixed winding. In such cases, the required 90° phase shift must be provided by the amplifier. The amplifier would be specially designed so that the V_c output is 90° out of phase with the V_e input. This can be done by making appropriate choices of coupling capacitors and other electronic circuit components. When this is done, the amplifier can be used only for a particular frequency, since the phase-shift angle would change if the frequency changed. This practice is more often seen in aircraft servo systems than in industrial servo systems.

Torque-Speed Characteristics of ac Servo Motors. As mentioned at the beginning of this section, ac servo motors are essentially the same as standard split-phase ac

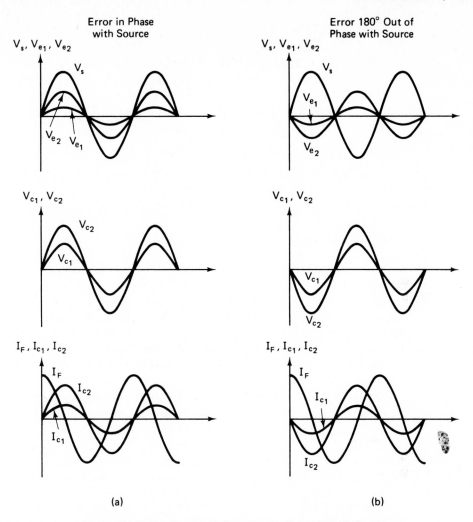

Figure 10-19. (a) Graphs of the various voltages and currents in an ac split-phase servo motor for the error voltage in phase with the source voltage. (b) The same graphs for the case in which the error voltage is out of phase with the source voltage. In the current graphs, the fixed winding current is shown exactly 90° out of phase with the control winding current. This is an idealization.

motors except that the rotor bar resistance is made higher. We have already seen one of the advantages of this, that it prevents single-phasing. Single-phasing would be disastrous in a servo control system because it would mean that the motor would not stop running when the position error was reduced to zero. The other important advantage of high rotor bar resistance is that it makes the torque-speed relationship of the motor better for servo applications. Figure 10-20 shows the torque-speed relationship of a standard split-phase motor compared to that for an

Figure 10-20. Torque vs. speed curves for a normal split-phase ac motor and for an ac servo motor. This curve assumes a constant control voltage applied to the servo motor.

ac servo motor. The curve of the servo motor is for a constant value of control voltage.

The curve for the normal split-phase motor is telling us that when the motor is running slow it has a certain amount of torque-producing ability. As the speed increases, the torque-producing ability of the motor also increases. This relationship holds true up to a certain point, called the *maximum torque point*. After that, any further increase in speed results in a reduction in the torque-producing ability of the motor.

The reason for this behavior is that the torque-producing ability of the motor is determined basically by the amount of current flowing in the rotor conductor bars. This current, though, is determined by two things:

a. The *amount* of voltage induced in the rotor conductor loops
b. The *frequency* of the voltage induced in the rotor conductor loops

Regarding effect **a**, the amount of voltage induced in the rotor conductor loops is large when the speed is slow and small when the speed is fast, because the rotating stator field is moving very fast *relative to the conductor bars* when the rotor is moving very slow. (In other words, the rotor is "slipping behind" drastically.) On the other hand, the rotating stator field is moving very slow relative to the conductor bars when the rotor is moving very fast. (The rotor is not "slipping behind" much because the slip is small.) Therefore the voltage is large when the rotor is moving slow, and the voltage is small when the rotor is moving fast. This effect tends to *reduce* rotor current at fast speeds.

However, regarding effect **b**, the frequency of the induced voltage is high when the rotor speed is slow, and the frequency is low when the rotor speed is fast. This is because the frequency of the induced voltage in a rotor conductor loop is equal to the number of times per second that the rotating field "laps" the rotor loop. That is, one cycle of ac voltage will be induced in a rotor conductor loop every time the rotating stator field gains one revolution on the rotor.*

The higher the frequency, the higher the inductive reactance of the one-turn rotor conductor loops ($X_L = 2\pi fL$). As you know, a higher inductive reactance causes lower ac current flow, and a lower inductive reactance allows higher ac current flow. This effect therefore tends to *increase* the rotor current at fast rotor speeds.

What we have here are two opposite effects. One effect, voltage, tends to *reduce* current and torque at fast speeds, and the other effect, frequency, tends to *increase* current and torque at high speeds. It's just a matter of which effect is stronger. For the normal split-phase motor, the frequency effect is stronger in the low-speed range, so torque-producing ability goes up as speed goes up. At a certain point (maximum torque point) the voltage effect takes over and predominates. Thereafter, the torque-producing ability goes down as speed goes up. This explains why normal split-phase motors exhibit the torque-speed relationship that they do, which is drawn in Fig. 10-20. In fact, this explains why virtually all induction motors have the torque-speed characteristics they do (including the workhorse of industry, the three-phase induction motor).

Unfortunately, though, these torque-speed characteristics are not good for a servomechanism. In servo control, it is desired to produce a lot of torque at slow speeds so that the motor can accelerate the positioned object rapidly. Besides, it is better if the torque-producing ability of the motor is reduced at high speed, because that makes it less likely that the motor will overshoot its mark. That is, it's less likely that the controlled object will shoot past the desired position and have to back up.

Servo motors are therefore built to have a large rotor conductor bar resistance. This is easily done by making the bars thinner and shallower. With a large rotor bar resistance, the inductive reactance of the rotor bar loops is swamped out, and the frequency effect is minimized. The voltage effect thus predominates. As seen above, the voltage effect tends to *reduce* torque-producing ability at high rotor speeds. The result is the torque-speed characteristic for an ac servo motor shown in Fig. 10-20.

The complete set of characteristic curves for an ac servo motor is shown in Fig. 10-21. The different curves are for different control voltages applied to the control field.

As Fig. 10-21 shows, the torque-producing ability of a servo motor is greatest

---

*You can understand this by picturing a fast race car carrying a magnet on its side and a slow race car carrying a coil of wire on its side. If both cars are going around the race track, the fast car's magnet will create one cycle of induced voltage in the slow car's coil every time the fast car laps the slow car. If the speed of the fast car is held constant (like a rotating stator field), then the speed of the slow race car will determine the frequency of the induced voltage. The slower the slow car goes, the more often it will be lapped, and the higher the frequency. The faster the slow car goes, the less often it will be lapped, and the lower the frequency.

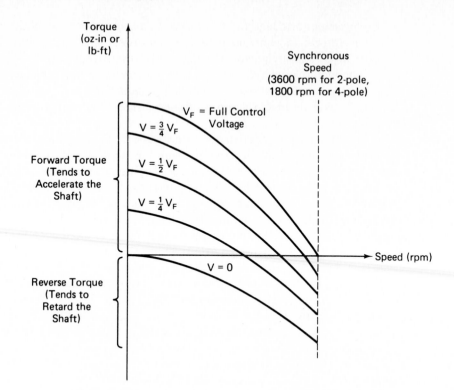

Figure 10-21. Torque vs. speed curves of an ac servo motor for various values of control voltage.

at low speed and declines as speed goes up. This is true for either a large or a small control winding voltage.

One interesting fact about servo motors emerges from these curves. At small control voltages (meaning that the controlled object is getting close to the desired position) the servo motor can actually exert a *reverse torque* on its shaft. This is helpful in preventing overshoot because if the object has a large inertia, it may tend to coast past the desired position even if the servomechanism exerts no motive force whatsoever. That is, in a case involving high inertia and low friction, the servo motor could be completely disconnected from the power lines when the controlled object was still some distance from the desired position, and the object would not be able to stop in time.

In a real servo system, the control winding is never disconnected from the amplifier. The amplifier output will be reduced to a very small voltage, though, as the object gets close to the desired position. If the inertia of the controlled object tries to make the rotor spin faster than it wants to, the interaction of the fixed winding magnetic field with the fast-turning rotor causes current to flow *backwards* in the control winding. Under these circumstances, the control winding is no longer being driven by the amplifier but is acting like a generator. Because of the reversed control winding current (180° different from what it normally is), the torque exerted on the rotor is in the opposite direction. This tends to bring the motor shaft to a screeching halt and minimizes overshoot of the desired position.

This is not to say that servo systems never overshoot the mark. Far from it. Usually the reversed torque applied by the servo motor is not sufficient by itself to stop the load in time. An obvious solution to this problem is to allow the motor to run only very slowly, by applying only very small control voltages. This would defeat one of the good features of a closed-loop control system, though, because it would lengthen the correction time. Instead, the usual solution is to provide *damping*, which is the tendency to exert a countertorque when the speed is fast. Many ingenious techniques have been devised to provide damping. Such things as eddy-current drag cups and tachometer feedback loops are examples of damping methods applied to servomechanisms. Any book dealing exclusively with servo systems will give an explanation of such methods.

Ac servo motors are generally preferred to dc servo motors for the same reasons that ac motors are always preferred to dc motors:

a. Their squirrel-cage rotors are simple and rugged compared to the complex armature windings found on dc machines.
b. They have no brush-to-commutator contacts which require frequent inspection and maintenance.
c. There is no insulation around the armature conductors as there is on a dc motor, so the armatures can dissipate heat much better.
d. Because the rotor has no complicated insulated windings on it, its diameter can be made quite small to reduce rotor inertia; this helps prevent overshoot in a servomechanism.

Nevertheless, there are some servo systems which require a dc motor as the final correction device. This usually happens when the positioned object is very large. When a very large object is being positioned, the motor must naturally have a high horsepower rating. High-horsepower ac servo motors suffer from the problem of their rotor conductors overheating, because of the necessity to make the rotor conductor resistance high. As a rough rule of thumb, if the servo motor must have a power rating in excess of 100 hp, a dc servo motor can do a better job than an ac servo motor. Therefore servo systems which deliver large amounts of power normally use dc motors. Dc servo motors will be discussed in Sec. 10-12.

10-11 SOLID-STATE AC SERVO AMPLIFIERS

An ac servo amplifier amplifies the position error voltage to produce the servo motor control voltage, as shown in Fig. 10-18. An ac servo amplifier has the same general requirements that all good amplifiers have:

a. It should have a high input impedance, so it doesn't load down its signal source (the error voltage source).

b. It should have a hefty voltage gain (A_V) which is fairly independent of temperature changes, component aging, and variations among components.

c. It should have a low output impedance, so it can drive a heavy current load without its output voltage sagging.

d. It should keep the signal distortion down to a reasonable level. An ac sine-wave input should yield an ac sine-wave output.

e. It should operate efficiently so its transistors run cool, especially its final output transistors.

One thing a servo amplifier does *not* need is a wide bandwidth. On the contrary, high-frequency roll-off of its frequency response curve is a blessing because it tends to eliminate high-frequency noise signals which sneak into the amplifier.

We will study three different ac servo amplifiers. Among the three of them, most of the important features and variations among servo amplifiers will be illustrated.

10-11-1 Servo amplifier 1:
A four-stage transistor
amplifier with push-pull output

Refer to Fig. 10-22, which is the schematic drawing of a four-stage servo amplifier.

Transistor Q_1 and its associated components comprise the input stage of the servo amplifier. The 60-Hz ac input signal (the error voltage) is fed to the base of Q_1 through coupling capacitor C_1. Q_1 is connected in an emitter-follower configuration to display a high impedance to the error voltage source. Because of the high impedance looking into the input stage, the reactance of C_1 does not need to be very low. Therefore, C_1 is quite a bit smaller than the other coupling capacitors in this circuit.

The input resistance (impedance) of the input stage is given by

$$R_{in_1} = R_1 \,/\!/\, R_2 \,/\!/\, r_{b\ in_1}, \tag{10-1}$$

where $r_{b\ in_1}$ stands for the ac resistance looking into the base of Q_1. The ac resistance looking into the base of Q_1 is specified by

$$r_{b\ in_1} = \beta_1(r_{e_{j_1}} + r_{e\ out_1}), \tag{10-2}$$

where β_1 is the current gain of Q_1, $r_{e_{j_1}}$ is the ac resistance of the base-emitter junction of Q_1, and $r_{e\ out_1}$ is the ac resistance looking out from the emitter of Q_1 to ground. The ac resistance across the base-emitter junction of a forward-biased transistor is rather low, usually less than 50 Ω, so $r_{e_{j_1}}$ is negligible compared to $r_{e\ out_1}$. Therefore, assuming the β_1 is about 100 (a conservative estimate), the input resistance of the input stage is given approximately by

$$R_{in_1} = R_1 \,/\!/\, R_2 \,/\!/\, 100(r_{e\ out_1})$$

Figure 10-22. A straightforward solid-state servo amplifier.

or

$$(R_{\text{in}_1} = 560 \text{ K} \mathbin{/\mkern-3mu/} 680 \text{ K} \mathbin{/\mkern-3mu/} 100(r_{e \text{ out}_1}) = 307 \text{ K} \mathbin{/\mkern-3mu/} 100(r_{e \text{ out}_1}). \tag{10-3}$$

Now the ac resistance looking out from the emitter of Q_1 to ground is equal to the parallel resistance of two current flow paths. Path 1 runs directly down through emitter resistor R_3 to ground. Path 2 runs through C_2, through R_4, through the gain-adjust pot, and into the Q_2 circuitry. The resistance of path 1 is simply 47 K, the size of R_3. The resistance of path 2 is equal to the sum of R_4, plus the resistance of the gain-adjust pot, plus the ac resistance looking into the second stage. Considering the worst case, with the gain-adjust pot dialed completely out for maximum amplifier gain, the ac resistance of path 2 is equal to R_4 (10 K) plus the input resistance of the second stage. The input resistance of the second stage is at least 11 K if β_2 is at least 100 [since $R_{\text{in}_2} = R_5 \mathbin{/\mkern-3mu/} R_6 \mathbin{/\mkern-3mu/} \beta(r_{b \text{ in}_2}) = 150 \text{ K} \mathbin{/\mkern-3mu/} 22 \text{ K} \mathbin{/\mkern-3mu/} 100(270 \ \Omega)$ $= 11 \text{ K}]$. Therefore

$$r_{e \text{ out}_1} = r_{\text{path 1}} \mathbin{/\mkern-3mu/} r_{\text{path 2}} = 47 \text{ K} \mathbin{/\mkern-3mu/} (10 \text{ K} + 11 \text{ K}) = 14.5 \text{ K}.$$

Going back to Eq. (10-3), we can now calculate the input resistance of the first stage as

$$R_{\text{in}_1} = 307 \text{ K} \mathbin{/\mkern-3mu/} 100(14.5 \text{ K}) = 307 \text{ K} \mathbin{/\mkern-3mu/} 1450 \text{ K} = 253 \text{ K}.$$

An input resistance of 253 K is adequate for most applications. That is, most error voltage sources would not be unduly loaded down by an amplifier whose input resistance was this high.

In an ac servo amplifier, as in most ac amplifiers, it is a good practice to keep the reactance of the coupling capacitors less than about 3% of the input resistance of the amplifier stage. In this particular case, for a 60-Hz signal,

$$X_{C_1} = \frac{1}{2\pi f C_1} = \frac{1}{2\pi(60)(1 \times 10^{-6})} = 2.65 \text{ K},$$

so

$$\frac{X_{C_1}}{R_{\text{in}_1}} = \frac{2.65 \text{ K}}{253 \text{ K}} \cong 0.01,$$

which means that X_{C_1} is only about 1% as large as R_{in_1}.

The Q_1 emitter follower can provide no voltage gain. Therefore the voltage appearing at coupling capacitor C_2 is slightly smaller than the input error voltage. The C_2 voltage is then further divided between the R_4/gain-adjust pot combination and the second stage of the servo amplifier. The second stage, comprised of transistor Q_2 and its associated components, is a voltage amplifier with a voltage gain of about 10. The emitter resistors R_8 and R_9 serve to stabilize the dc bias point against temperature changes and transistor batch variations. The emitter bypass capacitor C_3 allows ac current to bypass R_9. This tends to increase the voltage gain. R_8 is left unbypassed to help raise the input impedance of the second stage.

The output voltage from the second stage is taken from the collector of Q_2 and fed through C_4 into the third stage. The third stage amplifies the signal further and applies it across the primary winding of T_1. The primary winding of T_1 is tuned to 60 Hz by C_1. Together, the winding inductance and the capacitor C_1 form a parallel resonant circuit with a resonant frequency of around 60 Hz. This allows the third stage to provide maximum voltage amplification to the 60-Hz signal frequency and very little amplification to extraneous signals at other frequencies.

The ac output voltage from the third stage is stepped down by T_1 and appears across the center-tapped secondary winding. T_1 is a 10:1 step-down transformer; it reduces the third-stage output voltage down to less than 2 V while boosting the current capability by a factor of 10. This provides a proper match with power transistors Q_4 and Q_5, which need very little ac input voltage but do require a fair amount of input current. Power transistors Q_4 and Q_5 are wired in a common-emitter push-pull configuration. Here is how the push-pull power output stage works.

During the half cycle that the T_1 secondary voltage is positive on top, the top half of the winding delivers base current to Q_4. Current flows out the top terminal of the winding, through the base-emitter junction of Q_4, through R_{13} and R_{17}, and back into the center tap. Q_4 then carries collector current via this path: from the +48-V supply, into the center tap of the T_2 primary, through the top half of T_2, through the collector-to-emitter path of Q_4, through R_{13}, and into the ground.

While all this is happening, Q_5 is biased OFF. This is because the bottom terminal of the T_1 secondary is negative relative to ground, robbing Q_5 of the small base current which was supplied through R_{15}. With Q_5 turned OFF, the bottom half of the T_2 primary winding carries no current.

Thus, during the positive half cycle of the signal voltage, the T_2 primary winding carries a net current in the *up* direction. This induces a T_2 secondary voltage of a certain polarity, say positive on top. This voltage is applied to the motor control winding. Therefore one half cycle of an ac sine wave is created across the control winding.

During the other half cycle of the signal voltage, the T_1 secondary is positive on the bottom. This allows the bottom half of the T_1 secondary to deliver base current to Q_5, which turns on and carries collector current via this path: from the +48-V supply, into the center tap of the T_2 primary, through the bottom half of T_2, through the collector-emitter path of Q_5, through R_{13}, and into the ground. While all this is happening, Q_4 is biased OFF by the negative voltage applied to its base by the top half of the T_1 secondary. With Q_4 turned OFF, the top half of the T_2 primary carries no current.

Thus, during the negative half cycle of the signal voltage, the T_2 primary carries a net current in the *down* direction. This induces a T_2 secondary voltage of the opposite polarity from before, positive on the bottom. In this manner the other half cycle of the sine wave is applied to the motor control winding.

You may wonder why we would use such a complicated method to deliver a complete sine wave of voltage to the control winding. The reason is a very good one. Here it is: This method allows the output transistors to spend half of their time "resting" and cooling down. One transistor in the push-pull stage handles one half of the signal, and the other transistor in the push-pull stage handles the other half of the signal. *Neither transistor has to carry steady dc current at all times.*

This method has a great power advantage over the conventional transistor biasing method, in which a certain dc bias current is established and the ac signal current is superimposed on it. That dc bias current is of no use in driving the load because it is blocked out by the coupling capacitor or transformer. Nevertheless, it creates a power drain on the dc power supply and forces the transistor to dissipate that power. A push-pull output stage as shown in Fig. 10-22 eliminates such power waste. The output transistors still burn a lot of power because of the relatively large motor currents they carry. However, all the power they do burn is the result of ac current, which is at least useful for driving the load. By eliminating the dc power consumption problem, smaller power transistors can be used, and the size of their heat sinks can be reduced. Virtually all ac servo amplifiers use some variation of the push-pull configuration for their power output stage.

It is not hard to see that if the phase of the error voltage input changed by 180° relative to the fixed winding voltage, the lead-lag relationship would be reversed. That is, if the control winding current had been leading the fixed winding current by 90°, it would now be lagging the fixed winding current by 90°. This would cause the motor to turn in the opposite direction. Therefore the *phase* of the error voltage determines the direction of rotation of the servo motor, and the size of the error voltage determines its speed of rotation.

10-11-2 Servo amplifier 2: A four-stage chopper-stabilized transistor amplifier with negative feedback and unfiltered dc supply to the control winding

Figure 10-23 shows an amplifier design which illustrates some other popular features of ac servo amplifiers. First, notice that the error voltage input is a dc signal. Some error detectors are unable, for one reason or another, to provide an ac voltage to the servo amplifier. Instead, they deliver a dc voltage. Straightforward amplification of a dc error voltage is difficult because of the drift problems inherent in a dc amplifier. A popular alternative is to *chop* the dc signal to make it resemble an ac signal. The signal chopper is just a vibrating reed switch whose common terminal is alternately connected to the dc input and ground potential. The signal appearing at the common terminal of the chopper switch is a square wave whose height equals the magnitude of the dc input voltage.

The vibration of the mechanical switch is created by the chopper drive coil, an electromagnet which is powered by a 60-Hz supply. The chopper drive supply is snychronized with the fixed motor winding ac supply. On one half cycle of the 60-Hz ac line, the mechanical switch is pushed up to make contact with the input terminal; on the other half cycle it is pulled down to make contact with the ground terminal. The result is a square wave which is either in phase with or 180° out of phase with the fixed winding supply, depending on whether the dc error voltage is positive or negative.

Figure 10-23. Another transistor servo amplifier. This design illustrates several popular features of servo amplifiers, including chopped dc input, negative feedback from one stage to another, power supply decoupling between stages, and unfiltered dc supply to the output stage.

The square-wave input signal is applied to coupling capacitor C_1 and into the base of the Q_1 transistor. Q_1 is a *pnp* transistor connected as a common-emitter amplifier. Emitter resistor R_3 stabilizes the bias currents against temperature changes and transistor batch variations. R_3 also serves to give the Q_1 amplifier stage an inherently high input resistance.

The output of this amplifier is taken from the collector of Q_1 and is direct-coupled to the base of Q_2, which is another common-emitter amplifier. Part of the output voltage of Q_2 appears across R_3, which is the emitter resistor of Q_1. R_3 is used by both transistors Q_1 and Q_2. The portion of the Q_2 output voltage which appears across R_3 *in the emitter of* Q_1 is of such a polarity that it *opposes the original input signal* applied to Q_1. This is called *negative voltage feedback*. It serves to raise the input resistance of the Q_1 amplifier stage even higher than it would be naturally. Negative voltage feedback also further stabilizes the voltage gain of the Q_1-Q_2 combination against temperature change and transistor variations.

To understand why this arrangement provides negative feedback, consider what happens as the input signal at C_1 goes into its negative half cycle. The negative-going voltage tends to turn Q_1 on harder, causing the collector of Q_1 to go more positive relative to ground. This drives the base of Q_2 more positive and tends to turn Q_2 on harder. As Q_2 begins conducting greater collector current down through R_3 and R_6, it causes the emitter of Q_1 to become more negative (closer to ground, farther from the $+V_{CC}$ supply). The fact that the Q_1 emitter is forced to become more negative tends to cancel the original effect of the input signal going negative.

To explain in detail, if the Q_1 emitter had merely followed the input signal, a certain amount of current would have flowed through the base-emitter junction of Q_1; the feedback, however, tends to reverse-bias the Q_1 base-emitter junction, which tends to turn Q_1 off somewhat. The end result is that less current flows through the base-emitter junction than would have without the negative feedback. With less input current for the same input voltage, the input resistance of the amplifier is increased. The input resistance of this amplifier approaches 1 million ohms. The input resistance is also stabilized against temperature changes, etc.

The price paid is the reduction in voltage gain of the Q_1-Q_2 input stage, which becomes much lower than it would have been without the feedback. The voltage gain becomes stabilized though, as mentioned above.

To sum up, the effects of negative voltage feedback are to

a. Raise the input resistance and stabilize it against circuit variations.
b. Lower the voltage gain and stabilize it against circuit variations.

The output signal of the Q_1-Q_2 input stage is taken from the collector of Q_2 and coupled through C_4 into the base of Q_3. Transistor Q_3 is another common-emitter amplifier. Its output is taken from the Q_3 collector and direct-coupled to the base of Q_4. The Q_4 emitter resistor, R_{12}, serves to help stabilize the Q_4 bias current and also to provide *negative current feedback* to the base of Q_3. Negative current feedback serves to stabilize the amplifier gain against circuit variations but also, unfortunately, to lower the input resistance of the Q_3 stage. At this point in the circuit, however, input resistance has ceased to be so important, because the Q_3 stage is being driven by the Q_1-Q_2 stage, which has good current-delivering ability.

To understand why the R_{12}/gain adjust pot combination provides negative current feedback, consider what happens as the signal at C_4 enters its positive half cycle. The positive-going voltage tends to turn Q_3 on harder. This causes the voltage at the collector of Q_3 to go more negative (closer to ground). This negative-going voltage tends to turn Q_4 off, causing a reduction in current in the Q_4 emitter lead. The voltage across R_{12} is therefore going more negative at this instant. This negative-going voltage sucks current through C_6, robbing the base of Q_3 of some of the input current coming through C_4. What happens is that the attempt to increase current into Q_3 causes a reaction which tends to decrease the current into Q_3. This is the essence of negative current feedback.

The effectiveness of the negative current feedback depends on the ac resistance between the emitter of Q_4 and the base of Q_3. This ac resistance can be varied by adjusting the gain-adjust pot. The ac resistance between these two points is whatever portion of the pot is not shorted out by C_6. As the wiper is moved to the left, more of the total pot resistance is shorted out by C_6, and the ac resistance is lowered. Lower ac resistance causes the feedback effect to be stronger, thus lowering the voltage gain of the Q_3-Q_4 combination.

If the pot wiper is moved to the right, less of the gain pot would be shorted out by C_6, thereby raising the ac resistance between the Q_4 emitter and the Q_3 base. This would reduce the strength of the negative current feedback and raise the voltage gain.

The output from the Q_4 amplifier appears across the primary of transformer T_1, which is tuned to a resonant frequency of 60 Hz by C_7. This helps reduce the gain of the square-wave harmonic frequencies and causes the output to be more sinusoidal.

Notice that the +60-V dc power supply line to the Q_3 and Q_4 amplifiers is separated from the Q_1-Q_2 supply line by R_8. This is a *decoupling resistor*. It works in conjunction with *decoupling capacitor* C_2 to filter out any power supply line noise signals before they reach the Q_1-Q_2 input stage. That is, any voltage disturbances appearing on the +60-V dc supply line due to the current drawn by Q_3 and Q_4 are prevented from appearing at Q_1 and Q_2. The R_8 and C_2 decoupling components can accomplish this because the reactance of C_2 is much less than the resistance of R_8 at 60 Hz or higher frequencies.

In this example, the reactance of C_2 at 60 Hz is 53 Ω ($X_C = 1/2\pi fC$). This is only about 3.5% as large as the 1500-Ω resistance of R_8, so only about 3.5% of any noise disturbance will appear across C_2. Of course, keeping noise signals from appearing across C_2 is the same as keeping them from appearing across Q_1 and Q_2. It is important to keep power supply noise signals away from the input stage because if they get into the input stage, they are eligible for full amplification by the amplifier. Noise signals in the succeeding stages are not as harmful because they are not subject to the full amplification of the amplifier.

In the power output stage, the power transistors again are connected in a push-pull configuration, although this time they are wired as common-base amplifiers. Let us look first at the biasing arrangement. Refer to the waveforms of the power output stage in Fig. 10-24, specifically parts (a), (b), and (c). There is a small dc bias current supplied to Q_5 and Q_6 by the -60-Vp unfiltered supply. The path of bias current flow is this: from ground into the base of the transistor, out the emitter lead and through half the secondary winding, through R_{13}, and into the -60-Vp supply. This bias tendency is pulsating because the -60-Vp supply is full-wave pulsating dc. The use of unfiltered pulsating dc reduces the transistor's heat dissipation requirements slightly.

The T_1 secondary winding delivers emitter current to power transistors Q_5 and Q_6. The emitter current waveforms are drawn in Figs. 10-24(d) and (e). When the T_1 secondary winding goes positive on its top terminal, the bottom half of the secondary winding turns on Q_6. The current flow path is the loop comprised of the bottom half of the secondary winding, the base-emitter junction of Q_6, and the germanium diode. On the alternate half cycle, when the secondary voltage is positive on the bottom, the top half of the winding turns on Q_5 in the same way.

The positive supply voltage to the center tap of the motor control winding is also full-wave pulsating dc. This unfiltered pulsating supply voltage helps reduce transistor power consumption considerably, because there is less voltage across the main transistor terminals as the instantaneous collector current climbs toward its peak value. That is, as the collector current rises toward its peak, the power transistor stays just barely out of saturation because the collector supply voltage is following the current up. With the transistor just barely out of saturation, the voltage drop across its main terminals stays small, causing the power consumption to be small $[P_{\text{inst}} = (V_{\text{inst}})(I_{\text{inst}})]$. The waveforms of the positive voltage supply, the control winding current, and the collector-to-emitter voltages across the transistors are drawn in Figs. 10-24(f), (g), (h), and (i).

Besides cutting down on transistor power dissipation, using unfiltered pulsating dc to drive the push-pull power output stage also relieves the load on the *filtered* dc power supply. This reduces the ripple in that supply.

The servo motor control winding is a center-tapped winding, with only one half of the winding carrying current at any instant in time. This does not alter the fact that the magnetic field set up by the control winding changes directions every half cycle. In other words, when the current flows through the control winding from center to bottom (Q_6 conducting), the magnetic field it creates points in a certain direction relative to the stator of the motor. When the current flows through the winding from center to top (Q_5 conducting), it sets up a magnetic field pointing in the *other* direction relative to the stator. This alternation of magnetic field direction is all that is necessary to interact with the fixed winding field to create a net rotating field.

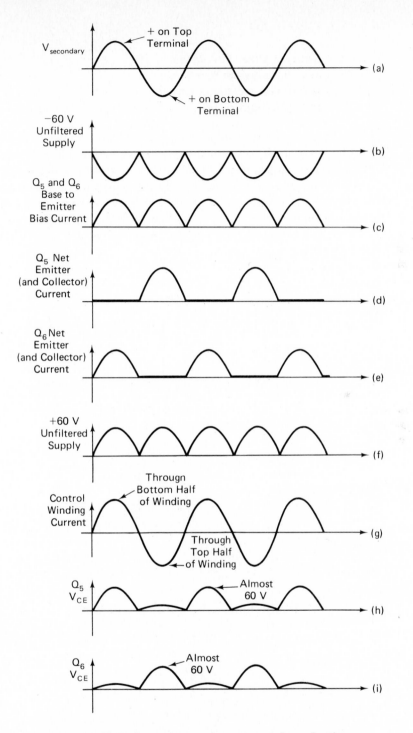

Figure 10-24. Various voltage and current waveforms for the amplifier of Fig. 10-23.

10-11-3 Servo amplifier 3: A hybrid amplifier using an IC op amp in the front end with a discrete push-pull output stage

A simple servo amplifier using an IC op amp is shown in Fig. 10-25. As we saw in Chapter 8, an IC op amp is a complete amplifier circuit packaged in a case no larger than that used for a single discrete transistor. Once the power supplies are hooked up and a few external components are attached, the op amp is ready to go.

In this application the op amp is being used as a noninverting ac amplifier. The voltage gain is set by the combination of R_1, R_2, and the gain-adjust pot. The voltage gain of the op amp noninverting amplifier can be adjusted over the range

Figure 10-25. A hybrid (having both IC and discrete components) servo amplifier.

392

of 5.7 to 55.7 in this circuit since

$$A_{V_{min}} = \frac{4.7\text{ K} + 1\text{ K}}{1\text{ K}} = 5.7,$$

and

$$A_{V_{max}} = \frac{54.7\text{ K} + 1\text{ K}}{1\text{ K}} = 55.7.$$

The input impedance of the noninverting amplifier is naturally very high, so no special techniques are needed to raise the input impedance.

The components C_1, C_2, and R_3 are connected to the proper op amp terminals to determine the frequency response characteristics. The op amp manufacturer's data sheets always give advice on the sizes of these components.

The ac error voltage is applied to the noninverting input terminal, and an amplified signal is taken from the op amp's output terminal. It is passed through R_4 and C_3 into transistor Q_1. This transistor is connected in common-emitter configuration to provide maximum power amplification. R_5, R_6, and R_7 set the bias current of Q_1, and that bias point is stabilized by R_7. C_4 is connected in parallel with the T_1 primary winding to resonate at the signal frequency.

The T_1 secondary drives the Q_2-Q_3 transistor pair, which is connected in push-pull configuration. The transistors are biased slightly on by the R_8-R_9 voltage divider attached to the $+60$-V dc supply.

The 100-μF capacitor C_6 provides the path for the ac signal current to flow through the base-emitter junctions of Q_2 and Q_3.

10-12 DC SERVO MOTORS

As mentioned in Sec. 10-10, ac servo motors are generally preferred to dc servo motors, except for use in very high-power systems. For very high-power systems, dc motors are preferred because they run more efficiently than comparable ac servo motors. This enables them to stay cooler. An efficient motor also prevents excessive power waste, although power waste is generally not a primary concern in servomechanisms.

A dc servo motor is no different from any other dc shunt motor for general-purpose use. It has two separate windings: they are the field winding, placed on the stator of the machine, and the armature winding, placed on the rotor of the machine. Both windings are connected to a dc voltage supply. In many dc shunt motor applications the windings might actually be connected in parallel (shunt) and driven by the same dc supply, but in a servo application, the windings are driven by *separate* dc supplies. This situation is illustrated schematically in Fig. 10-26(a).

The field winding of a dc motor is usually symbolized schematically as a coil. The field winding is connected to the dc voltage supply labeled V_F in Fig. 10-26(a). The armature winding of a dc motor is symbolized schematically as a circle in con-

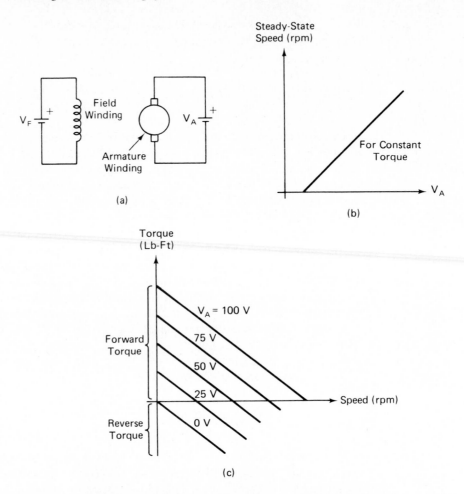

Figure 10-26. (a) Schematic of a dc servo motor. (b) Graph of speed vs. armature voltage for a constant shaft torque. (c) Torque vs. speed curves for various applied armature voltages.

tact with two squares. This suggests the physical appearance of a dc armature as a cylinder having brushes bearing against its surface. The armature winding is connected to the dc voltage supply labeled V_A in Fig. 10-26(a).

We will not get into a detailed physical description of dc motor speed control at this time. That subject is covered in Chapter 13. Suffice it to say that the steady-state speed can be controlled either by varying V_F or by varying V_A in Fig. 10-26(a). In virtually all modern servo systems, the adjustment is made to V_A, the armature supply voltage.

The relation between steady-state speed and applied armature voltage is shown in Fig. 10-26(b) for a constant torque. It is approximately a linear relation. However, in a real servo system, the motor torque is not constant. It varies as the controlled object approaches the desired position and the motor voltage is reduced. The more meaningful torque-speed curves at various armature voltages are shown in Fig. 10-26(c).

These curves tell us that the torque-producing ability of a dc servo motor is greater at low speeds than at high speeds for a given applied armature voltage V_A. This enables the motor to accelerate the load (the positioned object) rapidly from a standstill. Furthermore, the torque-speed curves show that as the positioned object approaches its desired position and V_A is reduced, the dc motor is capable of delivering a reverse torque to slow the load down if the speed of approach is high. This is possible because the motor armature winding starts to act like a generator under this condition. The current in the armature conductors reverses direction (assuming the armature voltage supply is capable of sinking current), and the reverse current creates reverse torque.

Thus, the standard dc shunt motor satisfies the requirements for a good servo motor. It has high torque at low speeds, and it has a built-in *damping effect*, in that a high approach speed results in an automatic slow-down tendency. Of course, as with any dc motor, it cannot be expected to operate with as little trouble as an ac induction motor. A dc motor is inherently a high-maintenance device, because of the complex insulated armature winding and because of the fact that current must be carried to and from the armature through brush-to-commutator contacts.

10-13 AMPLIFIERS FOR DC SERVO MOTORS

Theoretically, there is nothing wrong with the idea of using a dc amplifier to amplify a dc error voltage in a servo system. The amplifier would simply boost the dc error voltage up to a larger dc voltage to drive the armature of the servo motor. There are some practical problems with this idea, though. For instance, all coupling capacitors between the amplifier stages would have to be eliminated, since dc current cannot pass through a capacitor. Likewise, transformer coupling from one stage to another would have to be abandoned, since transformers will not respond to a dc current. In short, all dc amplifiers must be *direct-coupled*. This means that the output of one stage must be connected to the input of the next stage through resistors and/or wires only. Direct coupling is illustrated in the input stage of the amplifier in Fig. 10-23; the Q_1-Q_2 connection contains no capacitors or transformers. The other stages in that amplifier are not direct-coupled, though.

Direct coupling throughout a multistage amplifier causes certain practical problems which are just about insurmountable. In a direct-coupled amplifier, any change in transistor bias voltage due to power supply variation, temperature changes, or component aging is treated exactly like a real input signal. For example, if the dc bias voltage at the collector of a certain transistor were to change slightly due to a temperature rise, the other stages of the amplifier could not distinguish that change from a genuine voltage signal caused by a dc error input voltage.

In other words, direct-coupled amplifiers tend to change their dc output voltage due to *internal* bias voltage variations which have nothing to do with the *externally* applied input signal. As can easily be appreciated, this is very bad. This phenomenon is called *drift*. If this occurs in a servo system, the servo motor will run when it isn't supposed to run, and the controlled object will be improperly positioned.

There are various ways to minimize drift, but they all involve circuit complications, and they never work perfectly. Therefore servomechanism builders avoid using dc servo amplifiers. If the error input voltage must be a dc voltage for some reason, it can be chopped and amplified in an ac amplifier as shown in Fig. 10-23.

The fact remains that dc servo motors require rather large dc voltages capable

Figure 10-27. Control circuits for a dc servo motor. (a) A method using just one ac supply winding. (b) A method using two ac supply windings. (c) Waveforms of error voltage and output voltage showing that a larger error voltage causes earlier firing of an SCR.

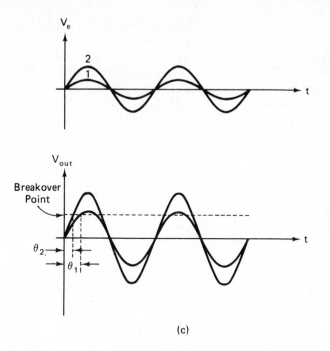

(c)

Figure 10-27. (Cont.)

of delivering large currents. What is needed is an ac amplifier capable of delivering a dc output. The SCR, with its *amplifying* and *rectifying* ability, is the ideal output device for such an amplifier.

Figure 10-27 shows two methods of using SCRs to control a dc servo motor. In Fig. 10-27(a) there are two SCRs connected, facing in opposite directions. SCR1 controls the armature current in one direction, from top to bottom. It thus controls the motor when it is rotating in one direction, say clockwise. SCR2 controls the armature current in the other direction, bottom to top. It controls the motor when it is rotating in the other direction, counterclockwise. Here are the details of operation.

The ac error signal (V_e) is amplified in a preamplifier of the same type as those presented in Sec. 10-11. The voltage at the output terminals of the preamplifier (V_{out}) is an amplified version of the error signal and is in phase with the error signal. At a certain point in the positive half cycle, the output voltage will reach the breakover point of the four-layer diode, causing a burst of current into the T_1 primary winding. This occurs during every positive half cycle as long as the input error voltage is above a certain minimum value. The specific point in the half cycle is determined by how large the error voltage is. For a large error voltage, the breakover will occur early in the half cycle; for a small error voltage, the breakover will occur later in the half cycle. This is shown clearly in the waveforms of Fig. 10-27(c). These waveforms apply to either SCR circuit.

For a small V_e, the curves labeled 1 show what happens. V_{out} reaches the breakover point relatively late in the half cycle, resulting in firing delay angle θ_1. For a larger V_e (bigger position error) the curves labeled 2 apply. V_{out} reaches the breakover point of the four-layer diode earlier in the cycle, resulting in a firing delay angle θ_2.

Every time the four-layer diode breaks over to deliver a burst of current into the T_1 primary winding, positive pulses occur at the two T_1 secondary windings in Fig. 10-27(a). Both of these positive pulses tend to fire the SCRs to which they are connected. That is, both SCR1 and SCR2 receive a gate current pulse telling them to turn ON. However, only one SCR actually does turn ON, because only one of them is forward biased at this instant. If V_s is positive on the top when the pulses arrive, SCR1 turns ON and SCR2 remains OFF because of the reverse bias across its main terminals. If V_s is positive on the bottom when the pulses arrive, SCR2 turns ON and SCR1 remains OFF. Thus the phase relationship between V_s and V_e determines which SCR will fire. If V_e is in phase with V_s, SCR1 fires; if V_e is 180° out of phase with V_s, SCR2 fires. The phase of the error voltage determines the direction of rotation of the motor and the direction of movement of the controlled object. The size of the error voltage determines the firing delay angle of the SCR and therefore the average current (and voltage) delivered to the armature. It therefore controls the servo motor's speed.

Figure 10-27(b) shows a somewhat different arrangement of SCRs and motor armature, but the action of the gate control circuitry is exactly the same as in Fig. 10-27(a). If V_e is in phase with V_s, then at the instant the gate pulses arrive at SCR1 and SCR2 the top T_1 secondary winding will be forward-biasing SCR1 and the bottom T_1 secondary winding will be reverse-biasing SCR2. Therefore SCR1 will fire and SCR2 will remain OFF, causing current to pass through the armature from top to bottom. To understand this, refer to the phasing marks on the transformer windings. If V_e and V_s are in phase, then when the dotted terminals of the T_2 windings are positive, the triangled terminals of the T_1 windings are also positive. During the negative half cycle, V_{out} will be the wrong polarity to fire the four-layer diode, so no gate pulses are delivered and no SCR turns ON.

If V_e is 180° out of phase with V_s, then at the instant the gate pulses arrive at the SCRs, the bottom T_1 secondary winding will be forward-biasing SCR2 and the top T_1 secondary winding will be reverse-biasing SCR1. Therefore SCR2 will fire and SCR1 will remain OFF, causing current to pass through the armature from bottom to top. To see this, recognize that if V_e and V_s are 180° out of phase, then when the dotted terminals of the T_2 windings are positive, the triangled terminals of the T_1 windings are negative.

As in the first circuit, if V_e and V_s are in phase, the motor spins in one direction, and if V_e and V_s are 180° out of phase, the motor spins in the other direction. The magnitude of the error input voltage again determines the firing delay angle of the SCR, thereby controlling the average voltage applied to the motor armature.

QUESTIONS AND PROBLEMS

1. For a solenoid-actuated valve, explain the function of each of the following parts: valve seat, valve plug, valve stem, armature, solenoid coil, spring.

2. Why is it that ac solenoid valves can have their coils damaged if the valve stem sticks, but dc solenoid valves don't have that problem?

3. In an electric motor-driven two-position valve, what does the motor gearing provide: high speed and low torque or low speed and high torque?

4. In the limit switch timing diagram of Fig. 10-2, LS2 is released slightly before LS1 is released. Why is this necessary?

5. In Fig. 10-2, why is it necessary that LS2 be actuated at the 180° mark?

6. About how long does it take for a two-position control valve to go full travel?

7. In Fig. 10-3, what is the purpose of the extreme position limit switches?

8. Make a distinction between a pneumatic valve *operator* and a pneumatic valve *positioner*.

9. Explain the difference between an air-to-open valve and an air-to-close valve.

10. Generally speaking, under what circumstances are pneumatic and hydraulic valve actuators preferred to electric motor actuators?

11. In Fig. 10-4, what causes the variable air pressure to increase and decrease?

12. Identify the forces which combine to balance the balance beam in Fig. 10-4.

13. In Fig. 10-4, if the feedback spring were to break, what would the valve do if a small electrical input signal were applied to the coil?

14. Repeat Question 13 for the problem of the baffle breaking.

15. Distinguish between the *blind end* and the *rod end* of a cylinder.

16. In Fig. 10-5, which way must the pilot cylinder move in order to cause the main power cylinder to stroke down?

17. In Fig. 10-5, what causes the power cylinder to stroke up: an increase in the input pressure signal or a decrease in that signal?

18. Give a detailed step-by-step explanation of what happens in the apparatus of Fig. 10-5 if the electrical input signal is decreased.

19. In Fig. 10-6, identify the two forces that combine to balance the vertical balance beam.

20. Repeat Question 18 for the apparatus of Fig. 10-6.

21. What is the difference between a relay and a contactor?

22. Define the term *pick-up current* for a relay.

23. Define the term *drop-out current* for a relay.

24. Explain why electromagnetic relays have hysteresis.

25. Is the hysteresis of a relay necessarily a bad thing? Explain.

26. Draw a three-phase wye connection. Repeat for a delta connection.

27. Which load connection, wye or delta, delivers the entire line voltage to each phase of the load?

28. Which load connection, wye or delta, delivers more electrical power to a resistive load? By what factor is the power greater?

29. Name some of the most important industrial uses of power thyristors. Which is the most important of all?

Questions 30–49 refer to split-phase ac motors.

30. How far out of phase should the two stator winding currents be, ideally?

31. In Fig. 10-13(a), how many rotations does the magnetic field make for one ac cycle? If the ac line frequency is 60 Hz, what is the rotational speed of the magnetic field, expressed in rpm?

32. Describe the construction of a squirrel-cage rotor. What materials are used to construct a modern squirrel-cage rotor?

33. Why does no current flow down the length of the core of a squirrel-cage rotor?

34. In a squirrel-cage rotor, is it true that neighboring rotor bars carry current in opposite directions? Explain.

35. Draw an end-on view of a squirrel-cage rotor like that in Figs. 10-15(b) and (c). Draw the stator magnetic field as coming out of the southwest (lower left) and pointing toward the northeast (upper right). Assuming the rotation is counterclockwise, show the direction of the current in every rotor bar. Show the direction of the force exerted on every bar. Will the rotor spin counterclockwise?

36. Repeat Question 35, but this time assume that the magnetic field rotation is *clockwise*.

37. What is the most common method of creating a phase shift between the two stator winding currents?

38. Is it ever possible for a split-phase ac motor to run on just a single stator winding?

39. Is it ever possible for a split-phase ac motor to start on just a single stator winding?

40. What is the difference in construction between a normal split-phase motor and an ac servo motor?

41. Distinguish between the control winding and the fixed winding in an ac servo motor.

42. In an ac servo motor, if the current in the control winding leads the current in the fixed winding, how could the direction of rotation be reversed?

43. What is it about the torque-speed characteristic of a normal split-phase motor that makes it unacceptable for use in a servo system?

44. Name the two effects which interact to produce the rising torque-speed behavior of the normal split-phase motor in Fig. 10-20. Which effect is accentuated to produce the drooping torque-speed behavior of the servo motor? How is this effect accentuated by the motor manufacturer?

45. List some of the reasons ac servo motors are preferred to dc servo motors.

46. Under what circumstances are dc servo motors used instead of ac servo motors?

47. What is *damping* in a servo system? What good does it do? Why is damping better than just running the servo motor at slow speed?

48. What is the function of the servo amplifier in a servo system?

49. List some of the desirable characteristics of an ac servo amplifier and explain why each one is desirable.

Questions 50–59 refer to servo amplifier 1, drawn in Fig. 10-22.

50. The input stage does not provide any voltage gain whatsoever, so what good is it?

51. Explain why resistors R_8 and R_9 are placed in the emitter lead of Q_2. What good do they accomplish?

52. What is the purpose of bypass capacitor C_3? What good does it accomplish?

53. What is the purpose of C_5 in the collector circuit of Q_3?

54. What do we call the coupling method used between Q_3 and the power output stage?

55. What is the purpose of resistors R_{14} and R_{15}?

56. About how much base current flows in Q_4 and Q_5 when the error signal is zero?

57. Describe the current flow paths in the push-pull power output stage when an error voltage is present.

58. What happens to the base current in the power transistors when the error voltage becomes nonzero?

59. Discuss the advantage of a push-pull amplifier stage over a conventional ac amplifier stage.

Questions 60–67 refer to servo amplifier 2, drawn in Fig. 10-23.

60. Explain the purpose of chopping a dc signal into an ac signal prior to amplification.

61. What do we call the circuit arrangement in which the Q_1 emitter resistor is part of the Q_2 collector resistance? What benefits are provided by this circuit arrangement?

62. What do we call the circuit arrangement of R_8 and C_2? What benefit is provided by this arrangement?

63. What do we call the circuit arrangement in which the base of Q_3 is connected through a resistance to the emitter of Q_4? What benefit does this provide?

64. Explain in intuitive terms why the gain of the servo amplifier is reduced as the gain adjust pot wiper is moved to the left.

65. Describe the flow path of the dc base bias current for power transistors Q_5 and Q_6.

66. Does the motor control winding current have to pass through the germanium diode in the push-pull stage? Discuss your answer carefully. Does the germanium diode have to be a power rectifier, or can it be a small-signal diode?

67. Explain why the power transistors tend to run cooler when unfiltered dc voltage is used to drive the push-pull stage.

Questions 68–70 refer to servo amplifier 3, drawn in Fig. 10-25.

68. What is the voltage gain of the op amp noninverting amplifier when the gain adjust pot is adjusted to its center position?

69. Why is there no provision made to correct the output offset of the op amp? Why isn't it necessary?

70. Do you think the C_1, C_2, and R_3 frequency response components would be sized to provide a low value of upper cutoff frequency or a high value of upper cutoff frequency? Obtain the frequency response curves of a 709 op amp and see what these particular component sizes would do.

71. In Fig. 10-27(a), what is it that determines whether SCR1 drives the armature or SCR2 drives the armature?

72. If SCR1 in Fig. 10-27(a) is driving the armature and correcting the position of the controlled object but the object overshoots its desired position, which SCR will cause it to back up to the proper position?

73. Figure 10-27(c) shows V_{out} to be exactly in phase with V_e. Is this absolutely necessary, or could V_{out} be out of phase with V_e?

11

Input Transducers—

Measuring Devices

All of industrial control depends on the ability to accurately and swiftly measure the value of the controlled variable. By and large, it has been found that the best way to measure the value of a controlled variable is to convert it into an electrical signal of some sort and to detect the electrical signal with an electrical measuring device. This approach is superior to converting the value of the controlled variable into a mechanical signal because electrical signals have certain advantages over mechanical signals:

a. Electrical signals can be transmitted from place to place much more easily than mechanical signals. (All you need is a pair of wires.)

b. Electrical signals are easier to amplify and filter than mechanical signals.

c. Electrical signals are easy to manipulate to find out such things as the rate of change of the variable, the time integral of the variable, whether the variable has exceeded some limit, etc.

Devices which convert the value of a controlled variable into an electrical signal are called *electrical transducers*. The number of different electrical transducers is very great. Electrical transducers have been invented to measure virtually every physical variable, no matter how obscure. Industrially, the most important physical variables that are encountered are position, speed, force, pressure, flow rate, temperature, light intensity, and humidity. Accordingly, in this chapter, we will concentrate our attention on electrical transducers which measure these particular variables.

OBJECTIVES

After completing this chapter you will be able to:

1. Explain the meaning of the terms linearity and resolution as applied to potentiometers
2. Explain the operation of a linear variable differential transformer (LVDT)
3. Describe the construction and operation of a Bourdon tube, and list the most popular shapes of Bourdon tubes
4. Describe the construction and operation of bellows used for pressure measurement
5. Explain the construction and operation of a thermocouple, and how it is compensated against variations in cold-junction temperature; state the temperature ranges in which thermocouples are applied
6. Describe the operation of a resistive temperature detector (an RTD), and state the temperature ranges in which RTDs are applied
7. Describe the operation of thermistors, and state the temperature ranges in which they are applied
8. Describe the behavior of photovoltaic cells and photoconductive cells, and state the relative advantages and disadvantages of these two devices
9. Describe how photocell detectors work; describe a chopped photocell detector and discuss its advantages
10. Describe the use of photocells for measuring a material's translucence, for automatic bridge balancing, and for dc signal chopping
11. Explain the operation of LEDs, and distinguish between visible LEDs and infrared LEDs
12. Explain the operation of a phototransistor, and discuss its advantages over a photoconductive cell
13. Describe the operation of optical coupler/isolators, and list some of their industrial uses
14. Describe the operation of strain gages, and show how they are stabilized against temperature variations
15. List the five main types of industrial tachometers; describe the operation of each type and state its relative advantages and disadvantages
16. Describe the operation of resistive hygrometers and psychrometers for measuring relative humidity, and take readings correctly from a psychrometer table

11-1 POTENTIOMETERS

The *potentiometer* is the most common electrical transducer. Potentiometers can be used alone, or they can be attached to a mechanical sensor to convert a mechanical motion into an electrical variation. A potentiometer is quite simple in conception. It consists of a resistive element and a movable contact that can be positioned anywhere along the length of the element. This movable contact is called by various names, including *tap*, *wiper*, and *slider*. We will use all three terms interchangeably.

Figure 11-1 shows two schematic representations of a potentiometer. In Fig. 11-1(a), the resistive element is drawn in circular form; this representation suggests the physical construction of most potentiometers, in which the resistive element really is circular and spans an angle of about 300°. The wiper position is then adjusted by turning the shaft to which the wiper is attached. The shaft can be turned by hand or with a screwdriver, depending on whether it has a knob attached or a screwdriver slot in its end.

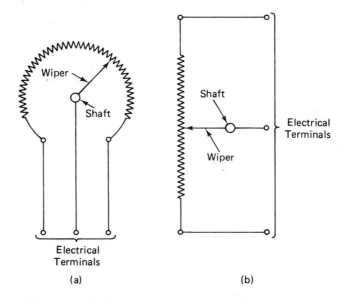

Figure 11-1. Schematic symbols of potentiometers. (a) Circular symbol, which suggests the pot's physical appearance. (b) Straight-line symbol.

The more popular electrical schematic representation is shown in Fig. 11-1(b). This is more popular only because it is easier to draw.

Potentiometer Linearity. The great majority of potentiometers are *linear*. The term linear means that a given mechanical movement of the wiper produces a given change in resistance, no matter where the wiper happens to be on the element. In other words, the resistance of the element is evenly distributed along the length of the element. The precise degree of linearity of a potentiometer is very important in some applications. Manufacturers therefore specify a *percent linearity* on the potentiometers they make. The meaning of percent linearity, or just linearity for short, can be understood by referring to Fig. 11-2.

Figure 11-2(a) shows a graph of resistance versus shaft angle for a perfectly

Figure 11-2. Graphs of resistance vs. shaft angle for a potentiometer. (a) Perfectly linear potentiometer. (b) Real potentiometer, with the resistance deviating from a perfect straight line. (c) Stepwise or noncontinuous variation in resistance.

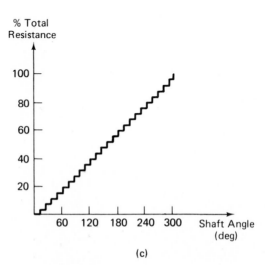

linear potentiometer. The resistance plotted on the vertical axis can be thought of as the resistance between the wiper terminal and one of the end terminals in Fig. 11-1. The shaft angle plotted on the horizontal axis is the angle through which the shaft has been rotated, with 0° being the position in which the wiper is in direct contact with the end terminal. As can be seen, a perfectly linear potentiometer yields a given amount of resistance change for a given number of degrees of shaft rotation, no matter where the shaft happens to be. That is, a shaft movement from 0° to 60° yields a resistance change of exactly 20% of total resistance; likewise, a shaft movement from 180° to 240°, a 60° rotation, produces a resistance change from 60% to 80% of the total resistance, also an exact 20% change.

It is, of course, impossible to manufacture potentiometers that have perfect linearity. The true state of affairs is shown in Fig. 11-2(b), in which the resistance deviates from the ideal straight line. The point of worst deviation from the ideal straight line determines the *percent linearity* of the potentiometer. For example, in the graph of Fig. 11-2(b), at the worst point, the actual resistance deviates from the ideal straight line by 10%. This means that the actual resistance differs from the expected resistance by an amount which is 10% of the total resistance. The linearity of this potentiometer is therefore 10%.

When a manufacturer specifies a 10% linearity for his potentiometer, he is guaranteeing that the resistance will deviate from the straight-line resistance by *no more than* 10% of total resistance. Thus, a 500-Ω potentiometer with a 10% linearity would have a graph of resistance versus shaft angle in which the actual resistance strayed from an ideal straight line by no more than 50 Ω.

While 10% linearity might be adequate for many industrial potentiometer applications, it almost certainly would not be adequate for a *measuring* application. Usually, potentiometers used as transducers have linearities of less than 1% and sometimes as low as 0.1%. For a 500-Ω pot with a 0.1% linearity, the actual resistance would deviate from the expected straight-line resistance by no more than 0.5 Ω.

Potentiometer Resolution. Many potentiometers are of the *wire-wound* variety. In a wire-wound pot, a thin piece of wire is wound many times around an insulating core. The wiper then moves from one turn of wire to the next as the pot is adjusted. The result is that the wiper resistance does not vary in a perfectly smooth fashion, but varies in steps. This phenomenon is shown, greatly exaggerated, in Fig. 11-2(c).

The important point here is that there is a limit on the minimum resistance change possible. The smallest possible resistance change is equal to the resistance of one turn of wire (for a wiper that does not bridge across turns). For example, a 500-Ω wire-wound pot having 200 turns would have a resistance per turn of 500 Ω/ 200 = 2.5 Ω. The smallest possible pot adjustment would move the wiper from one turn to the neighboring turn, so the smallest possible resistance change would be 2.5 Ω. This smallest possible change in resistance determines the potentiometer *resolution*.

Resolution of a potentiometer can be considered to be the minimum possible resistance variation, expressed as a percentage of the total resistance.* For the pot described in the above paragraph, the resolution would be 2.5 Ω/500 Ω = 0.5%.

As a general rule, pots which have inherently good resolution have inherently bad linearity and vice versa. Of course, by taking great pains in the manufacturing process, it is possible to make potentiometers which have both good resolution and good linearity. Potentiometers used for measuring purposes are usually of this type. They have good resolution, good linearity, and good environmental characteristics (temperature and humidity don't affect them). They are somewhat expensive, costing as much as 20 times what a simple control pot would cost.

Very often, a potentiometer is installed in a circuit with a voltage applied between its end terminals, as illustrated in Fig. 11-3(a). The rotation of the shaft then creates a *voltage* variation between the terminals, instead of just a *resistance* variation between terminals. If the shaft position represents the value of some measured variable, the potentiometer establishes a correspondence between the measured variable and V_{out}.

Another common potentiometer hookup is shown in Fig. 11-3(b). Resistors R_1 and R_2 are equal, and the measuring apparatus is arranged so that the pot wiper is exactly centered for some neutral or reference value of measured variable. This is a bridge circuit. If the bridge is driven by a dc voltage source, the magnitude of V_{out} corresponds to the *amount* by which the measured variable differs from its reference value, and the polarity of V_{out} corresponds to the *direction* of the difference, greater than or less than the reference value. If the bridge is driven by an ac source, the magnitude of V_{out} corresponds to the amount of deviation from the reference value, and the *phase* of V_{out} corresponds to the direction of the deviation. If the measured value is greater than the reference value, the pot wiper moves *up* in Fig. 11-3(b). Then V_{out} will be *in phase* with the ac source. If the measured value is less than the reference value, the pot wiper moves *down* in Fig. 11-3(b), and V_{out} will be 180° *out of phase* with the ac source.

Another common arrangement of a potentiometer in a bridge circuit is shown in Fig. 11-3(c). Recall that the basic idea of bridge circuits is that the bridge will be balanced when the ratio of resistances on the left equals the ratio of resistances on the right. In other words, $V_{out} = 0$ if

$$\frac{R_1}{R_2} = \frac{R_3}{R_4}.$$

A bridge of this sort can be used in either of two ways:

a. The measured variable can be used to position the pot shaft, and then the output voltage (V_{out}) from the bridge represents the value of the measured variable.

b. The measured variable can be used to cause one of the resistors, say R_4, to vary.

---

*This definition is a little weak because the minimum possible variation at one shaft angle might be different from that at another. This occurs because not all the wire turns have exactly the same resistance. For a formal and precise definition of resolution, consult a book devoted entirely to transducers.

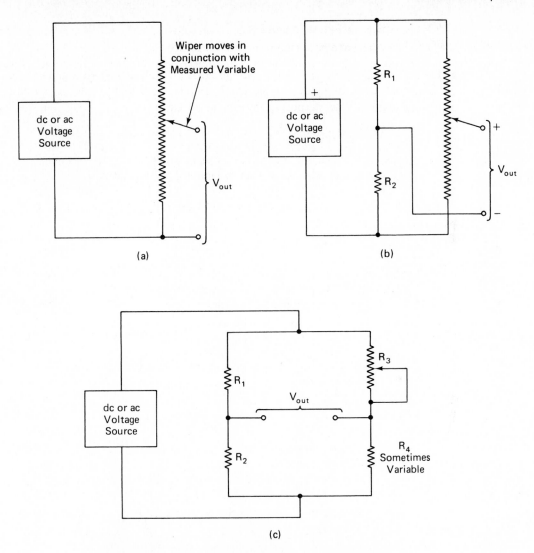

Figure 11-3. Potentiometers used in measuring circuits. (a) Potentiometer as a simple voltage divider. (b) Bridge circuit, with the pot comprising one side of the bridge. (c) Bridge circuit with the pot comprising one leg of the bridge.

R_4 may be a potentiometer itself, or it may be a resistor that varies in response to some stimulus, such as temperature. R_3 is then adjusted either manually or automatically until V_{out} equals zero, meaning that the bridge is balanced. The position of the R_3 pot shaft then represents the value of the variable. The shaft can be attached to some indicating device to read out the value of the measured variable.

11-2 LINEAR VARIABLE DIFFERENTIAL TRANSFORMERS (LVDTs)

A *linear variable differential transformer* gives an ac output voltage signal which is proportional to a physical displacement. Figure 11-4 shows the construction, schematic symbol, and output waveforms of an LVDT.

Figure 11-4. (a) The physical construction of an LVDT. (b) Schematic drawing of an LVDT. (c) When the LVDT core is perfectly centered, V_{out} equals zero. (d) When the core moves up, V_{out} is in phase with V_{in}. (e) When the core moves down, V_{out} is out of phase with V_{in}.

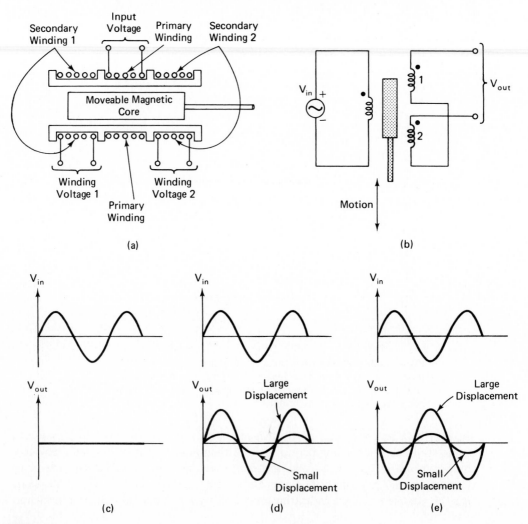

(a)

(b)

(c)

(d)

(e)

Figure 11-4(a) shows that an LVDT has one primary winding and two secondary windings all wrapped on the same form. The form itself is hollow and contains a magnetic core which is free to slide inside the form. As long as the magnetic core is perfectly centered in the form, the magnetic field linkage will be the same for secondary winding 1 and secondary winding 2. Therefore both secondary winding voltages will be equal. If the core moves to the left in Fig. 11-4(a), the magnetic linkage will be greater to secondary winding 1 because more of the core is inside that winding than secondary winding 2. Therefore winding voltage 1 will be greater than winding voltage 2. On the other hand, if the core moves to the right in Fig. 11-4(a), winding voltage 2 will be greater than winding voltage 1, because secondary winding 2 will have more of the core inside it. The LVDT is built so that the *difference* between the two secondary winding voltages is proportional to core displacement.

When the LVDT is in use as a measuring device, the secondary windings are connected together in series opposition, as indicated in Fig. 11-4(b). Therefore, if the core is centered and winding voltage 1 equals winding voltage 2, the net output voltage (V_{out}) is zero. This is shown in Fig. 11-4(c). If the core moves up in Fig. 11-4(b), winding voltage 1 becomes larger than winding voltage 2, so V_{out} becomes nonzero. The farther the core moves, the greater V_{out} becomes. This is shown in Fig. 11-4(d). Also, V_{out} is *in phase* with V_{in} because of the way the output voltage phase is defined in Fig. 11-4(b).

If the core moves down below its center position in Fig. 11-4(b), winding voltage 2 becomes larger than winding voltage 1, and V_{out} again becomes nonzero. This time V_{out} is 180°; *out of phase* with V_{in}, as shown in Fig. 11-4(e). Thus the size of V_{out} represents the amount of displacement from center, and the phase of V_{out} represents the direction of the displacement.

Most LVDTs have a displacement range of about plus or minus 1 in. That is, the core can move up 1 in. from center or down 1 in. from center. If the LVDT is to be used to measure mechanical displacements much greater than 1 in., an appropriate mechanical ratioing apparatus (gearing) must be used.

As far as voltage sizes are concerned, most LVDTs are designed to operate on an input voltage of less than 10 V ac. Full-scale output voltages fall in the same general range. That is, full-scale output voltages can range from about 0.5 V ac to about 10 V ac for different LVDT models.

11-3 PRESSURE TRANSDUCERS

The different approaches to industrial pressure measurement are numerous. We will concentrate our attention on just two common classes of pressure-sensing devices, *Bourdon tubes* and *bellows*. These devices detect the measured pressure and convert it into a mechanical movement. The mechanical movement is then transduced into an electrical signal by either a pot or an LVDT.

11-3-1 Bourdon tubes

A *Bourdon tube* is a deformed metal tube with an oval cross section. It is open at one end and sealed at the other end. The whole tube is elastic because of the elasticity of the metal used in its construction. The fluid whose pressure is being measured is admitted to the inside of the tube at the open end, which is mechanically anchored. The tube then deflects by an amount proportional to the magnitude of the pressure. This deflection is mechanically transmitted to the wiper of a potentiometer or to the core of an LVDT to provide an electrical signal. Figures 11-5(a) through (d) show the various shapes of Bourdon tubes and the motions which they produce.

Figure 11-5(e) shows how a C-shaped Bourdon tube could be linked to a potentiometer. Figure 11-5(f) shows how a C-shaped tube could be linked to an LVDT. Spiral and helical Bourdon tubes are often preferable to C-shaped Bourdon tubes because they produce greater motion of the sealed tip per amount of pressure.

Bourdon tubes are most often used to measure pressures in the range from 10 to 300 psi.

11-3-2 Bellows

A *bellows* is essentially a series of metal diaphragms connected together. When subjected to fluid pressure, a metal diaphragm will distort slightly because of the elasticity of the material used to construct it. When several diaphragms are soldered together in series, the total movement of the end diaphragm can be considerable. Figure 11-6(a) shows a cutaway view of a bellows. With the pressure inlet port anchored, the bellows will expand as the fluid pressure rises, and the output boss will move to the right. As the fluid pressure falls, the bellows contracts, and the output boss moves to the left. The contraction force can be provided by the springiness of the bellows diaphragms themselves or by a combination of diaphragm springiness with an external spring.

Figures 11-6(b) and (c) show two common bellows arrangements. In Fig. 11-6(b), the pressure is applied to the inside of the bellows and tends to expand the bellows against the pull of the tension spring. As the bellows expands, it actuates a mechanical linkage which moves the wiper of a potentiometer to furnish an electrical output signal.

In Fig. 11-6(c), the measured pressure is applied to the outside of a bellows, forcing it to contract against the push of a compression spring. As it moves, it actuates a mechanical linkage which moves the core of an LVDT to furnish an electrical output signal. These pressure transducers are calibrated by adjusting the initial tension or compression of the return spring. An adjustment nut, not shown in Fig. 11-6, is provided for this purpose.

Bellows-type pressure transducers find their main usefulness in measuring pressures in the range of 0.5 to 20 psi.

Figure 11-5. Bourdon tubes. (a) C-shaped Bourdon tube, the most common type. (b) Helix-shaped Bourdon tube. (c) Spiral Bourdon tube. (d) Twisted Bourdon tube. (e) C-shaped Bourdon tube linked to a potentiometer. (f) C-shaped Bourdon tube linked to an LVDT.

(a)

(b)

(c)

Figure 11-6. (a) The basic construction of a bellows. (b) Bellows arrangement in which the input pressure is applied to the inside of the bellows. (c) Bellows arrangement in which the input pressure is applied to the outside of the bellows.

11-4 THERMOCOUPLES

By far the most common device for measuring industrial process temperatures is the *thermocouple*. A thermocouple is a pair of dissimilar metal wires joined together in a complete loop, as shown in Fig. 11-7(a). The dissimilar wires have two junction points, one at each end of the loop. One junction, called the hot junction, is subjected to a high temperature, and the other junction, the cold junction, is subjected to a low temperature. When this is done, a small net voltage is created in the loop; this voltage is proportional to the *difference* between the two junction temperatures.

What happens in a thermocouple loop is that a small voltage is produced at each junction of the dissimilar metals, due to an obscure phenomenon called the Seebeck effect. The higher the temperature at the junction, the greater the voltage produced by that junction. Furthermore, the relationship between voltage and temperature is approximately linear; that is, a given increase in temperature produces a given increase in voltage. The proportionality constant between voltage and temperature depends on which two metals are being used. Since a complete loop always has two junctions, two voltages are produced. These voltages oppose each other in the loop, as Fig. 11-7(b) shows. The net voltage available to drive current through the resistance of the loop is the difference between the two individual junction voltages, which depends on the difference between the two junction temperatures.

To measure the temperature difference, it is only necessary to break the loop open at a convenient point (at some cool location) and insert a voltmeter. The voltmeter must be rather sensitive since the voltage produced by a thermocouple loop is in the millivolt range. The voltage reading can then be converted into a temperature measurement by referring to standard tables or graphs which relate these two variables. Graphs of voltage versus temperature difference for several popular industrial thermocouples are given in Fig. 11-8. In each case the first metal or metal alloy mentioned in the thermocouple is the positive lead, and the second metal or metal alloy is the negative lead.

To avoid the problem of identifying thermocouples by proprietary registered trade names, a letter code for thermocouple types has been adopted. Thus, type *J* thermocouples have the response shown in Fig. 11-8 no matter what particular name is used to identify the metal alloy. The same is true for type *K* and type *R* thermocouples and for other types not graphed in Fig. 11-8.

When a voltmeter is inserted into the thermocouple loop, it is usually most convenient to insert it as shown in Fig. 11-7(c). In that figure, metal *A* and metal *B* don't actually touch each other at the cold junction. Instead, both metals are placed in contact with standard copper conducting wires. The connections are normally made on a terminal strip. The copper wires then lead off and connect to the sensitive voltmeter. It may seem that this would disrupt the total net voltage generated by the thermocouple loop, but that does not happen. The net loop voltage remains the same because there are now *two* cold junctions, one between metal *A* and copper

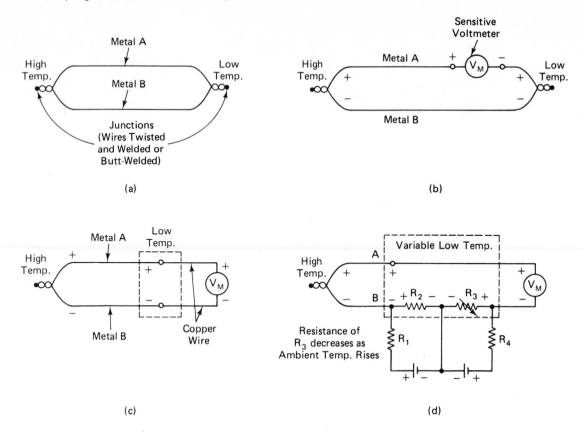

Figure 11-7. (a) Basic thermocouple. (b) Thermocouple with a voltmeter inserted in the loop. (c) Thermocouple loop with no cold junction between metal *A* and metal *B*. (d) Thermocouple loop which is compensated against variations in the cold-junction temperature.

and the other between metal *B* and copper. The sum of the two junction voltages produced by these cold junctions equals the voltage that would have been produced by the single cold junction of metal *A* with metal *B*. Of course, the two cold junctions must be maintained at the same temperature that the single junction would have felt. This is no problem, since the copper wires and terminals are always inside some enclosure which is thermally insulated from the process being measured, and which is subjected to the same temperature to which a single junction would have been subjected, namely the ambient temperature in the industrial location. Therefore the circuit in Fig. 11-7(c) will yield the same reading as the circuit in Fig. 11-7(b).

One further matter is important in the use of thermocouples in industry. It regards the variation in ambient temperature at cold junctions. Here is the situation: If we *knew beforehand* the temperature of the cold junctions, then instead of

Figure 11-8. Voltage vs. temperature curves for types *E, J, K,* and *R* thermocouples. The words *chromel, constantan,* and *alumel* are proprietary trade names of individual manufacturers of thermocouple wire.

relating the voltmeter reading to the temperature *difference*, we could relate it to the hot-junction temperature itself. This would be possible because we could construct the tables of temperature versus voltage to reflect the fact that the cold junctions are at a certain known *reference temperature,* as it is called.

As an example, consider the type *J* thermocouple in Fig. 11-8. The graph shows that at a temperature of 400°F, the thermocouple loop voltage is 12 mV. If we knew that the cold junction was always at 75°F, say, then we could conclude that a 12-mV loop voltage represented a hot-junction temperature of 475°F (475°F − 75°F = 400°F). As long as the cold junction was constantly maintained at the reference temperature of 75°F, we could go right through the thermocouple table and add 75°F to every temperature difference reading. The resulting temperature value would then represent the temperature at the hot junction.

As a matter of fact, this is exactly what is done in industrial thermocouple tables. The figure of 75°F has been chosen because it represents a fairly reasonable guess at the average ambient temperature in an industrial setting. (In thermocouple tables for *laboratory use,* the reference temperature is usually considered to be 32°F, the freezing point of water.)

For the above approach to work accurately, the cold junction must be constantly maintained at the 75°F reference temperature. This is usually impractical unless the temperature-measuring instrument happens to be located in an air-conditioned control room. In all probability, though, the measuring instrument is located right out with the industrial equipment and machinery. The ambient temperature may easily vary from about 50°F in the winter to about 100°F in the summer; even wider seasonal swings in ambient temperature are common. Because of this variation in cold-junction temperature, industrial thermocouple loops must be *compensated.*

A simple automatic compensation method is illustrated in Fig. 11-7(d). The two dc voltage supplies and the four resistors are arranged so that the voltages across R_2 and R_3 are in opposition. The voltage polarities across R_1 and R_4 do not matter, since R_1 and R_4 are outside the thermocouple loop. R_3 is a temperature-sensitive resistor, having a negative temperature coefficient. This means that its resistance goes down as its temperature goes up. The circuit is designed so that at 75°F the small voltage across R_3 equals the small voltage across R_2. The voltages across the two resistors exactly cancel each other, and the voltmeter reading is unaffected. Now if the cold-junction temperature should rise above 75°F, the voltmeter reading would tend to decrease because of a smaller difference between the hot and cold junctions. This would tend to yield a measured temperature reading which is *lower* than the actual temperature at the hot junction. However, the resistance of R_3 decreases as the cold-junction temperature rises, resulting in a smaller voltage across its terminals. The R_3 voltage no longer equals the R_2 voltage. Therefore the R_2-R_3 combination introduces a *net* voltage into the loop which tends to increase the voltmeter reading. A careful check of the polarities of the voltages in Fig. 11-7(d) will prove that this is so. Because of the design of the compensating circuit, the net voltage introduced by the R_2-R_3 combination exactly cancels the decrease in loop voltage introduced by the temperature rise at the cold junction.

If the cold-junction temperature should fall below 75°F, the R_2-R_3 combination introduces a net voltage in the opposite direction. This offsets the increase in loop voltage introduced by the larger temperature difference between hot and cold junctions. Verify this to yourself by carefully checking the polarities in Fig. 11-7(d).

Many industrial temperature-measuring/recording instruments use an automatically balancing bridge to indicate temperatures. The thermocouple loop voltage is balanced by moving the wiper of a potentiometer in a Wheatstone bridge circuit. The potentiometer shaft is ganged to another shaft-which operates the temperature-indicating needle. Therefore for every value of thermocouple loop voltage there is a corresponding position of the temperature-indicating needle. A temperature scale is then marked off behind the needle.

11-5 THERMISTORS AND RESISTIVE TEMPERATURE DETECTORS (RTDs)

Besides using thermocouple *voltage* to electrically measure temperature, it is also possible to utilize the *resistance change* which occurs in many materials as their temperature changes. Materials used for this purpose fall into two classes, pure metals and metallic oxides.

Pure metals have a fairly constant *positive temperature coefficient of resistance.*

The temperature coefficient of resistance, usually just called the *temperature coefficient*, is the ratio of change in resistance to change in temperature. A positive coefficient means that the resistance gets larger as the temperature increases. If the coefficient is constant, it means that the proportionality factor between resistance and temperature is constant and that resistance and temperature will graph as a straight line. Resistance versus temperature graphs for several popular metals are given in Fig. 11-9(a). The *resistance factor* in this graph means the factor by which the actual resistance is greater than the reference resistance at 0°F. For example, a factor of 2 indicates that the resistance is twice as great as it was at 0°F. When a pure metal wire is used for temperature measurement, it is referred to as a *resistive temperature detector*, or RTD.

Figure 11-9. (a) Resistance vs. temperature curves for pure metal RTDs. (b) Resistance vs. temperature curves for typical thermistors.

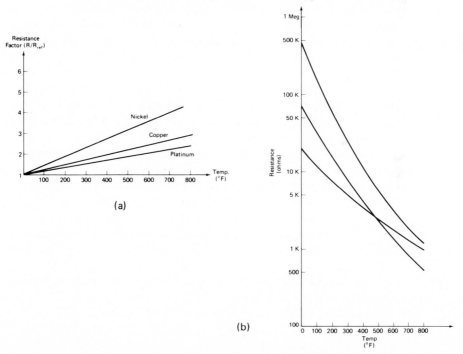

When metallic *oxides* are used for temperature measurement, the metallic oxide material is formed into shapes which resemble small bulbs or small capacitors. The formed device is then called a *thermistor*. Thermistors have large negative temperature coefficients which are nonconstant. In other words, the resistance change per unit of temperature change is much greater than for a pure metal, but the change is in the other direction—the resistance gets smaller as the temperature increases. The fact that the coefficient is nonconstant means that the change in resistance per unit temperature change is different for different temperatures. Figure 11-9(b) shows graphs of resistance versus temperature for three typical industrial thermistors. Note that the vertical scale is logarithmic to allow the great range of resistances to be shown. The temperature-sensitive resistor which compensated the thermocouple in Sec. 11-4 would be a thermistor.

Figure 11-10 shows three circuits for utilizing thermistors and/or RTDs. In schematic drawings, temperature-sensitive resistors are symbolized by a circled resistor with an arrow drawn through, and the letter *T* outside the circle. A resistor with a positive temperature coefficient can be indicated by the arrow pointing toward the top of the circle, and a resistor with a negative temperature coefficient can be symbolized by an arrow pointing toward the bottom of the circle. These are not universally accepted rules, but we will use them in this book.

In Fig. 11-10(a), the temperature transducer is shown in series with an ammeter and a stable voltage supply. As the temperature rises, the resistance goes down, and the current increases. If the specific characteristics of the thermistor are known, it is possible to relate the current measurement to the actual temperature. The supply voltage must not change or the current-to-temperature correspondence will become invalid.

In Fig. 11-10(b), the temperature transducer increases its resistance as the temperature rises. This causes a larger portion of the stable supply voltage to appear across its terminals. Thus the voltmeter reading can be related to temperature. If desired, the voltmeter scale can be marked in temperature units instead of volts for a direct temperature readout.

In Fig. 11-10(c), a bridge circuit is used. In precision, bridge circuit measurements are inherently superior to other measurements because the meter which detects bridge imbalance can be made very sensitive. Therefore even a slight imbalance in the bridge can be detected and adjusted out. The bridge detecting meter can be made very sensitive because when the bridge is *close* to being balanced, the voltage across the bridge is near zero; since the detecting meter does not have to measure a large voltage, it can be made to respond vigorously to a small voltage. In other words, it can be made very sensitive. By contrast, the meters in Figs. 11-10(a) and (b) cannot be made very sensitive because they must be able to read (relatively) large values of voltage or current.

The bridge circuit works like this: As the temperature of the thermistor rises, its resistance drops. This changes the ratio of resistances on the right-hand side and throws the bridge out of balance (assuming it was balanced to start with). Either manually or automatically, R_3 is adjusted until the ratio of resistances on the right-hand side is once again equal to the ratio on the left-hand side, bringing the bridge back into balance. The position of the R_3 pot shaft then represents temperature, since for every possible value of R_4 resistance there is only one value of R_3 resistance which will balance the bridge. The R_3 shaft is mechanically linked to another shaft which positions the temperature pointer.

When using the balanced bridge measurement method, the temperature readout

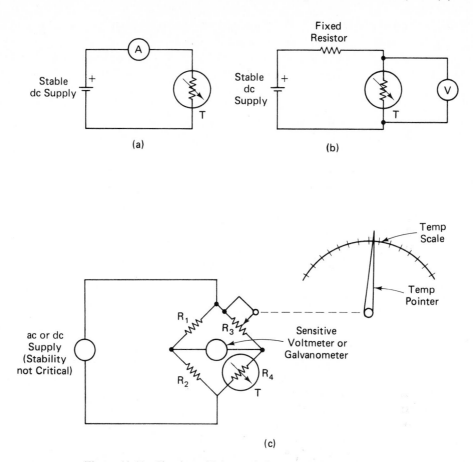

Figure 11-10. Circuits utilizing resistive temperature transducers. (a) The ammeter reading corresponds to the measured temperature. (b) The voltmeter reading corresponds to the measured temperature. (c) A bridge arrangement. When the bridge is balanced, the position of the pot wiper corresponds to the measured temperature.

scale will be linear if the transducer is linear. A linear readout scale means that equal *distances* on the scale represent equal *differences* in temperature, or, said another way, that the scale temperature marks are all equally spaced. Since we have seen that a thermistor is very nonlinear, we would expect that the temperature scale in Fig. 11-10(c) would also be nonlinear.* The extreme nonlinearity of thermistors makes them poorly suited for measuring temperatures over wide ranges. However, for measuring temperatures with narrow bands, they are very well

---

*For an example of a nonlinear scale, look at the ohms scale on a VOM. The marks representing a given ohms difference are far apart on the right but close together on the left.

suited, because they give such a great response for a small temperature change. This great response is also what recommends thermistors in applications like that described in Fig. 11-7(d), compensating a thermocouple loop over a fairly narrow band of cold-junction temperatures. The hearty response of the thermistor makes it easy to generate sufficient compensation.

The natural nonlinearity of thermistors can be partially corrected by connecting several matched thermistors together in a series-parallel combination. The resulting circuit is called a *thermistor composite network*. These networks are quite linear over a fairly wide temperature range (about 200°F), but they are naturally more expensive than plain thermistors.

As a general rule, thermistors are preferable when the expected temperature band is narrow, and RTDs are preferable when the expected temperature band is wide. Most thermistors are manufactured for use somewhere between −150°F and +800°F, although special thermistors have been developed for use at extremely low temperatures, near absolute zero. RTD thermometers are available for use at temperatures from −400°F to +2000°F.

Besides their uses as measurers of temperature in an external medium, thermistors also have applications which make use of the *internal* heat generated as they carry current. In any external temperature-measuring application, it is important to eliminate the effect of the thermistor's internally generated heat; this is done by making the thermistor current very small. In some applications, however, the ability of a thermistor to change its own resistance as it generates I^2R heat energy can be very useful. For example, thermistor self-heating can be used for establishing time delays, protecting delicate components from surge currents, detecting the presence or absence of a thermally conductive material, etc.

11-6 PHOTOCELLS AND PHOTOELECTRIC DEVICES

Photocells are small devices which produce an electrical variation in response to a change in light intensity. Photocells can be classified as either *photovoltaic* or *photoconductive*.

A photovoltaic cell is an energy source whose output voltage varies in relation to the light intensity at its surface. A photoconductive cell is a passive device, not capable of producing energy; its resistance varies in relation to the light intensity at its surface.

Industrially, applications of photocells fall into two general categories:

a. Sensing the presence of an opaque object:

1. The sensing may be done on an all-or-nothing basis, in which the photocell circuit has only two output states, representing either the presence of an object or

the absence of an object. This is the kind of sensing used to count pieces traveling down a conveyor or to prevent a mechanism from operating if an operator's hands are not safely out of the way.

2. The sensing may be done on a continuous basis, with the photocell circuit having a continuously variable output, representing the variable position of the object. This is the kind of sensing used to "watch" the edge of a moving strip of material to prevent it from straying too far from its proper position.

b. Sensing the degree of translucence (ability to *pass* light) or the degree of luminescence (ability to *generate* light) of a fluid or solid. In these applications the process has always been arranged so that the translucence or luminescence represents some important process variable. Some examples of variables which could be measured this way are density, temperature, and concentration of some specific chemical compound (carbon monoxide, carbon dioxide, water, etc.).

The outstanding advantage of photocells over other sensing devices is that no physical contact is necessary with the object being sensed.

11-6-1 Photovoltaic cells

The symbols often used for photovoltaic cells are shown in Fig. 11-11(a). The two wavy arrows pointing toward the circled battery suggest that external light energy produces the battery action. Since wavy arrows are inconvenient to draw, the Greek letter λ is often used to suggest light activation.

The open-circuit output voltage versus light intensity is graphed in Fig. 11-11(b) for a typical photovoltaic cell. Notice that the graph is logarithmic on the light intensity axis. This graph indicates that the cell is more sensitive at low light levels, since a small *change* in intensity (say from 1 to 10 fc) can produce the same boost in output voltage as a larger change in intensity (say from 100 to 1000 fc) at a higher light intensity level.

The output current characteristics of a photovoltaic cell operating into a load are graphed in Fig. 11-11(c) for various load resistances; as can be seen, a single photovoltaic cell cannot deliver very much current. The output currents are measured in microamps in this example. Photocells can be stacked in parallel, however, for increased current capability.

An example of a photovoltaic cell furnishing all-or-nothing-type information to a logic circuit is illustrated in Fig. 11-12. In Fig. 11-12(a), the light from the light source is gathered and focused at the photovoltaic cell, which is mounted some distance away. Distances 10 ft or more are not uncommon in industrial situations. When the photovoltaic cell is activated by the light, it picks up sensitive relay R,

(a)

(b)

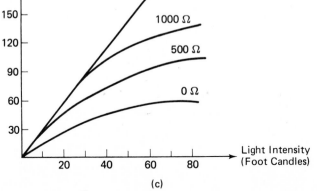

(c)

Figure 11-11. (a) Schematic symbols used for photovoltaic cells. (b) Graph of voltage vs. illumination for a typical photovoltaic cell. (c) Graphs of current vs. illumination for several different load resistances.

whose contact passes the input signal to the logic circuit. If an object blocks the light path, the photocell deenergizes the relay, and the logic circuit receives no input.

The object blocking the light path could be anything. It could be a moving object whose passage was to be counted by an electronic or mechanical counter;

it could be a moving object whose passage alerts machinery farther down the line to prepare for its arrival; it could be a workpiece or machine member which is supposed to get out of the way before the logic circuitry will allow some other motion to occur.

Figure 11-12. (a) Photovoltaic cell energizing a relay directly. (b) Photovoltaic cell energizing a relay through a transistor switch. (c) All photocell components contained in a single enclosure.

If the photovoltaic cell has trouble picking up the relay directly, it can operate through a transistor amplifier, as shown in Fig. 11-12(b). It is a good idea to do this anyway, since photovoltaic cells are subject to *fatigue* when they deliver near their full current for any length of time. The output voltage and current decline when a photovoltaic cell suffers from fatigue.

Sometimes the light source, focusing device, photocell, amplifier, and relay are all included in the same package, as shown in Fig. 11-12(c). Light leaves the package, passes some distance through space, is reflected by a reflecting surface, and reenters through the same aperture. It is then reflected off the one-way mirror and strikes the photocell. The amplifier, relay, and contacts are all contained in the package, so the final output is the switching of the relay contacts to indicate whether or not an object has blocked the light path.

Often the problem arises that the signal light cannot be distinguished from ambient light. The photoelectric system may then be unreliable, because the photovoltaic cell can deliver an output due just to the ambient light. The system then would indicate that there was no object present blocking the path when in fact there *was* an object present. There is a solution to this problem. Instead of just passing the light directly out through the focusing apparatus, the light beam is "chopped." That is, the beam is periodically interrupted at some specific frequency by a moving object inside the package between the light source and the outlet.

The best way to do this is to install a rotating disc between the light source and the one-way mirror in Fig. 11-12(c). Part of the disc is translucent and part of it is opaque, so the light beam is alternately passed and blocked at some constant frequency, usually several hundred hertz. Let us assume for purposes of illustration that the light beam is chopped at a frequency of 400 Hz.

It is now quite easy to distinguish between ambient light and true signal light simply by *tuning the amplifier to 400 Hz*. That is, design the amplifier so that it will not amplify dc signals at all and will give very little amplification to other frequencies which might sneak in the light aperture (such as 60- and 120-Hz light pulsations from mercury lamps). The amplifier will then respond only to voltage signals from the photovoltaic cell at the frequency of 400 Hz. The only way such an unusual light pulsation frequency could get to the cell is from the true light signal. All extraneous light signals are ignored.

Figure 11-13 shows an application of photovoltaic cells to measure the translucence of a liquid being passed through a sample cell. Suppose that the translucence is known to be a sure indication of the concentration of some impurity in the liquid. The semitransparent mirror passes half of the light from the source to the liquid, and the other half of the light is reflected to PCI. Only part of the light sent to the liquid can pass through it and strike PC2. Therefore the voltages generated by PC1 and PC2 will be different, with the PC1 voltage greater.

Photovoltaic cells 1 and 2 are connected in a bridge as shown in Fig. 11-13(b). The bridge is manually or automatically balanced by adjusting R_2. The final posi-

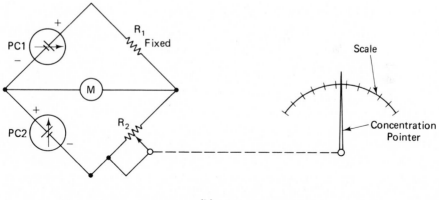

Figure 11-13. Photocell bridge circuit for measuring a liquid's translucence. (a) Arrangement of the light source, the photocells, the semitransparent mirror, etc. (b) Electrical schematic of the bridge circuit.

tion of the R_2 wiper will depend on the voltage difference between PC1 and PC2, which in turn depends on the concentration of the impurity. Thus, once the bridge has been balanced, every value of R_2 corresponds to some certain value of impurity concentration. The R_2 shaft is mechanically linked to a pointer shaft, which has a scale of concentrations marked beneath it for direct readout.

This measurement arrangement has some stabilizing features that deserve comment. First, both photocells are excited by the same light source. This eliminates the possibility of an error due to one light source changing in intensity more than another. In Fig. 11-13, if the light source changes in intensity due to aging of the bulb or supply voltage variations, both photocells are affected equally. These equal changes are cancelled out by the action of the bridge.

Second, photovoltaic cells are somewhat temperature sensitive. That is, their output voltage depends slightly on their temperature. However, if PC1 and PC2 are physically close together, they experience the same temperature changes, so any temperature errors are also cancelled out by the bridge.

11-6-2 Photoconductive cells

Photoconductive cells change resistance in response to a change in light intensity (the formal term is *illumination*), as mentioned earlier. As the illumination goes up, resistance goes down. The schematic symbols often used for photoconductive cells are shown in Fig. 11-14(a). A graph of resistance versus illumination for a typical photoconductive cell is given in Fig. 11-14(b). Note that both scales are logarithmic to cover the wide ranges of resistance and illumination that are possible.

The chief virtue of the modern photoconductive cell is its sensitivity. As Fig. 11-14(b) illustrates, the cell's resistance can change from over 1 million Ω to less

Figure 11-14. (a) Schematic symbols for a photoconductive cell. (b) Curve of resistance vs. illumination for a typical photoconductive cell.

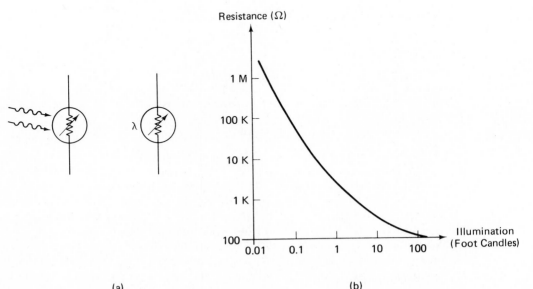

(a)

(b)

than 1000 Ω as the light intensity changes from darkness (illumination is less than 0.01 fc) to average room brightness (10 to 100 fc).

Photoconductive cells can be used for many of the same purposes that photovoltaic cells are used, except, of course, that they cannot act as energy sources. Photoconductive cells are preferred to photovoltaic cells when very sensitive response to changing light conditions is needed.

When fast response is necessary, photovoltaic cells are preferable to photoconductive cells. Likewise, if a photocell is to be rapidly switched on and off, as suggested in Sec. 11-6-1, photovoltaic cells are preferred because they can be switched at higher frequencies than photoconductive cells. As a rough rule of thumb, photoconductive cells cannot be successfully switched at frequencies higher than 1 kHz, whereas photovoltaic cells can be switched successfully at frequencies up to about 100 kHz and sometimes higher.

Photochoppers. One interesting application of photoconductive cells is in chopping a dc voltage signal for insertion into an ac amplifier. Dc signal chopping was mentioned in Sec. 10-11-2 in conjunction with servo amplifiers. The photoconductive cell is a good alternative to the vibrating mechanical switch chopping method. This is illustrated in Fig. 11-15.

In Fig. 11-15(a) the square-wave driving voltage is applied to two neon bulb-rectifier diode combinations. When V_{drive} is positive, rectifier diode A is forward biased, and rectifier diode B is reverse biased. Therefore neon bulb A is turned on, and neon bulb B is turned off. A neon bulb is capable of quickly turning on and off as voltage is applied and removed. A normal incandescent bulb cannot turn on and off quickly since it depends on the heating of its filament to emit light.

When V_{drive} goes negative, rectifier diode B is forward biased, and rectifier diode A is reverse biased. Therefore neon bulb B turns on, and neon bulb A turns off. Resistor R is inserted to limit the current flow through the neon bulbs.

Photoconductive cells PCA and PCB are exposed to neon bulbs A and B, respectively, inside light-tight enclosures. In such an enclosure, no external light can get in to affect the photocells. The photocells are specially chosen to have a large resistance change between light and dark conditions. In this case, let us assume the resistance changes from about 10 megohms to below 1 kilohm. The ratio of resistances is therefore about 10,000 to 1 (10 megohms/1 kilohm = 10,000). The cells are said to have a *light-to-dark ratio* of 10,000.

Photoconductive cell A is placed in *series* with the amplifier input lead, and photoconductive cell B is placed in *shunt* across the amplifier, as shown in Fig. 11-15(b). Thus when neon bulb A is turned on, the amplifier sees a low resistance in series with its input and a very high resistance in shunt. These low and high resistances can be thought of as closed and open switches, as shown in Fig. 11-15(c). Therefore at the instant shown in Fig. 11-15(c), the dc input voltage gets through to the coupling capacitor practically unattenuated (if the amplifier input impedance is much higher than 1 K).

(b)

Figure 11-15. Photoconductive cells used to chop a dc signal. (a) Circuitry for alternately switching the photocells from light to dark. (b) The photocells wired into the amplifier input. (c) Equivalent circuit when PC*A* is light and PC*B* is dark. (d) Equivalent circuit when PC*B* is light and PC*A* is dark.

When V_{drive} goes negative, neon bulb B turns on, and the situation at the amplifier is as shown in Fig. 11-15(d). The amplifier sees an open switch in series and a closed switch in shunt. The dc input signal cannot get through to the coupling capacitor at this instant. The dc input voltage is thus being chopped just as it was in Sec. 10-11-2. This chopping method has the advantage of electronic reliability (no moving parts), and it would be less expensive than a vibrating mechanical switch.

Photocells for Automatic Bridge Balancing. Figure 11-16 shows another popular use for photoconductive cells. The bridge circuit in Fig. 11-16(a) balances the measurement potentiometer against the valve position potentiometer to bring about proportional control. The galvanometer and photocell arrangement presented in this figure is a cheap and reliable method for accomplishing automatic balancing of the bridge. Here is how it works.

The galvanometer is a zero-center meter. That is, if there is no current flowing through it, the needle returns to the center of the scale. If current flows through it from left to right, the needle moves to the right of center; if current flows through it from right to left, the needle moves to the left of center. Attached to the needle is a lightweight vane which is opaque. Two photoconductive cells are mounted a slight distance from the vane on one side, and two light sources are mounted a slight distance from the vane on the other side. Refer to the expanded drawing in Fig. 11-16(a). If the galvanometer needle is centered, the vane covers both photocells, making both of their resistances high. If the needle moves off center, either photocell 1 or photocell 2 will be uncovered, depending on the direction of needle movement. When a photocell is uncovered, its resistance drops drastically because of the light striking its surface. The lowered resistance turns ON one of the transistor switches in Fig. 11-16(b), picking up one of the relays. The relay contacts then drive the valve motor either open or closed, moving the valve position pot until the bridge is back in balance. When the bridge is rebalanced, the galvanometer current drops to zero, and the needle returns to the center of the scale. Both photocells again go dark, turning OFF whichever transistor was turned ON. The relay which was picked drops out, and the valve freezes in that particular position.

11-6-3 Optical coupling and isolation: Phototransistors, light-emitting diodes

Figure 11-17 shows two ways of building an *optical isolator*. Figure 11-17(a) shows a standard incandescent light source and a photoconductive cell to accomplish the isolation, and Fig. 11-17(b) shows a *light-emitting diode* (LED) and *phototransistor* to accomplish the isolation. We will look at some industrial uses of optical isolators and then explain the operation of both of these designs. The design in Fig. 11-17(b) has certain advantages over that in Fig. 11-17(a), and these advantages will be pointed out.

An optical isolator is basically an interface between two circuits which operate at (usually) different voltage levels. The most common industrial use of the optical isolator is as a signal converter between high-voltage pilot devices (limit switches, etc.) and low-voltage solid-state logic circuits. Optical isolators can be used in any situation where a signal must be passed between two circuits which are electrically isolated from each other. Recall from Chapters 2 and 3 that complete electrical isolation between circuits (meaning that the circuits have no conductors in common) is often necessary to prevent noise generated in one circuit from being passed to the other circuit. This is especially necessary for the coupling between high-voltage

(a)

(b)

Figure 11-16. Photoconductive cells used to automatically balance a bridge. (a) Bridge circuit, with an opaque vane attached to the galvanometer needle. (b) Photocell circuitry. If both photocells are dark, neither relay picks, and the valve motor does not move. If either photocell goes light, the proper relay will pick up to apply power to the valve motor and bring the bridge back into balance.

(a)

(b)

Figure 11-17. (a) An optical isolator using an incandescent bulb and a photoconductive cell. (b) Another optical isolator using an LED and a phototransistor.

information-gathering circuits and low-voltage digital logic circuits. The information circuits are almost always badly exposed to noise sources, and the logic circuits cannot tolerate noise signals.

The optical coupling method eliminates the need for a relay-controlled contact or an isolating transformer, which are the usual methods of providing electrical isolation between circuits. Refer to Sec. 2-7-3 and Fig. 2-14 for a review of these methods.

The Incandescent Bulb-Photoconductive Cell Optical Isolator/Coupler. The optical isolator in Fig. 11-17(a) has an incandescent bulb wired in series with a protective resistor. This series combination is connected through a pilot device to a 115-V signal. If the pilot device is open, there will be no power applied to the incandescent bulb, so it will be extinguished. The photoconductive cell, being shielded from outside light, will go to a very high resistance, allowing the voltage at the transistor base to rise. The transistor switch turns ON, pulling V_{out} down to ground voltage, a logical 0.

If the pilot device closes, power will be applied to the bulb, causing it to glow. The resistance of the photoconductive cell will decrease, pulling the base voltage down below 0.6 V. The transistor turns OFF and allows the collector to rise to $+V_{cc}$, a logical 1. Therefore if a 115-V input signal is present, the circuit yields a logical HI. If there is no 115-V signal present, the circuit yields a logical LO.

The optical coupling method is superior in many applications, because it gets rid of some of the less desirable features of relays and transformers. Relays and transformers have certain shortcomings as couplers and isolators, namely:

a. They are fairly expensive.

b. They are bulkier and heavier than optical devices.

c. They create magnetic fields and switching transients which may be a source of troublesome electrical noise.

d. Relay contacts may create sparks, which are very undesirable in certain industrial situations.

The optical coupler works equally well on either ac or dc high-voltage signals. For this reason, signal converters using optical coupling are sometimes referred to as *universal* signal converters.

The LED-Phototransistor Optical Isolator/Coupler. Figure 11-17(b) shows an optical isolator/coupler using a light-emitting diode and phototransistor instead of an incandescent bulb and photoconductive cell. A light-emitting diode, usually called an LED, is a semiconductor diode which emits light when it carries current in the forward direction. The forward breakdown voltage of an LED is higher than 0.6 V since LEDs are not made from silicon as rectifier diodes are. They usually have forward breakdown voltages in the range of 1.0 to 2.2 V. Also, LEDs have reverse breakdown voltages which are much lower than those of silicon rectifier diodes. Figure 11-18(a) shows the current-voltage characteristic of a typical LED. Figure 11-18(b) shows the relationship between light power output and forward current for an LED.

A visible LED is not very bright compared to a #44, 6-V bulb, for example. Some LEDs do not even emit a visible light, but emit infrared light invisible to the human eye. Of course, such LEDs must then be used with photodetectors which are sensitive to infrared radiation. In commercially built optical couplers this is standard practice, since no human has to see the light anyway. Also, infrared

Figure 11-18. (a) Current vs. voltage characteristic curve of a typical LED. (b) Curve of light output vs. forward current for a typical LED.

LEDs are more efficient than visible LEDs, converting more of their electrical energy into light* and less into heat.

A phototransistor is a semiconductor transistor which responds to the intensity of light on its lens, instead of to base current. Some phototransistors can respond to both incident light and base current. The phototransistor in Fig. 11-17(b) doesn't have a base lead, so it responds to light only. The wavy arrows pointing toward the base location symbolize that the transistor is a phototransistor.

Figure 11-19 shows the characteristic curves of a typical phototransistor. Notice that the family of curves represents different values of light power density (the formal term is *irradiance*), not different values of base current. Phototransistors do not have a response as linear as that of a junction transistor. Notice the inconsistent spacing of the curves, indicating a nonlinear relationship between collector current and light intensity.

The LED-phototransistor combination in Fig. 11-17(b) has some important advantages over the bulb-photoconductive cell combination in Fig. 11-17(a):

a. An LED has an extremely long life compared to a bulb of any kind. An LED will emit light forever if operated at the proper current; an incandescent bulb does well if it lasts 10,000 hr.

---

*Strictly speaking, the term *light* refers to electromagnetic radiation which is visible to the human eye. Infrared radiation, which is at too low a frequency to be visible, is not really light. However, popular usage has blurred this distinction, and we hear the phrases "infrared light" and "ultraviolet light." We adopt the less rigorous usage of the term *light*, and we will refer to infrared radiation as light.

Figure 11-19. Characteristic curves of a typical phototransistor.

b. An LED can withstand the mechanical vibrations and shocks in industrial environments much better than a filament lamp, thus providing greater reliability.
c. The LED and phototransistor have faster response speed than a lamp and photoconductive cell. This may be an advantage for certain high-frequency switching applications.

Of course, there is no reason an optical coupler/isolator could not combine an LED with a standard photoconductive cell, and this is sometimes done. Usually, though, an LED light source is combined with a phototransistor light detector, because of the better match between their operating speeds and between their light emission and detection wavelengths.

11-7 STRAIN GAGES

Strain gages are used in industry to accurately measure large forces, especially large weights. There are also strain gages designed to measure small forces, but they are not as common. A strain gage is basically a resistance wire which is firmly cemented to a surface of a strong object which then receives a force. When force is applied to the object, it distorts very slightly. That is, the object either stretches slightly or compresses slightly, depending on whether it feels a tension or a compression force. The resistance wire, being cemented to the surface of the object, also distorts slightly. The distortion of the wire changes its resistance, which is detected and related to the value of the force.

What a strain gage really measures is *strain*, which is the change in the length of the strong object as a percent of its original length. The strain of the strong object represents the force applied to the object through Hooke's law, which says that

$$\frac{F}{A} = Y\epsilon = Y\left(\frac{\Delta L}{L_o}\right), \tag{11-1}$$

where F stands for the force applied to the object (in the direction of distortion); A is the cross-sectional area of the object; Y is Young's modulus, which depends on the particular material of which the object is made; and ϵ stands for strain, the change in length per unit of original length ($\epsilon = \Delta L/L_o$). The important point is that the change in the length of the object depends on the force applied to the object and can be related to that force on a one-to-one basis.

The resistance of the wire which comprises the strain gage depends on the length and cross section of the wire, as shown by

$$R = \frac{\rho L}{A}. \tag{11-2}$$

In Eq. (11-2), R stands for the resistance of the wire in ohms; ρ is the *resistivity* of the wire material, a property of the particular material used; L is the length of the wire; and A is the cross-sectional area of the wire. It can be seen that if the wire is stretched slightly, the resistance R will increase because the length will increase and the area A will decrease. On the other hand, if the wire is compressed slightly, R will decrease because the length L will decrease and the cross-sectional area will increase. Therefore, the resistance of the wire depends on the change in length of the wire and can be related to that change in length on a one-to-one basis.

To sum up, the wire resistance depends on the length and cross-sectional area of the wire, and the wire length depends on the length of the strong object, since they are cemented together. The length of the object depends on the applied force, so the end result is that the resistance of the wire depends on the applied force. By precisely measuring the resistance change, we can measure the force.

Figure 11-20(a) shows a top view of a strain gage, looking at right angles to the mounting surface. The resistance wire is usually a copper-nickel alloy with a diameter of about one-thousandth of an inch (0.001 in.). The wire is placed in a zigzag pattern on a very thin paper backing, called the *base*. The wire is zigzagged in order to increase the effective length which comes under the influence of the strain. The entire zigzag pattern is called the *grid*. Copper lead wires are attached at the ends of the grid.

Figure 11-20(b) shows a strain gage mounted on the surface which is to undergo the strain. The base is placed flat on the surface [the surface may be curved, as in Fig. 11-20(b)], and the entire strain gage is completely covered with special bonding cement. The bonding cement establishes an intimate contact between the wire grid and the strain surface of the strong object. Because of this intimate contact, and because the wire has practically no strength of its own to resist elongation or compression, it elongates or compresses exactly the same distance as the strong object. The strain of the wire grid is therefore exactly the same as the strain of the strong object.

Figure 11-20. (a) Physical appearance of a strain gage. (b) Strain gage bonded to a cylindrical object. (c) How the strain gage would be wired into a bridge circuit. (d) Dummy gage bonded to the cylindrical object along with an active gage. (e) How the dummy and active gages would be wired into the bridge to provide temperature stability.

The percent change in resistance for a given percent change in length is called the *gage factor* of the strain gage. In formula form,

$$GF \text{ (gage factor)} = \frac{\Delta R/R}{\Delta L/L} = \frac{\Delta R/R}{\epsilon}. \tag{11-3}$$

Most industrial strain gages have a gage factor of about 2. This means that if the length of the object changes by 1% ($\epsilon = 0.01$), the resistance of the strain gage changes by 2%.

A strain gage wired into a bridge circuit is shown in Fig. 11-20(c). The bridge is usually designed to be in balance when the force exerted on the strong object equals zero. A trimmer pot may be added to one of the legs of the bridge to adjust for exact balance at zero force. As a force is applied, the bridge becomes unbalanced, and the voltage across the bridge can be related to the amount of force. A greater force creates a greater change in gage resistance and a larger voltage output from the bridge.

To compensate for temperature effects, a second strain gage, identical to the first one, can be mounted at right angles to the application line of the force. The force has no effect on this strain gage, since the gage is not aligned with the force. The gage is wired into the bridge as shown in Fig. 11-20(e) and is labeled as a *dummy gage*; the force-sensing gage is then called the *active gage*. The purpose of the dummy gage is to cancel out any temperature-related resistance change in the active gage. With both gages experiencing the same temperature, any resistance change in the active gage due to temperature variations appears in the dummy gage too. Since the error appears on both sides of the bridge, it is cancelled out.

For precise weight measurement, a carefully shaped and machined object, containing several strain gages, is used. The gages are strategically placed on the machined surfaces at various angles to yield the utmost temperature stability. The gages themselves are designed to provide a linear relationship between bridge output voltage and force (weight) on the machined object. The object, in combination with its strain gages, is then called a *load cell*. Accurate weighing scales measuring large weights almost always have load cells as transducers.

11-8 TACHOMETERS

A *tachometer* is a device which measures the angular speed of a rotating shaft. The most common units for expressing angular speed are revolutions per minute (rpm) and radians per second. A radian is equal to $\frac{1}{2}\pi$ revolutions, or approximately 57 mechanical degrees. We will use the units of rpm exclusively.

Tachometers in industry use one of two basic measuring methods:

a. The angular speed is represented by the *magnitude* of a generated voltage.
b. The angular speed is represented by the *frequency* of a generated voltage.

In the domain of magnitude tachometers there are two principal types. They are the *dc generator tachometer* and the *drag cup tachometer*.

In the domain of frequency tachometers, there are three principal types. They are the *rotating field ac tachometer*, the *toothed-rotor tachometer*, and the *photocell pickup tachometer*. These names are not universally accepted, but they describe the action of the various tachometers rather well, and we will adopt them in this book.

11-8-1 dc generator tachometers

The dc generator tachometer is a dc generator, pure and simple. The field is established either by a permanent magnet mounted on the stator or by a separately excited electromagnet on the stator. The output voltage is generated in a conventional dc armature winding with a commutator and brushes. The equation for true generated voltage in a dc generator is

$$V_G = kB(rpm),$$

where V_G represents the true generated voltage, k is some proportionality constant which depends on the construction details (rotor length, rotor diameter, etc.), B is the strength of the magnetic field, and *rpm* is the angular speed measured in revolutions per minute.

With the field strength held constant, the generated voltage is proportional to the angular speed of the shaft. It is therefore possible to connect the tachometer shaft to the shaft being measured, apply the generated voltage to a voltmeter, and calibrate the meter in terms of rpm. One nice feature of a dc generator tachometer is that the polarity of the generated voltage reverses if the direction of rotation reverses. Therefore this type of tachometer can indicate rotational direction as well as speed.

11-8-2 Drag cup tachometers

A drag cup tachometer has two sets of windings on its stator at right angles to each other, just like an ac servo motor. Refer to Fig. 10-14(a). The rotor is not a squirrel-cage rotor, however. It is a hollow copper cylinder, called a *cup*, with a laminated iron inner core, which does not contact the cup. The cup attaches to the tachometer input shaft and rotates at the measured speed.

One of the stator windings, called the exciting winding, is driven by a stable ac voltage source. The other stator winding is the output winding. The exciting winding sets up an alternating magnetic field which induces eddy currents in the copper cup. The eddy currents set up an *armature reaction field* at right angles to the field from the exciting winding. The right-angled field will then induce an ac voltage in the output winding whose magnitude depends on the speed of rotation of the cup. The result is an ac output voltage which varies linearly with shaft speed.

The output voltage frequency is equal to the exciting frequency (usually 60 Hz), and it is 90° out of phase with the exciting voltage. The direction of shaft rotation determines whether the output voltage leads or lags the exciting voltage. Therefore this tachometer also can indicate direction as well as speed of rotation.

All tachometers which rely on a voltage magnitude to represent speed are subject to errors caused by three things:

a. Signal loading
b. Temperature variation
c. Shaft vibration

Regarding problem **a**, the voltage delivered by any kind of generator will vary slightly as the current load on the output winding varies. This is because the *IR* voltage drop in an output winding varies as its current varies.

Regarding problem **b**, as temperature changes, the magnetic properties of the core change, causing variations in the magnetic field strength. As the magnetic field strength varies so does the generated voltage.

Regarding problem **c**, as the shaft vibrates, the precise spacing between the field and armature windings changes. This change in spacing causes variations in generated voltage.

Modern tachometer designs have minimized these errors and have produced tachometers in which the voltage-speed linearity is better than 0.5%. This is quite adequate for the majority of industrial applications.

11-8-3 Rotating field ac tachometers

The rotating field ac tachometer is a rotating field alternator, pure and simple. The field is usually created by permanent magnets mounted on the rotor. The rotor shaft is connected to the measured shaft, and the rotating magnetic field then induces an ac voltage in the stator output windings. The equation for the frequency of the generated voltage in an ac alternator is

$$f = \frac{P(rpm)}{120}, \tag{11-4}$$

where f is the frequency in hertz, P is the number of magnetic poles on the rotor, and *rpm* is the rotational speed. It can be seen that the output frequency is an exact measure of the angular speed of the shaft.

11-8-4 Toothed-rotor tachometers

The toothed-rotor tachometer is the most popular of the frequency tachometers. This tachometer has several ferromagnetic teeth on its rotor. On its stator it has a permanent magnet with a coil of wire wrapped around the magnet. This arrangement is illustrated in Fig. 11-21 for a rotor with six teeth.

As the rotor rotates, the teeth come into close proximity with the magnet and then pass by. When a tooth is close to the magnet, the magnetic circuit reluctance is low, so the field strength in the magnet's core increases. When no tooth is close, the magnetic circuit reluctance is high, so the field strength in the magnet's core decreases. Therefore one cycle of field strength is produced every time a tooth passes by. This variation in magnetic field strength induces a voltage in the coil wrapped on the permanent magnet. One voltage pulse is produced by each tooth. This is shown in Fig. 11-21(b).

The relation between pulse frequency and speed is given by

$$\text{rev/sec} = \text{pulses/sec} \div 6$$

(a)

(b)

Figure 11-21. (a) Arrangement of a toothed-rotor tachometer. (b) Output voltage waveform of a toothed-rotor tachometer.

because it takes six pulses to represent one revolution. The number of revolutions per minute equals the number of revolutions per second multiplied by 60, or

$$\text{rpm} = 60 \ (\text{rev/sec}) = 60 \ (\text{pulses/sec} \div 6),$$

$$\text{rpm} = 10f, \tag{11-5}$$

where f represents the pulse frequency; whatever readout method is used would have to reflect Eq. (11-5), namely that the rotational speed equals the measured frequency multiplied by the factor 10. For a different number of teeth on the rotor, the factor would be different.

11-8-5 Photocell pickup tachometers

A photocell pickup tachometer is basically the same device that was suggested in Sec. 11-6-1 for chopping a light beam. A rotating disc is placed between a light source and a photovoltaic cell. Part of the disc passes the light beam, and part of the disc blocks the light beam. Therefore the photovoltaic cell is constantly being turned on and off, at a frequency which depends on the angular speed of the disc. By connecting the disc shaft to the measured shaft, a voltage waveform will be generated by the photocell. The frequency of the waveform will then be a measure of the angular speed of the shaft.

For example, if the disc had four light areas and four dark areas, the speed would be given by

$$rpm = 15f,$$

where f is the frequency of the photocell output waveform. You should justify this equation to yourself. Use the same derivation approach that was used in Sec. 11-8-4.

11-8-6 Frequency tachometers versus magnitude tachometers

The main advantage of frequency-measuring tachometers is that they are not subject to errors due to output loading, temperature variation, and shaft vibration. Also, their linearity is perfect. However, frequency tachometers all have the disadvantage of awkwardness in reading out the speed. It takes a lot more effort to convert a frequency into a readable form than it does to convert a voltage magnitude into readable form. It is true that frequency measurements lend themselves to digital detection and readout, but digital measurement and readout are much more complex than a simple analog meter readout. The digital circuits must repeatedly go through the cycle of (a) count, (b) store, (c) display, (d) reset.

On the other hand, to a human being digital readout is more intelligible than analog readout, because the person taking the reading doesn't have to figure out the value of each meter marking. Thus, as far as the readout is concerned, frequency tachometers and magnitude tachometers both have their advantages and disadvantages.

Many times in industrial control the measured speed is to be used as a *feedback* signal. This is certainly the case in a closed-loop speed control system, where the measured speed is compared with the set point to find the speed error signal. Measured speed is also used as feedback in a servo control system, where the speed of approach is used to subtract from the position error signal to prevent overshoot. This process, called *error rate damping*, is common in servo systems. In cases like these, the speed signal must be expressed as an analog voltage instead of a digital number. Therefore in feedback systems, the magnitude tachometers have the advantage over frequency tachometers because the magnitude tachometers automatically provide an analog voltage signal.

Frequency tachometers *could* provide an analog voltage signal, but only by adding an extra signal-processing circuit (a D-to-A converter or a frequency demodulator). Overall, magnitude tachometers are preferred to frequency tachometers for feedback applications.

11-9 HUMIDITY TRANSDUCERS

There are many industrial operations which must be carried out under specific and controlled conditions of moisture content. In some cases the moisture contained in the ambient air is important; in other cases the moisture contained in the product itself is more important to the success of the industrial process. We will discuss two common methods of measuring the moisture content of ambient air and one method of measuring the moisture content of a sheet product. The most common scale for measuring the moisture content of air is the *relative humidity* (*RH*) scale. Formally, relative humidity is the ratio of the water vapor (moisture) actually present in the air to the maximum amount of water vapor that the air could possibly hold.

11-9-1 Resistive hygrometers

A resistive hygrometer is an element whose resistance changes with changes in the relative humidity of the air in contact with the element. Resistive hygrometers usually consist of two electrodes of metal foil on a plastic form. The electrodes do not touch each other, and they are electrically insulated from each other by the plastic form. A solution of lithium-chloride is then used to thoroughly coat the entire device. This construction is illustrated in Fig. 11-22(a).

Figure 11-22. (a) Physical appearance of a resistive hygrometer. (b) Curve of resistance vs. relative humidity for a resistive hygrometer.

As the relative humidity of the surrounding air increases, the lithium-chloride film absorbs more water vapor from the air. This causes its resistance to decrease markedly. Since the lithium-chloride film is in intimate contact with the two metal electrodes, the resistance between the electrode terminals also decreases markedly. The terminal resistance can then be related to the relative humidity. A typical characteristic curve of resistance versus relative humidity for a resistive hygrometer is shown in Fig. 11-22(b).

Resistive hygrometer transducers cannot be used over the full range of relative humidities, from 0% to 100%. Most of them have an upper safe operating limit of about 90% *RH*. Exposure to air with a relative humidity higher than 90% can result in excess water absorption by the lithium-chloride film. Once that happens, the resistance characteristics of the hygrometer are permanently altered.

11-9-2 Psychrometers

A psychrometer is a relative humidity-measuring device which has two temperature transducers (thermometers). One of the thermometers measures the temperature of an element which is simply located in the ambient air. This element is called the *dry bulb*. The second thermometer measures the temperature of an element which is surrounded by a fibrous material saturated with pure liquid water. This element is called the *wet bulb*. The ambient air is forced to flow over both the dry bulb and the wet bulb by a blower of some sort. This arrangement is illustrated in Fig. 11-23(a).

The temperature transducers shown in Fig. 11-23(a) are platinum wire RTDs. The dry bulb stays at the temperature of the flowing ambient air, so the dry bulb temperature is simply equal to the ambient air temperature, regardless of its humidity. The wet bulb, however, is colder than the dry bulb because of the evaporation of the liquid water contained in the fibrous material which surrounds the wet bulb. The greater the evaporation rate of the water, the greater the cooling effect on the wet bulb, and the lower the wet bulb temperature reading. The evaporation rate depends on the relative humidity of the moving air. If the air is dry (low relative humidity), the evaporation rate will be great, and the wet bulb will be *much* colder than the dry bulb. If the air is moist (high relative humidity), the evaporation rate will not be so great, and the wet bulb will be only a *little bit* colder than the dry bulb. The *difference* in temperatures is therefore an indication of the relative humidity of the air.

To understand why the evaporation rate out of the water-soaked fibrous material depends on relative humidity, think of it this way: If the ambient air were at 100% *RH*, it would not be able to absorb any more water at all, because it would already be saturated. Therefore the water would not evaporate from the fibrous material at all. It is easy to reason back from this extreme condition to understand that the drier the air is, the better it is at accepting extra water (causing evaporation).

Therefore, the lower the relative humidity, the faster the water evaporates from the fibrous material.

The percent relative humidity can be read from a psychrometer table by knowing two things:

a. The dry bulb temperature
b. The difference in temperatures between the two bulbs

(a)

Difference Between Dry Bulb and
Wet Bulb Temperatures (°F)

	1	5	10	15	20	25
40	92	60	—	—	—	—
50	93	68	38	12	—	—
60	94	73	49	26	6	—
70	95	77	55	37	20	3
80	96	79	61	44	29	16
90	96	81	65	50	36	24
100	96	83	68	54	42	31

Dry Bulb Temp. (°F)

Relative Humidities (%)

(b)

Figure 11-23. (a) The arrangement of the dry bulb and the wet bulb in a psychrometer. The example shows RTDs as the temperature detectors. (b) A psychrometer table, which relates the three variables of dry bulb temperature, temperature difference between bulbs, and percent relative humidity.

An abbreviated psychrometer table of this type is shown in Fig. 11-23(b). More precise psychrometer tables, marked in 0.5°F graduations, are available in psychrometer handbooks.

The psychrometer can be used to take manual readings of percent relative humidity, or it can be used in a control application to automatically maintain a certain desired humidity. We will explore an automatic humidity control system using a psychrometer in Chapter 12.

11-9-3 Detecting moisture conditions in a solid material

Hygrometers and psychrometers are devices which are capable of measuring the moisture content of air as expressed on a universal well-accepted measurement scale, the relative humidity scale. Given the proper calibration data, they can measure the relative humidity of *any* mixture of gases (not only air) on this well-known scale. Sometimes, though, it is not important to know moisture content as expressed on some universal scientific scale. Sometimes it is only important to know whether the moisture content is above or below a certain desired moisture content, or *reference* content. This is especially true in industrial situations where a continuous strip of material (paper, textiles, etc.) must be handled and processed. When moisture content of a solid strip of material in relation to some reference content is all that is needed, the usual method of taking the measurement is shown in Fig. 11-24(a). In this figure the strip of material is passed between two rollers which contact opposite sides of the strip. Each roll has a lead wire attached to its stationary bearing support. The resistance measured between the leads is then an indication of the moisture content of the material in question. Virtually all materials will have a low resistance when wet and a higher resistance when dry. The user deter-

Figure 11-24. (a) A method for detecting the moisture content of a sheet material. (b) The bridge circuit establishes a correspondence between the output voltage and the moisture content.

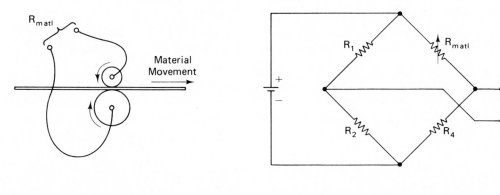

(a)

(b)

mines the resistance between the lead wires when the material has the optimum moisture content for his purposes. The system then detects any deviation from that resistance and attempts to correct that deviation.

A circuit for detecting moisture/resistance variations is the familiar Wheatstone bridge drawn in Fig. 11-24(b). Suppose the resistance of the material (R_{matl}) is 3250 Ω when the moisture content is just right. The bridge could then be built with $R_1 = R_2 = R_4 = 3250 \ \Omega$. Then if the material were at the right moisture level, R_{matl} would equal 3250 Ω, and the bridge would balance. V_{out} would equal 0. If the moisture content were to change because of some upset in the system, the bridge would go out of balance. The magnitude of V_{out} would indicate the amount of deviation between actual and desired conditions, and the polarity of V_{out} would indicate the direction of deviation. The V_{out} signal could then be applied to some sort of controller to restore the moisture content to the proper level.

In this type of control application, there is no measurement taken per se. That is, there is no measurement result obtained which could be expressed on a universal measurement scale. There is simply a comparison of an actual moisture condition with a desired moisture condition, and no attempt is made to express these moisture conditions on a numerical scale.

QUESTIONS AND PROBLEMS

1. What are the advantages of electrical signal transducers over mechanical signal transducers?

2. If a 10,000-Ω potentiometer has a linearity of 2%, what is the most the actual resistance can differ from the ideal resistance at any point?

3. Roughly speaking, what is considered an acceptable linearity for a potentiometer used as a measuring transducer?

4. If a 1000-Ω wire-wound potentiometer has 50 turns, what is its percent resolution (assuming the wiper contact cannot bridge across turns)?

5. In Fig. 11-3(c), if $R_1 = 5$ K, $R_2 = 12$ K, and $R_4 = 15$ K, what value of R_3 will cause the bridge to be balanced?

6. The discussion in Sec. 11-2 explained why the words *variable differential* are used in the name of the LVDT. Why do you suppose the word *linear* is used in its name?

7. About how far can the cores of LVDTs move?

8. Roughly, how much signal voltage can you expect from an LVDT?

9. What are Bourdon tubes made of?

10. Are Bourdon tubes used to measure liquid pressure, or gas pressure, or both?

11. What is the advantage of a spiral or helical Bourdon tube over a C-shaped Bourdon tube?

12. Are Bourdon tubes useful for measuring low pressures? Explain.

13. A Bourdon tube itself is a mechanical transducer. How are Bourdon tubes used to provide an electrical measurement signal?

14. Generally speaking, what range of pressures can a bellows transducer handle?

15. Is a pressure bellows itself a mechanical transducer or an electrical transducer? Explain.

16. What returns a bellows to its original position when its pressure is relieved?

17. Strictly speaking, does a thermocouple loop measure the hot-junction temperature, or does it measure the difference between hot- and cold-junction temperatures?

18. In industrial thermocouple tables, what cold-junction temperature is usually assumed?

19. What is the purpose of thermocouple compensating circuits?

20. If the cold junction of a thermocouple loop were located in a heated and air-conditioned control room, would a compensating circuit be necessary? Explain.

21. Of the common thermocouple types, which type is the most sensitive?

22. Of the common thermocouple types, which type is best suited to temperatures above 2000°F?

23. Of the common thermocouple types, which type do you suppose is the most expensive?

24. What is the difference between a positive temperature coefficient of resistance and a negative temperature coefficient of resistance?

25. Does an RTD have a positive or a negative temperature coefficient of resistance? Repeat the question for a thermistor.

26. Which device is more sensitive, a thermistor or an RTD?

27. Which device is more linear, a thermistor or an RTD?

28. Of the common RTD materials, which is the most sensitive?

29. Over what range of temperatures do RTDs find their greatest industrial use? Repeat for thermistors.

30. Draw the schematic symbol for a temperature-sensitive resistor which has a positive temperature coefficient. Repeat for a negative temperature coefficient.

31. What is the difference between a linear measurement scale and a nonlinear measurement scale?

32. What is a logarithmic scale? Why are they used in graphs?

33. When being used to measure the temperature of an external medium, is thermistor self-heating good or bad? Explain.

34. Explain how the self-heating of a thermistor could be used to detect whether a liquid was above or below a certain height in a tank.

35. What is the difference between a photovoltaic cell and a photoconductive cell?

36. What is the main operating advantage of photoconductive over photovoltaic cells?

37. What is the main operating advantage of photovoltaic cells over photoconductive cells?

38. Do photoconductive cells have a positive or negative illumination coefficient of resistance? Explain.

39. What does the term *light-to-dark ratio* mean?

40. Roughly speaking, about how much voltage can a typical photovoltaic cell deliver? Repeat the question for current.

41. What is photovoltaic cell fatigue?

42. Can photovoltaic cells be stacked in parallel? Can they be stacked in series?

43. Explain the operation of a *chopped* photodetector. What is its advantage over a plain photodetector?

44. Which chopping frequencies should not be used for chopped photodetectors? Why?

45. Are photocells at all sensitive to temperature?

46. What are the units of illumination in the English system? (There are two names for the same unit; see if you can track down the other name.)

47. What are the units of illumination in the metric system? What is the conversion factor between the English and metric units?

48. What illumination level is considered pitch dark?

49. What is the illumination level outdoors on a bright sunny day?

50. Why can't incandescent bulbs be used for the photochopper of Fig. 11-15?

51. Name some of the industrial uses of optical coupler/isolators.

52. What are the advantages of optical couplers over the couplers discussed in Chapter 2?

53. What are the advantages of LEDs over conventional light sources?

54. What is the difference between visible LEDs and infrared LEDs? State the relative advantages of each.

55. Write Hooke's law and explain what it means.

56. Upon what three things does the resistance of a piece of wire depend? State the relationship in formula form.

57. Combining your answers to Questions 55 and 56, explain how a strain gage works.

58. Define *gage factor* of a strain gage. About how great is the gage factor of an industrial strain gage?

59. Describe how a dummy gage can eliminate temperature-related errors in the use of strain gages.

60. What are the five basic types of industrial tachometers? Classify each type as either a magnitude tachometer or as a frequency tachometer.

61. Describe the principles of operation of each of the five types in Question 60.

62. What are the three chief sources of error in industrial tachometers?

63. Which class of tachometers has the greater linearity, magnitude or frequency?

64. What are the advantages of magnitude tachometers over frequency tachometers?

65. What is a resistive hygrometer? Explain its construction and principle of operation.

66. Are resistive hygrometers linear? Explain.

67. What is the relative humidity limit for a typical resistive hygrometer? What happens if that limit is exceeded?

68. Describe a wet bulb-dry bulb psychrometer's construction and operating principles.

69. Explain why a greater temperature difference between wet and dry bulb temperatures means a lower relative humidity.

70. If the dry bulb temperature is 60°F and the wet bulb temperature is 45°F, what is the % *RH*?

71. If the dry bulb temperature is 75°F and the wet bulb temperature is 67.5°F, estimate the % *RH*.

12

Nine Examples of Closed-Loop Industrial Systems

In this chapter, we will look at nine different industrial closed-loop control systems in detail. Among them, these nine systems contain nearly all of the final correcting devices and input transducers studied in Chapters 10 and 11. The control modes represented in these systems include On-Off, proportional, and proportional plus integral.

OBJECTIVES

After completing this chapter you will be able to:

1. Discuss and explain the process of controlling temperature in a quench oil tank used for quenching heat-treated metal parts
2. Discuss the operation of soaking pits used to heat steel ingots prior to hot-rolling, and discuss and explain a system for controlling the pressure in a soaking pit recuperator
3. Explain the operation of an all-solid-state proportional plus reset temperature controller with thermocouple input

4. Discuss and explain the process of maintaining a constant tension in a strip handling system

5. Discuss the process of recoiling moving strip, and explain how an edge-sensing system is used to ensure that the strip is wound up straight

6. Discuss the operation of an automatic powder weighing system with load cell input, and explain how a servomechanism is used to position an optical shaft encoder to read out the weight

7. Discuss the steel carburizing process, and explain the operation of a system which controls the carbon case depth by controlling the CO_2 content in the carburizing atmosphere

8. Discuss and explain the control of relative humidity in a textile moistening process

9. Discuss and explain the control of relative humidity in granaries and in explosives-storage warehouses

12-1 THERMISTOR CONTROL OF QUENCH OIL TEMPERATURE

Very often, heat-treated metal parts must be quenched in either oil or water in order to impart the proper metallurgical qualities to the metal. In most such processes, the parts are immersed in a bath of quench oil as soon as they leave the heat-treating chamber. Naturally, the temperature of the quench oil tends to rise due to the continual dunking of the hot metal parts. To accomplish the desired quenching results, the quench oil must be maintained within a certain temperature range; this is done by cooling the oil in a heat exchanger. The situation is illustrated schematically in Fig. 12-1(a). The hot parts slide down a chute into the quench tank, landing on a link belt which catches them on its spurs. The moving belt carries them horizontally through the quench oil and then up and out of the tank.

An oil outlet pipe allows oil to flow out of the tank and into the fixed displacement recirculating pump. A fixed displacement pump moves a fixed volume of liquid on each revolution, so the speed of rotation of the pump determines how much oil recirculates through the cooling system. The outlet of the pump feeds into a water-cooled heat exchanger. From the heat exchanger, the recirculating oil passes back into the quench tank.

The motor driving the recirculating pump is a series universal motor, capable of operating on either dc or ac. In this system, it is operated on ac voltage. The average voltage applied to the motor terminals determines its speed of rotation. Since the motor drives the pump, the rotational speed of the motor determines how much oil recirculates and thereby determines the amount of cooling that takes place. As the motor speeds up, more oil recirculates, and the oil in the tank tends to cool down. As the motor slows down, less oil recirculates, and the oil in the tank tends to warm up.

The quench oil temperature is sensed by a thermistor mounted inside a *probe*, which is a protective shield. A thermistor is an ideal temperature transducer for

Figure 12-1. Quench oil temperature controller. (a) Physical layout of the quench tank and the cooling apparatus. (b) Circuit for controlling the recirculating pump.

this application because it produces a large response for small temperature changes, and because it is suited to the fairly low temperatures encountered in quenching processes (usually less than 200°F). The thermistor is connected into the control circuitry as shown in Fig. 12-1(b). Here is how the control circuit works.

The bridge rectifier, in conjunction with the clipping circuit comprised of R_1 and ZD1, supplies an approximate square wave across the Q_1 circuitry. This square wave has a peak value of 20 V and is synchronized with the ac line pulsations, as we have seen before. At the instant the 20-V supply appears, the R_2-R_{TH} series combination divides it up. The voltage available for driving the base-emitter circuit of Q_1 depends on exactly how the R_2-R_{TH} voltage divider divides the 20 V. If the thermistor resistance R_{TH} is high, a small voltage will appear across R_2, and the base-emitter drive will be small. If the thermistor resistance is low, a larger voltage will appear across R_2 due to voltage divider action, and the base-emitter drive will be large.

The voltage available to drive the base-emitter circuit determines the Q_1 emitter current, according to

$$I_E = \frac{V_{R_2} - 0.6\text{V}}{5.6\text{ K}}, \tag{12-1}$$

where V_{R_2} stands for the voltage appearing across resistor R_2.

Equation (12-1) is just Ohm's law applied to the emitter resistor. It shows that an increase in V_{R_2} causes an increase in emitter current.

The Q_1 collector current is virtually the same as the emitter current. As the diagram shows, the Q_1 collector current charges capacitor C_1. When C_1 charges to the peak point of the UJT, the UJT fires. The resulting current pulse is delivered to the triac gate. The triac then turns ON and applies power to the motor terminals.

To summarize the behavior of this circuit, the greater the voltage across R_2, the greater the charging current to C_1. If the C_1 charging current is greater, the UJT will fire sooner in the half cycle, and the power delivered to the motor will be greater. This causes the motor and pump to spin faster.

If the voltage across R_2 is small, the Q_1 collector current will charge C_1 slowly. This causes late firing of the UJT and of the triac, and reduced motor speed.

Now let us see how the measured oil temperature affects the circuit action. An increase in oil temperature results in a lowering of the thermistor resistance R_{TH}, due to the thermistor's negative temperature coefficient. Lowering R_{TH} causes an increase in V_{R_2} by voltage divider action. As we have seen, an increase in V_{R_2} causes the pump to run faster. This recirculates more oil through the heat exchanger and tends to drive the tank temperature back down.

This particular circuit is designed to start recirculation when the tank oil temperature reaches 140°F. Below 140°F the motor does not run at all. Above 140°F the UJT and triac start firing, and the motor begins running. Therefore for a temperature of exactly 140°F, the triac should be on the verge of firing. This is equivalent to saying that the firing delay angle should be 180° when the temperature is 140°F. Then any slight increase in temperature beyond that point will reduce the firing delay angle to less than 180° and cause the motor and pump to start running.

The thermistor characteristic is such that at 140°F, $R_{TH} = 30$ K, so

$$\frac{V_{R_2}}{20\text{ V}} = \frac{R_2}{R_2 + R_{TH}} = \frac{8.2\text{ K}}{8.2\text{ K} + 30\text{ K}},$$

$$V_{R_2} = 4.3 \text{ V}.$$

The emitter current is given by Eq. (12-1):

$$I_E = \frac{4.3 \text{ V} - 0.6 \text{ V}}{3.3 \text{ K}} = 1.12 \text{ mA}.$$

Therefore I_C, the capacitor charging current, equals 1.12 mA also. Assuming that the UJT has a standoff ratio η equal to 0.64, the UJT peak voltage is given by

$$V_P = (0.64)(20 \text{ V}) + 0.6 \text{ V} = 13.4 \text{ V}.$$

Therefore the capacitor must charge to 13.4 V to fire the UJT and triac. The time required to do this can be found from

$$\frac{\Delta V}{\Delta t} = \frac{I}{C},$$

which expresses the voltage buildup rate for a capacitor. Rearranging, we obtain

$$\Delta t = \frac{C}{I}(\Delta V) = \frac{(0.68 \text{ } \mu\text{F})(13.4 \text{ V})}{1.12 \text{ mA}} = 8.14 \text{ msec}.$$

Therefore the UJT should fire at about 8.14 msec after the start of the cycle. This time can be expressed as an angle by saying that

$$\frac{\theta}{360°} = \frac{8.14 \text{ msec}}{16.67 \text{ msec}},$$

$$\theta = 176°,$$

where 16.67 msec is the period of the 60-Hz ac line. Thus the firing delay angle is calculated as 176° when the oil temperature is 140°F.

This means that the triac is just barely firing and is supplying a very small average voltage to the motor. Any further temperature rise from this point will cause the firing delay angle to be reduced, and consequently, the motor and pump to begin running faster. The pump is then able to hold the oil temperature very close to 140°F.

If it were desired for some reason to have a variable temperature set point, this could be easily accomplished. The R_2 resistor could be replaced by a potentiometer. Then as the pot resistance was increased, the temperature set point would be lowered. As the pot resistance was decreased, the temperature set point would be raised.

12-2 A PROPORTIONAL MODE PRESSURE CONTROL SYSTEM

12-2-1 Soaking pits for steel ingots

In the steel industry, a *soaking pit* is an underground pit used to heat steel ingots to about 2400°F prior to rolling. The ingots are placed in the pit by a crane; the pit cover is placed on top, also by a crane, and the gas burners are turned on to bring the pit temperature up to 2400°F. The combustion of natural gas with air creates waste gases, which leave the pit through an exhaust duct. Some of the heat energy contained in the hot waste gases is recovered and used to preheat the fresh combustion air coming into the burners. The preheating takes place in a heat exchanger called a *recuperator*. This process is illustrated schematically in Fig. 12-2(a).

The recuperator in Fig. 12-2 is just a large-diameter duct. It has hot waste gases entering on the left at a temperature of about 2400°F and leaving on the right at about 1800°F. The reduction in waste gas temperature represents the fact that some of the heat energy has been recovered and transferred to the cold combustion air, thus making the soaking process more energy-efficient. Incoming combustion air for the burners is drawn into the cold air intake by the combustion air blower, a large powerful fan. The incoming air passes through a duct to a butterfly valve which opens to the proper position to maintain the correct upstream air pressure ahead of the recuperator. The cold air makes two or three passes through the recuperator, picking up more heat energy on each pass. It finally emerges from the recuperator at a temperature of about 900°F. From there it travels to the burner air control valve, which passes the air to the burners as it is called for by the tem-

Figure 12-2. A soaking pit with a recuperator. (a) Physical layout. (b) Electronic circuit for positioning the control damper.

(a)

Figure 12-2. (Cont.)

(b)

457

perature controller. The temperature controller is not shown since we are concentrating on the pressure control system. When air arrives at the burners it is mixed with natural gas, the process fuel. Unused preheated air can be exhausted through a restriction in the supply header, as shown in Fig. 12-2(a). Sometimes part of the unused preheated air is routed back to the blower input to mix with the new air entering the cold air intake.

It is important to maintain the proper value of pressure in the cold air duct just ahead of the recuperator to prevent the recuperator tubes from overheating. The desired pressure may vary from one set of operating conditions to the next. Therefore, the pressure controller must have provision for adjusting the pressure set point. The controller opens or closes the air butterfly valve to correct any deviation of the measured pressure from set point. If the controller sees that the measured pressure is below set point, it opens the butterfly valve further to raise the air pressure upstream of the recuperator. If the measured pressure is above set-point pressure, the controller closes the butterfly valve further.

The input pressure transducer is a bellows-potentiometer transducer of the type shown in Fig. 11-6(b). The pressure signal for the transducer is taken from a pressure tap in the air duct upstream of the recuperator, as shown in Fig. 12-2(a). The transducer potentiometer has -15 V dc and ground voltage applied to its two end terminals, so the transducer output is a dc voltage which varies between 0 and -15 V. As the measured pressure increases, the pot wiper moves closer to the -15-V terminal. Thus higher pressures are represented by more negative voltages. This is indicated on the left of Fig. 12-2(b).

The final correcting device is an electrohydraulic positioner driving the shaft of the butterfly valve. The positioner is the same kind as that shown in Fig. 10-6. The position of the cylinder rod is controlled by the amount of current flowing through the sensing coil. In this positioner, the coil has a resistance of 2000 Ω. A coil current of 0 mA causes the cylinder to fully retract, completely closing the butterfly valve. A current of 5 mA causes the cylinder to fully extend, driving the butterfly valve wide open. A hydraulically operated positioner is needed for this application because of the great imbalancing forces exerted on the butterfly valve by the large volume of combustion air.

The electronic pressure control circuit is shown in Fig. 12-2(b). The set point and the measured pressure are the two electrical inputs to this circuit. These inputs are compared, and the difference between them, the error, causes control action. The output of the controller is the dc current delivered to the 2000-Ω sensing coil on the far right of the drawing. The control mode is strictly proportional. That is, the sensing coil current is varied in proportion to the error between set point and measured pressure.

12-2-2 The electronic comparer/controller

To begin understanding the operation of op amp 1 and its input circuits, let us temporarily make two simplifying assumptions:

a. Assume the span-adjust pot is turned all the way to the top. This will allow the entire wiper voltage from the set-point pot to be applied to R_4.

b. Assume the set point zero pot wiper is adjusted to exactly 0 V. This effectively eliminates this pot and R_2 and R_3 from consideration, since they supply only a 0-V signal to op amp 1.

In a little while, we will come back and see why these two potentiometers are necessary.

With these assumptions made, we can simplify the op amp 1 circuitry as shown in Fig. 12-3. It is not difficult to see that this is an unweighted summing circuit. The set point voltage, which is positive, is added to the measurement voltage, which is negative. The sum of these voltages appears in inverted form at V_{out_1}. In equation form,

$$-V_{out_1} = \frac{R_6}{R_4}(V_{st\,pt}) + \frac{R_6}{R_5}(V_{meas}) = \frac{100\,K}{100\,K}(V_{st\,pt}) + \frac{100\,K}{100\,K}(V_{meas}),$$

$$-V_{out_1} = V_{st\,pt} + V_{meas}. \tag{12-2}$$

Keep in mind that V_{meas} is a negative voltage.

Figure 12-3. Simplified schematic diagram of the op amp 1 circuitry of Fig. 12-2(b).

Equation (12-2) tells us that if the measured pressure agrees exactly with the set-point pressure, the output of op amp 1 equals 0 V. If the measured pressure is less than the set point, the output of op amp 1 is negative. If the measured pressure exceeds the set point, the output of op amp 1 is positive. The magnitude of V_{out_1} represents the amount of deviation of the measured pressure from the set-point pressure, and the polarity of V_{out_1} tells us the direction of the error.

Now let us go back and look at the set-point zero circuitry: as can be seen from Fig. 12-2(b), the voltage from the zero pot wiper is a third input to the op amp 1 summing circuit; but it carries very little weight, since R_3 is so large compared to R_4 and R_5. This third input is necessary to compensate for the fact that the set-point pot and the measurement pot cannot go all the way to 0 Ω. Even when these pots are turned down as far as they can go, there will be some *end resistance* left. Therefore the wiper voltages will not be absolutely zero. This would be no problem if we could be sure that the two voltages were the same. However, we cannot be sure of that. Since we definitely want V_{out_1} to be 0 V when both input pots are turned all the way down, we arrange for that to be so by injecting the lightly weighted signal from the zero pot. This cancels out any imbalance between the low ends of the two main input potentiometers.

Now let us consider the span adjust pot. One reason the span pot is necessary is that the range of the pressure input transducer might exceed the desired range of the set point. For example, suppose the pressure input transducer is designed to move its pot wiper from bottom to top as the bellows inlet pressure varies from 0 to 3 psig.* If the recuperator never under any circumstances needs an inlet pressure greater than 2 psig, then we would like the top of the set-point pot to represent a set-point pressure of 2 psig: that is, when the set-point pot is all the way to the top the measurement pot should be able to balance it by moving to the 2-psig point, which is only two thirds of its total distance. We can cause this to happen by delivering less than the full set-point wiper voltage to R_4. Instead of full wiper voltage, we arrange for only two thirds of the set-point wiper voltage to be delivered to R_4. This is done by turning the span pot down until the span pot wiper voltage is only two thirds of the set-point wiper voltage.

To summarize, the span pot reduces the set-point pot wiper voltage so that it can be balanced by the measurement pot wiper moving *less than its full range*. It is called a span pot because it determines the "span" of pressure values which can be set on the set-point pot. It would be adjusted by the system operators to provide whatever set-point span they desired.

Now consider op amp 2. It is wired as a noninverting amplifier with a variable gain. The output of this amplifier, V_{out_2}, drives the Q_1-Q_2 discrete circuitry which supplies current to the sensing coil of the butterfly valve positioner. Therefore the value of V_{out_2} determines the final position of the butterfly control valve. Since this is so, the voltage gain of the noninverting amplifier determines the controller's proportional band. If the voltage gain is high, it takes only a small error (small V_{out_1}) to cause a large change in V_{out_2}, and consequently a large change in valve position. This means the proportional band is narrow. If the voltage gain is low, it takes a larger error to cause a given change in V_{out_2}. Therefore it takes a large error (large V_{out_1}) to cause a given change in valve position, which makes the proportional band wider.

The maximum voltage gain occurs when the pot is adjusted to the far left. At

*The "g" following the psi units stands for *gage* pressure, or pressure above atmospheric pressure.

that point, $R_F = 1$ K and $R_{IN} = 22\ \Omega$, so

$$A_{V_{max}} = \frac{R_F}{R_{IN}} + 1 = \frac{1000}{22} + 1 = 46.5.$$

The minimum voltage gain occurs when the pot is moved all the way to the right. At that point, $R_F = 0$, so

$$A_{V_{min}} = \frac{0}{R_{IN}} + 1 = 1.$$

The voltage gain of the noninverting amplifier can thus be varied from 1 to 46.5.

Although V_{out_2} can vary from approximately $+12.0$ to -12.0 V, the voltage at the junction point of R_8 and R_9 is limited to a range of $+10$ to -10 V. The diode network below the junction point ensures this. Here is how it works.

The D_1-ZD1 pair prevents the R_8-R_9 junction from rising above $+10$ V. If the junction tries to rise above $+10$ V, the breakdown voltage of the D_1-ZD1 combination will be exceeded, causing that diode path to short out any excess over 10 V. This occurs because D_1 is forward biased by the positive output voltage, and zener diode ZD1 will have reached its breakdown point ($10\ \text{V} - 0.6\ \text{V} = 9.4\ \text{V}$). Any difference between V_{out_2} and $+10$ V is then dropped across 100-Ω resistor R_8.

The D_2-ZD2 diode pair prevents the R_8-R_9 junction voltage from going more negative than -10 V. If V_{out_2} goes below -10 V, D_2 will be forward biased by the negative output, and zener diode ZD2 will have reached its breakdown point. Any difference between V_{out_2} and -10 V will again be dropped across R_8.

The voltage appearing at the left end of R_9 can thus take on any value between $+10$ and -10 V, but it cannot exceed that range. Positive voltages mean that the measured pressure is above set point, and negative voltages mean that the measured pressure is below set point. Zero voltage means that the measured pressure agrees with the set point.

Let us consider the action of the discrete circuitry when zero volts appear at the left of R_9. The R_9-R_{10} voltage divider determines the voltage at the base of Q_1. The voltage dropped across R_9 can be found by the voltage division formula

$$\frac{V_{R_9}}{V_T} = \frac{R_9}{R_T} = \frac{R_9}{R_9 + R_{10}},$$

where V_{R_9} symbolizes the voltage drop across R_9 and V_T refers to the total voltage drop from the left end of R_9 to the -15-V supply. With 0 V on the left end of R_9, the total voltage drop is simply -15 V. Therefore

$$V_{R_9} = (15\ \text{V})\frac{3.9\ \text{K}}{3.9\ \text{K} + 4.7\ \text{K}} = 6.8\ \text{V}.$$

With 6.8 V dropped across R_9, the base voltage relative to ground is simply -6.8 V. This forward-biases the base-emitter junction of Q_1, causing Q_1 to conduct. The Q_1 emitter voltage will be 0.6 V below its base voltage, so $V_{E_1} = -6.8\ \text{V} - 0.6\ \text{V} = -7.4\ \text{V}$ relative to ground. Since the Q_1 emitter voltage is -7.4 V relative to ground, the Q_1 collector voltage must be $+7.4$ V relative to ground, since the Q_1 circuit is perfectly symmetrical. (You should prove this to yourself.)

Now let us assume for a moment that the maximum current adjust pot is dialed completely out. This will simplify the explanation. With this pot shorted out, the full $+7.4$ V appears at the top of R_{13}. It is then further divided by the R_{14}-R_{15} voltage divider to determine the voltage at the base of Q_2. By the voltage division formula, we can say that

$$\frac{V_{R_{14}}}{V_T} = \frac{R_{14}}{R_{14} + R_{15}},$$

where $V_{R_{14}}$ stands for the voltage drop across R_{14} and V_T symbolizes the total voltage drop between the left side of R_{14} and the -15-V supply. The total voltage drop is given by

$$V_T = +7.4 \text{ V} - (-15 \text{ V}) = 22.4 \text{ V},$$

so

$$\frac{V_{R_{14}}}{22.4 \text{ V}} = \frac{1 \text{ K}}{1 \text{ K} + 39 \text{ K}},$$

$$V_{R_{14}} = 0.56 \text{ V}.$$

Therefore the voltage at the Q_1 base is given by

$$V_{B_1} = +7.4 \text{ V} - 0.56 \text{ V} = +6.84 \text{ V}.$$

This voltage forward-biases transistor Q_2, causing it to turn on and conduct. Q_2 is wired as an emitter follower. The base-emitter voltage drop is 0.6 V, so $V_{E_2} = 6.84 \text{ V} - 0.6 \text{ V} = 6.24 \text{ V}$. Therefore transistor Q_2 will force enough current through the load coil to cause a voltage drop of 6.24 V across it. The current required to do this is given by

$$I_{\text{coil}} = \frac{V_{\text{coil}}}{R_{\text{coil}}} = \frac{6.24 \text{ V}}{2 \text{ K}} = 3.12 \text{ mA}.$$

The final conclusion of this derivation is that a 0-V input signal coming from op amp 2 causes a load coil current of 3.12 mA. Keep in mind that the entire range of currents necessary to stroke the positioner from fully closed to fully open is only 0 to 5 mA. A coil current of 3.12 mA would cause the positioner to drive the butterfly valve about 62% open, since 3.12 mA/5.0 mA = 0.62.

As the output voltage of op amp 2 takes on values other than 0 V, it causes the positioner to stroke the valve open or closed.

Now consider what happens if V_{out_2} goes positive. A positive V_{out_2} causes the voltage at the base of Q_1 to go more positive, thereby turning Q_1 on harder. This causes collector voltage V_{C_1} to become smaller, which tends to turn off Q_2 and reduce the load coil current. Thus positive values of V_{out_2} cause the butterfly valve to close.

A negative V_{out_2} causes the base voltage V_{B_1} to go further negative, thereby reducing the Q_1 collector current. Collector voltage V_{C_1} therefore rises, which tends to turn on Q_2 to increase the load coil current. Therefore negative values of V_{out_2} cause the butterfly valve to open.

The Q_1-Q_2 circuitry is designed so that -10 V at the left end of R_9 will cause the load current to equal 5 mA and so that $+10$ V at the left end of R_9 will cause the load current to equal nearly 0 mA. The exact adjustment of this response is done with the 10-K maximum current adjust pot. Variations in load coil resistance and in circuit component values may cause the actual load current to differ from the proper 5-mA value when $V_{out_2} = -10$ V. Such discrepancies are adjusted out with the maximum current adjust pot.

The purpose and operating principles of the pressure control system should now be clear. Any tendency of the measured pressure to drop below the set-point pressure causes the controller to deliver more current to the load coil. This opens the butterfly valve and admits more combustion air to bring the measured pressure back up toward the set point. Conversely, any tendency of the measured pressure to rise above set point causes the controller to reduce the current to the load coil. This closes the butterfly valve to bring the measured pressure back down toward set point.

In some soaking pit systems the control of combustion air is done on the basis of flow rate instead of recuperator upstream pressure. The control method is exactly the same as explained here, except that the pressure transducer becomes a *differential pressure* transducer, responding not to a single gage pressure but to the pressure *drop* across an orifice in the air duct. This pressure drop across an orifice is proportional to air flow rate, so the controlled variable becomes flow rate instead of recuperator pressure.

12-3 A PROPORTIONAL PLUS RESET TEMPERATURE CONTROLLER WITH THERMOCOUPLE INPUT

In industrial processes the most frequently controlled variable is temperature. In drying processes, melting processes, heat-treating processes, chemical reaction processes, etc., temperature is of prime importance. When the process temperature is above a few hundred degrees Fahrenheit the preferred transducer is usually a thermocouple. One of the most common temperature control schemes is a thermocouple input into a proportional plus reset electronic temperature controller, with the final correcting device being a variable position fuel valve. In this section we will investigate such a control scheme in detail.

12-3-1 The thermocouple-set point bridge circuit

On the left in Fig. 12-4(a) is the thermocouple bridge measurement circuit. This circuit combines the thermocouple millivolt signal with the temperature set-point signal to generate an error signal. The magnitude of the error signal represents the deviation between measured temperature and desired temperature. The desired temperature is represented by the position of the set-point potentiometer. Here is how the circuit works.

Figure 12-4. A thermocouple temperature control circuit. (a) The thermocouple bridge input, chopper, preamplifier, and demodulator. (b) The proportional plus integral control circuit which positions the valve.

(a)

Figure 12-4. (Cont.)

The 6.2-V zener diode, ZD1, provides the stable dc supply voltage to the bridge. The bridge is designed so the voltage at the top of R_3 equals the voltage at the top of R_5; both of these voltages are measured relative to ground, the bottom of the bridge. The 50-K zero pot, P_2, is adjusted to accomplish this. Causing these two voltages to be equal ensures that an extreme setting on the set-point pot P_1 exactly balances a 0-V signal from the thermocouple. That is, when P_1 is dialed all the way down with its wiper touching the junction of P_1 and R_5, the error signal will be zero when the thermocouple puts out a zero signal. This will happen only if the voltage across R_3 equals the voltage across the series combination of P_2 and R_5.

The system operator selects the desired set-point temperature by adjusting the position of P_1, the set-point pot. This pot has a pointer attached to its shaft which points to a marked temperature scale. The marked temperature scale is not shown in an electronic schematic diagram.

Once the set-point pot has been adjusted, the system seeks to bring the measured temperature into agreement with the set point. When the two are in agreement, the thermocouple signal voltage exactly equals the voltage between the wiper and the bottom terminal of P_1. If the measured temperature should rise above set point, the thermocouple (T/C) signal will be greater than the set-point signal, and the error voltage will be positive, as marked in Fig. 12-4(a). If the measured temperature should drop below set point, the T/C signal will be smaller than the set-point signal, and the error voltage will be negative. Its polarity would then be the opposite of the polarity shown in Fig. 12-4(a). The greater the deviation between measured temperature and set-point temperature, the greater the difference between these two voltages, and the greater the magnitude of the error signal.

12-3-2 The preamplifier, chopper, and demodulator

The rest of the electronic circuitry in Fig. 12-4(a) serves the purpose of amplifying the tiny dc error signal. Remember from Sec. 11-4 that thermocouples generate a very small-signal voltage, no more than a few tens of millivolts. The error signal, being the *difference* between a thermocouple signal and another signal in the millivolt range, is much smaller yet. The error signal is only a fraction of a millivolt when the measured temperature is close to set point. It is virtually impossible to build a dc amplifier that is so drift-free that it can reliably handle a dc signal this small.

For this reason, the preamplifier in this temperature control system is chopper stabilized. That is, the dc error signal is converted into an ac signal by chopping, is then amplified in an ac amplifier where drift is of no consequence, and is then converted back into a dc signal at the output of the amplifier. Before beginning the discussion of the preamplifier itself, let us deal with the chopping circuitry.

The chopping circuitry consists of an FET (field-effect-transistor) chopper and the chopper drive generator. The chopper drive generator is an astable clock, which we have seen before in Sec. 3-11. The astable clock develops a square-wave signal at the collector of transistor Q_7, which is then applied to the *gate* lead (marked G) of the FET chopper. The frequency of this square-wave signal is given by Eq. (3-1):

$$f = \frac{0.7}{R_B C} = \frac{0.7}{(160 \text{ K})(0.022 \text{ } \mu\text{F})} = 200 \text{ Hz}.$$

As this 200-Hz square wave is applied to the gate terminal of the FET chopper the FET behaves as follows:

a. When the square wave goes positive, the gate goes positive relative to the *source* terminal (marked S). This turns the FET OFF and causes it to exhibit a high resistance between the *drain* terminal (marked D) and source terminal. It can be thought of as an open switch.

b. When the square wave goes down, it removes the positive bias from the gate terminal, allowing the FET to turn ON. The FET then exhibits a low resistance from source to drain terminal and can be considered a closed switch.

The FET therefore alternates between an open switch and a closed switch.* When it is an open switch, the top line of the error signal (the T/C voltage) is connected to the preamp coupling capacitor, C_1. When the FET is a closed switch, it effectively applies the bottom of the error signal (the set-point voltage) to the preamp input. This is true because the bottom line is coupled to the preamp input through a Thevenin resistance of about 2 K, whereas the top line (T/C signal) is coupled through a Thevenin resistance of 18 K. Therefore the bottom line signal overwhelms the top line signal when the FET is ON (when it is a closed switch from source to drain).

The signal delivered to C_1 is therefore a square waveform whose peak-to-peak value equals the magnitude of the dc error signal.

The chopped error signal is coupled through C_1 and R_9 into the base of transistor Q_2. Transistors Q_2, Q_3, and Q_4 constitute a high-gain ac amplifier. There are no stabilizing resistors in the emitter leads of Q_2, Q_3, and Q_4. Instead, bias stability is provided by negative dc feedback from the collector of Q_4 into the base divider (R_{10} and R_{11}) of Q_2. The absence of degeneration resistors in the emitter leads provides a high voltage gain in this three-stage amplifier.

Notice that the Q_2 and Q_3 stages are decoupled from the Q_4 stage by R_{19} and C_6. This technique minimizes the appearance of dc supply line noise in the initial stages of the preamplifier, which is where it could do the most harm. This technique was discussed in Sec. 10-11-2.

---

*An FET is unlike a bipolar transistor in that it is a normally ON device; you must deliver an external signal to the gate in order to turn it OFF. This is the opposite of a bipolar transistor, which is normally OFF and requires an external base signal to turn it ON. An FET is superior to a bipolar transistor in a chopping application of this sort. Its superiority is due to the fact that there are no *pn* junctions between the drain and source of an FET, as there are between the collector and emitter of a bipolar transistor. The field-effect transistor can thus duplicate the action of a closed mechanical switch better than can a bipolar transistor.

The ac signal appearing at the collector of Q_4 rides on a dc level of about 6.5 V, which is the Q_4 collector bias voltage. The dc component is removed by output coupling capacitor C_2. The ac output signal is therefore centered on ground when it appears at R_{15}. The ac signal is converted to a dc signal by the action of Q_5, an FET *demodulator*. This FET is also being used in a switching mode. Its gate is driven by the square wave at the collector of Q_8, which is 180° out of phase with the square waveform at Q_7 which drives the gate of the FET chopper (Q_1). As Q_5 alternately switches open and closed, it removes one half of the ac signal appearing at the left end of R_{15}. When it turns ON, Q_5 shorts the R_{15}-R_{16} junction to ground, causing the instantaneous voltage on the left of R_{15} to be removed. (It is dropped across R_{15}.) When Q_5 turns OFF, it disconnects the R_{15}-R_{16} junction from ground and allows the instantaneous voltage on the left of R_{15} to come through R_{15} only a little bit diminished. (R_{16} is quite a bit larger than R_{15}.)

The initial polarity of the dc error signal from the bridge determines whether the FET demodulator removes the negative half cycle or the positive half cycle of the ac waveform. If the dc error signal is positive as shown, the FET demodulator removes the *negative* half cycle of the ac output. If the dc error signal is negative (measured temperature is below set point), then the FET demodulator removes the *positive* half cycle of the ac output. Try to reason these last statements out for yourself.

The voltage waveform which appears at point A is filtered by R_{16} and C_3. This low-pass filter converts the square waveform at point A into a dc voltage with just a slight ripple component. This dc voltage is called the *amplified error signal*. The amplified error signal may be positive or it may be negative, depending on the polarity of the original dc error signal. It will have the same polarity as the original dc error.

The amplified error signal is applied to Q_6, an emitter follower, which furnishes a high input impedance. The voltage appearing at the emitter of Q_6 is 0.6-V more negative than the amplified error signal because of the voltage drop across the base-emitter junction. This 0.6 V is restored by the R_{18}-D2 combination; there is a 0.6 V rise across silicon diode D2. The purpose of passing the amplified error signal through Q_6 and D2 is to buffer it from the demodulator. This results in an amplified error signal which can deliver a fairly large current into the circuit that it drives without disturbing the demodulator.

12-3-3 Proportional plus reset control

The amplified dc error signal is brought into the op amp summing circuit on the left in Fig. 12-4(b). Let us first concentrate on the proportional aspect of the control. We will then investigate the circuitry which furnishes the reset control action.

Proportional Action. The amplified error signal is applied to the op amp 1 summer through R_{28}. Assume for a moment that this signal is positive and has just now suddenly increased in magnitude because of a load disturbance. Here is what will happen.

The positive voltage at the left of R_{28} will tend to drive the inverting input of op amp 1 positive. This will cause the output to go negative. When V_{out_1} goes negative, it applies a negative voltage to the noninverting input of op amp 2, which is a high-gain noninverting amplifier. The noninverting amplifier output, V_{out_2}, becomes a large negative voltage and will forward-bias Q_{11} and Q_{12}, which are connected as a *Darlington pair*. Because of the very large current gain of a Darlington pair (the total current gain is the product of the two individual transistor current gains), a small trickle of electron current through R_{33} into the base of Q_{11} will cause Q_{12} to turn ON and saturate. When Q_{12} switches ON, it completes a circuit to relay R2, causing that relay to pick up. The R2 contacts change state in the 115-V motor control circuit, causing the valve motor to run. The motor runs in the proper direction to close the fuel valve, since a positive error signal from the preamp means that measured temperature is too high (above set point). As the valve closes, the valve position pot P_6 moves downward into its negative potential region. The negative voltage appearing at the wiper of P_6 is applied to P_3, the proportional band adjust pot. A portion of the P_6 negative voltage is picked off by P_3 and fed back to R_{27} and into the summing circuit. This negative voltage tends to cancel the positive error voltage applied to R_{28}. Eventually, if the valve position pot moves far enough, the negative signal applied to R_{27} will cause the summing circuit output to return to zero. At this point, V_{out_2} also goes to zero, so it can no longer keep Q_{11} and Q_{12} turned ON, and relay R2 will drop out. This stops the valve motor and causes the fuel valve to freeze in that position. The reduction in fuel flow should drive the measured temperature back down toward set point.

Considering *only* R_{27} and R_{28} inputs to the summing circuit, the general equation which describes the summing circuit is

$$-V_{\text{out}_1} = \frac{10\,\text{K}}{10\,\text{K}}(V_{\text{error}}) + \frac{10\,\text{K}}{68\,\text{K}}(V_{vp}),$$

where V_{vp} stands for *valve position voltage*, which is the voltage fed back from P_6 and P_3 into R_{27}. Any nonzero V_{out_1} will cause one of the Darlington pairs to turn ON. If V_{out_1} is negative, V_{out_2} is also negative, and Q_{11} and Q_{12} turn ON, as we have seen. If V_{out_1} had been positive, Q_9 and Q_{10} would have turned ON.

Whichever Darlington pair turns ON picks its associated relay, either R1 or R2. The relay contacts then cause the motor to run the fuel valve and the valve position pot in whatever direction tends to reduce V_{out_1} to zero. When V_{out_1} reaches zero, the energized relay drops out, and the valve motor stops.

Knowing that the circuit always acts to bring V_{out_1} to zero, we can rewrite the above equation as

$$0 = \frac{10\,\text{K}}{10\,\text{K}}(V_{\text{error}}) + \frac{10\,\text{K}}{68\,\text{K}}(V_{vp}),$$

$$-V_{vp} = 6.8(V_{\text{error}}).$$

This equation expresses the proportional nature of the control. It tells us that the greater the magnitude of V_{error}, the greater the magnitude of V_{vp}. Thus the valve correction is *proportional* to the amount of deviation from set point. This is the essence of proportional control.

If P_3 is adjusted toward the top, so that a large portion of the P_6 wiper voltage is fed back to R_{27}, the proportional band is wide. If P_3 is adjusted toward the bottom, so that only a small portion of the P_6 wiper voltage is fed back, the proportional band is narrow. This can be understood as follows.

If a large portion of the P_6 signal is fed back, it will be relatively easy for the valve position pot to cancel the effect of V_{error} and thereby return V_{out_1} to zero. Therefore the valve will not have to move very far. This being the case, it would take a large V_{error} to force the valve to go to an extreme position (fully open or fully closed). When it requires a large error to drive the final correcting device between its limits, the proportional band is wide.

On the other hand, if P_3 feeds only a small portion of the P_6 signal back to R_{27}, it will be difficult for the valve position pot to cancel V_{error}. In other words, a small V_{error} will cause a large change in valve position. Thus the error required to drive the valve from one limit to the other is not as great as before, and the proportional band is narrower.

Reset Action. As with any purely proportional controller, the correction imparted to the valve will never quite return the temperature to set point. All it will do is drive the temperature back toward set point. To get the actual measured temperature back to set point, the reset mode of control must be added. In Fig. 12-4(b), the reset control action is supplied by op amp 3 and op amp 4 and their associated components.

To understand how this circuitry works, consider the positive V_{error} which appeared before. The appearance of the positive V_{error} resulted in a readjustment of the fuel valve in the closed direction; the corresponding reduction in fuel flow caused the measured temperature to return to the neighborhood of the set point. However, a small positive V_{error} will persist.

This small V_{error} is applied to R_{35}, which is part of the input resistance of the op amp 3 integrator. Recall from Sec. 8-10 that the output of an integrator is proportional to *how long* the input has been present. In this particular case,

$$-V_{\text{out}_3} = \frac{1}{R_{\text{IN}}C_4} V_{\text{error}}\, t.$$

R_{IN} is the sum of the R_{35} and P_4 resistances, and t is the time in seconds that V_{error} has been present. Op amp 4 is simply an inverter with a gain of 1, so its output, V_{reset}, is the opposite polarity of V_{out_3}, or

$$V_{\text{reset}} = \frac{1}{R_{\text{IN}}C_4} V_{\text{error}}\, t. \tag{12-3}$$

V_{reset} is applied to the summing circuit through R_{29}, as Fig. 12-4(b) shows. It causes the inverting input of op amp 1 to go positive, even though V_{error} and V_{vp} have nearly cancelled each other. V_{out_1} then goes negative, causing V_{out_2} to go negative and turn ON the Q_{11}-Q_{12} pair *again*. Thus the motor runs the fuel valve a little further closed, tending to reduce the measured temperature some more, bringing it into better agreement with the set point. If V_{error} continues to persist, V_{reset} will continue to increase as time passes. As it increases, it causes more and more correction in fuel valve position. Eventually, V_{error} will reach zero. At that point, the input to the integrator will equal zero, and the integrator will stop building up (stop integrating). V_{reset} will stop growing, and no further fuel valve correction will be made. Therefore the reset corrective action depends on how much time an error has been present, which is the essence of reset control.

The reset time constant is adjusted by 10-M potentiometer P_4. When P_4 resistance is dialed out, the time constant is short, and the integrator builds up quickly. The reset circuitry is then quick to make its effects felt. When P_4 resistance is dialed in (higher resistance), the reset time constant is lengthened, and the integrator builds up slowly, as Eq. (12-3) shows. The reset circuitry is then slow to make its effects felt.

As mentioned in Chapter 9, on commercial temperature control instruments the reset adjustment is called *reset rate*. The high numbers on the *reset rate* scale mean quick reset action (small P_4 resistance), and the low numbers mean slow reset action (large P_4 resistance). The amount of reset action used depends on the nature of the specific temperature process, as explained in Chapter 9. The same holds true for the proportional band adjustment.

12-4 A STRIP TENSION CONTROLLER

Many industrial processes involve the handling of moving sheets or strips of material. One example of this is the textile process, which will be described in Sec. 12-8. Other examples are the heat treating, galvanizing, or pickling of steel strip; the finishing of plastic strip; and the drying of paper strip. In all of these applications, it is important to maintain the proper amount of *tension* in the strip. Too great a tension will cause the strip to stretch and deform and possibly break. Too low a tension will cause the strip material to sag. This may cause it to tangle up in the handling machinery.

The tension of a strip can be controlled by making slight adjustments to the relative speeds of the *leading roll* and *following roll* in the strip handling apparatus. This is shown in Fig. 12-5(a). With the following roll spinning at a given speed, the strip tension can be increased by increasing the speed of the leading roll. Strip tension can be decreased by decreasing the speed of the leading roll.

A popular way of making these slight speed adjustments is to change the position of a drive belt on two conical pulleys. This is illustrated in Fig. 12-5(b).

The prime mover, either a constant-speed or an adjustable-speed electric motor, has a double-ended shaft. One end attaches to the following roll, causing that roll to spin at some reference speed. The other end of the shaft goes to conical pulley *A*, causing it to spin at the same reference speed.

The drive belt transmits power from conical pulley *A* to conical pulley *B*, which

Figure 12-5. A strip tension controller. (a) The LVDT rides on the moving strip between the leading roll and the following roll. (b) The mechanism for controlling the speed of the leading roll relative to the following roll. (c) Closeup view of the LVDT sensor.

then drives the leading roll. If the drive belt is centered on pulleys *A* and *B*, then pulley *B* spins at the same speed as pulley *A*. However, by moving the *draw pulley* to the left or right the drive belt can be moved to the left or the right on the conical pulleys. As the drive belt is moved to the left the *A* diameter decreases while the *B* diameter increases. This causes the leading roll to slow down. As the drive belt is moved to the right, the *A* diameter increases and the *B* diameter decreases, causing the leading roll to speed up.

The drive belt follows the draw pulley, which is an idling pulley. Its shaft is supported in bearings which are mounted on a movable base. The movable base is controlled by a rack and pinion, driven by the *draw motor*. The draw motor is a dc shunt motor whose shaft is geared down to run at a slow speed.

Control is accomplished by running the draw motor in short bursts. In this way, the movable base can be shifted to the left or right a little bit at a time. The base carries the draw pulley, which then positions the drive belt on the conical pulleys. This imparts slight speed adjustments to the leading roll to adjust the strip tension.

The transducer, which supplies tension information to the control circuitry, is an LVDT, shown in Figs. 12-5(a) and (c). The core of the LVDT is attached to a springy metal arm, which is moved up and down by a roller which rides on the strip. If the strip tension increases, the strip of material rises a little bit, causing the core of the LVDT to rise. If the strip tension decreases, the strip drops down a little. The roller also moves down due to the spring action of the arm. This causes the core of the LVDT to move down. Therefore the output voltages from the secondary windings of the LVDT are an indication of the strip tension.

Refer to Fig. 12-6, which is a schematic diagram of the electronic control circuitry. The frame of the LVDT is situated so that when the strip tension is in the middle of the acceptable range, the LVDT core is centered. In this condition the two secondary voltages are both equal ($V_{s1} = V_{s2}$). These secondary voltages are rectified and filtered and applied to the inputs of differential amplifier op amp 1. D1 and D2 are germanium small-signal diodes with a low forward bias voltage. Thus the dc voltages appearing at the top of C_1 and C_2 are very nearly equal to the peak values of V_{s1} and V_{s2}.

If the strip tension is somewhat tighter than the midpoint of the acceptable range, the C_1 voltage will be larger than the C_2 voltage. If the tension is somewhat looser than the midpoint of the tension range, the C_2 voltage will be larger than the C_1 voltage. This can be seen by looking at the direction markings by the LVDT core. Resistors R_1 and R_2 are bleeder resistors to allow C_1 and C_2 to discharge to continually reflect the peak values of V_{s1} and V_{s2}.

The voltages across C_1 and C_2 are applied to R_3 and R_5, which are the input resistors of a differential amplifier having a gain of 4. In equation form,

$$V_{\text{out}_1} = \frac{20 \text{ K}}{5 \text{ K}}(V_{C_2} - V_{C_1}) = 4(V_{C_2} - V_{C_1}).$$

If strip tension is tighter than the midrange value, V_{out_1} is a negative dc voltage. If tension is looser than the midrange value, V_{out_2} is a positive voltage. V_{out_1} is applied to two voltage comparers, op amp 2 and op amp 3. These comparers have the function of determining if the measured tension is *too* tight or *too* loose. In other words, a certain deviation from the tension midrange will be tolerated, but beyond

Figure 12-6. Schematic diagram of the strip tension control circuitry.

a certain point, corrective action will be taken. The op amp 2 comparer checks for tension exceeding the limit for looseness, while the op amp 3 comparer checks for tension exceeding the limit for tightness. The limits themselves are adjustable and are set by potentiometers P_1 (looseness) and P_2 (tightness).

For purposes of discussion, suppose that P_1 and P_2 are set at $+8$ and -8 V, respectively. Then if V_{out_1} goes more positive than $+8$ V, it means the measured tension has exceeded the looseness limit. When this happens, the $+$ input of op amp 2 goes more positive than the $-$ input, so V_{out_2} switches from negative saturation to positive saturation (from about -13 to $+13$ V). This causes a $+5$-V signal to appear at the top of zener diode ZD3. Therefore the appearance of $+5$ V at D3 indicates that tension is too loose and that corrective action is necessary.

Whenever tension does not exceed the looseness limit, V_{out_2} is -13 V, which forward-biases ZD1, causing a -0.6-V signal at the cathode terminal of ZD1.

If strip tension exceeds the tightness limit, V_{out_1} will go more negative than the P_2 setting of -8 V. When this happens, the $-$ input of op amp 3 is more negative than the $+$ input, so V_{out_3} switches from -13 to $+13$ V. This causes the same result at zener diode ZD2 that was seen above for ZD1. That is, the cathode level changes from -0.6 to $+5$ V. This $+5$-V signal represents the fact that tension is too tight and that corrective action is necessary.

The "too loose" and "too tight" signal lines are applied to logic gates NAND1 and NAND2. Assume that the logic family used here operates on a $+5$-V supply level. These NANDs also receive the pulse train output of the astable pulse generator on their bottom inputs. The pulses have a duration of about 400 msec and occur approximately every 10 seconds. When the positive edge of the pulse arrives, it partially enables both NAND gates. If either the "too loose" or the "too tight" line is HI at this time, the corresponding NAND gate will change states, delivering a negative-going triggering edge to the T terminal of one of the one-shots.

If the "too loose" line is HI when the pulse arrives, one-shot OS1 is fired, signaling the leading roll to speed up. If the "too tight" line is HI, one-shot OS2 is fired, signaling the leading roll to slow down. Here is how the firing of the one-shots brings about the speed adjustment of the leading roll.

Assume that OS1 fires. Its Q output will go HI for a pulse duration of 2 seconds, during which time it will turn ON the LED in optical coupler 1. The detecting device in the optical coupler is a phototransistor, which switches ON during the presence of the one-shot's output pulse. This effectively connects the gate of SCR1 to its anode through a 10-K resistor. When the anode to cathode voltage enters the positive half cycle, the SCR will be supplied with enough gate current to fire. The anode voltage itself supplies the gate current, through the 10-K resistor. Therefore SCR1 will fire just after the zero crossover and will continue doing so for the full 2-sec duration of the one-shot pulse. With SCR1 firing on every positive-going zero crossover, the draw motor armature will be supplied with dc current. The flow path is as follows: from the top 115-V supply line, through SCR1, through the bridge cross-connection wire, through D4, and through the motor armature from left to right. During this 2-sec period, the draw motor will run and will shift the movable base to the *right* in Fig. 12-5(b). At the end of 2 sec, the draw motor will stop running because the one-shot output pulse will terminate. This will disable the optical coupler and break the path of SCR gate current.

Having moved the drive belt on the conical pulleys, the system now waits for the next pulse from the astable pulse generator, which will arrive about 10 sec later. During this interval, the speed adjustment to the leading roll has the opportunity to effect an increase in strip tension. This should bring the tension above the loose limit and remove the $+5$-V signal from ZD1. If it does accomplish that job, the next pulse from the pulse generator will have no effect on NAND1, since the top input will then be LO. On the other hand, if the speed increase was not enough to bring strip tension above the loose limit, another correction will be made when the next pulse arrives. OS1 will fire again, allowing SCR1 to turn ON and drive the draw motor for another 2-sec burst. This action is repeated until the tension is brought back into the acceptable range.

Of course, if the original tension error had been a "too tight" condition, zener diode ZD2 would have been HI and NAND2 would have been enabled when the pulse arrived from the pulse generator. OS2 would then have fired for 2 sec, instead of OS1. This would energize optical coupler 2, thereby turning on SCR2. The draw motor armature current would then be reversed, passing through the armature from right to left, through SCR2, through the bridge cross-connection wire, and through D3. This would shift the movable base and draw pulley to the *left* in Fig. 12-5(b). The leading roll would slow down and loosen the strip tension. As before, as many corrections as are needed would be made by the system.

The draw motor field winding is driven from the 115 V-ac line, through half-wave rectifier D5. D6 allows current to continue circulating through the field winding during the negative half cycle. As the field current declines when the ac line passes its positive peak, the induced voltage in the field winding is the proper polarity to forward-bias D6. Therefore, there is no abrupt halt to the field current as the ac line reverses. Diode D6 allows current to continue flowing in the field winding throughout the negative half cycle.

12-5 EDGE GUIDE CONTROL FOR A STRIP RECOILER

When a strip material has completed its processing, it is wound up into a coil for subsequent handling and shipping. This operation is called *recoiling* and is illustrated in Fig. 12-7. The compass directions indicated in Fig. 12-7(a) will be used to specify directions of movement in the discussion of the recoiling system. This will help avoid confusion.

The moving strip passes under a fixed idler roll and is recoiled on the *windup* reel. If the recoiling operation consisted merely of rotating the windup reel and feeding the strip onto it, the coils produced would almost certainly be crooked. That is, the individual coil laps would not be aligned with each other; the sides

Figure 12-7. A strip recoiler. (a) Physical layout of the recoiling mechanism. (b) Closeup view of the edge-sensing apparatus.

would either be wavy or "telescoped." This would make handling and shipping the coil more difficult and would increase the likelihood of edge damage.

To produce straight coils, some type of control is necessary to ensure that every coil lap is aligned with every other lap. There are two ways to do this:

a. Guide the traveling strip to correct any tendency it has to move laterally (back and forth) relative to a fixed windup reel.

b. Shift the windup reel back and forth to follow any lateral movement by the traveling strip.

Of these two methods, the second is preferred for most strip materials, especially metal strip.

The position of the windup reel is usually controlled by a hydraulic cylinder. The cylinder rod is attached to the mounting base of the reel, as shown in Fig. 12-7(a). The position-sensing device is a large photoconductive cell, with a diameter of about 1 inch. A view of the photoelectric edge-sensing assembly is presented in Fig. 12-7(b); this is the view seen looking toward the east along the top of the traveling strip. If the strip is properly positioned, it will block exactly half of the broad light beam radiated by the lamp. The other half of the light beam will strike the photocell.

If the traveling strip should wander to the south, deeper into the photoelectric assembly, a smaller amount of light will strike the photocell. This will be detected electrically and will initiate control action to move the windup reel to the south, to keep it aligned with the traveling strip.

If the strip should wander to the north, out of the photoelectric assembly, the increasing illumination of the photocell will cause the windup reel to move to the north. Figure 12-7(a) shows that the edge sensor is mounted on an arm which is attached to the windup reel base. The edge sensor thus moves with the windup reel, always maintaining a fixed strip edge position relative to the reel. In this way the windup reel is kept constantly aligned with the strip. Every coil lap edge aligns with every other coil lap edge, and the coil winds straight.

The electrohydraulic circuit which accomplishes this control is shown in Fig. 12-8. Here is how it works. The photoconductive cell in the edge sensor has an approximate resistance of 5 K when the traveling strip blocks half of the light beam. The bias adjustment pot, P_1, is adjusted to turn on transistor Q_1 enough to bring its collector voltage to about 2 V. Q_1 collector voltage is applied to the base of power transistor Q_2. This causes Q_2 to conduct, establishing a current flow in the 320-Ω actuator coil that swings the hydraulic jet pipe. Refer to Figs. 12-8(a) and (b).

The jet pipe assembly is designed so that a 10-mA current through the actuator coil causes the jet pipe to be perfectly centered. Adjustment of the center position is done with the balancing spring adjustment. The mechanical force exerted by the balancing spring is equal and opposite to the force created by the permanent magnet and actuator coil.

In Fig. 12-8(a), bias pot P_1 is manually adjusted to provide exactly 10 mA of actuator coil current when the light beam is exactly half blocked. Lowering the P_1 resistance increases the current in the actuator coil, and raising the P_1 resistance decreases the coil current. The adjustment of P_1 would be done by the system users prior to putting the system in service.

Temperature stability is provided to the electronic circuit by negative feedback resistor R_2, connected between the emitter of Q_2 and the base of Q_1. R_2 provides negative current feedback, which lowers the overall circuit gain, but at the same time stabilizes the gain and bias point. A stable bias point is very important in this circuit, since drift in bias point will change the actuator coil current and move the jet pipe.

(a)

(b)

Figure 12-8. (a) The edge-sensing and controlling circuit. (b) The electrohydraulic actuator for moving the base of the windup reel.

By adjusting the resistance of potentiometer P_2, the amount of feedback can be varied, and the gain of the circuit is varied. If the P_2 resistance is increased, the feedback increases, and the gain decreases. That is, a given change in photocell resistance will result in only a small change in actuator coil current. If the P_2 resistance is decreased, the feedback is decreased, and the gain is increased. This makes the circuit more sensitive to changes in photocell resistance.

Let us imagine that all the electronic adjustments have been made and that the recoiling system is in operation. We will trace out the sequence of actions as the system causes the windup reel to follow the lateral movements of the traveling strip. Refer to Figs. 12-8(a) and (b) and keep in mind the physical layout of the system as shown in Fig. 12-7(a).

If the edge of the traveling strip is passing directly through the middle of the edge sensor, 10 mA of current will flow through the actuator coil, and the jet pipe will be exactly centered. Neither distribution pipe will have a higher hydraulic pressure than the other, so the pilot cylinder in Fig. 12-8(b) will be centered by its spring. The pilot cylinder rod is attached to the spool of the main hydraulic control valve. With the pilot cylinder centered, the main control valve does not pass oil to either end of the main cylinder; therefore, the windup reel base remains stationary. As long as the edge of the traveling strip remains centered in the light beam, the system does not move the windup reel, and the coil winds straight.

Now suppose the strip edge wanders to the north. This will tend to move the edge *out* of the center, thus exposing more of the light beam and lowering the photocell's resistance. As the photocell resistance drops, Q_1 turns on harder, and the Q_1 collector voltage decreases. This reduces the Q_2 conduction and causes the actuator coil current to fall below 10 mA. The jet pipe cannot remain centered, but moves up and creates a pressure imbalance in the distribution pipes. In this case the top pipe has a higher pressure than the bottom pipe, so in Fig. 12-8(b) the pilot cylinder moves to the left. The pilot cylinder rod shifts the control valve spool to the left, thereby connecting the high-pressure hydraulic supply port to the blind end of the main cylinder, while allowing the rod end of the main cylinder to drain to the reservoir. This causes the main cylinder to extend, moving the windup reel to the north. Thus the windup reel follows the wandering strip edge. As soon as the edge is back in the center of the sensor, the coil current will return to 10 mA, and all corrective action will stop. The windup reel will continue rewinding the coil in this new position as long as the strip edge remains centered in the sensor.

If the traveling strip edge wanders to the south, it will move *into* the sensor and block more of the light beam. The photocell resistance will increase, causing Q_1 to conduct less current. This will raise the Q_1 collector voltage and cause Q_2 to conduct more heavily, thus increasing the actuator coil current above 10 mA.

The magnetic force on the jet pipe now exceeds the mechanical spring force, so the jet pipe moves down. This time the bottom distribution pipe receives the higher pressure, so the pilot cylinder moves to the right; this shifts the main control valve spool to the right and applies high-pressure oil to the rod end of the main hydraulic cylinder. This cylinder strokes to the south, driving the windup reel base to the south. The windup reel thus follows the wandering strip until the sensor is once again centered on the traveling edge. At that point, the actuator coil current returns to 10 mA, and the hydraulic control devices return to center. The main cylinder freezes in that particular position, and the coil continues winding straight.

12-6 AN AUTOMATIC WEIGHING SYSTEM

Automatic weighing systems are frequently used in industry. These systems automatically transfer a preset weight of material into a hopper or container of some type. They are used for the manufacture of products which require a number of carefully weighed ingredients.

12-6-1 The mechanical layout

A system for weighing one single ingredient is shown in Fig. 12-9(a). The ingredient being handled and weighed is a powder. In an industrial process, the best way to handle powders is with a *screw conveyor*. A screw conveyor is a large pipe, perhaps 1 ft in diameter, with a wide-pitch internal screw. The screw thread diameter is only slightly less than the inner diameter of the pipe, so there is close clearance between the threads and the inner wall of the pipe. As the screw shaft is rotated, powdered material is forced down the pipe. The greater the speed of the screw shaft, the greater the flow rate of material. The conveyor pipe discharges into the weighing hopper, as illustrated in Fig. 12-9(a).

The weighing hopper is supported from above by steel cables which attach to a strain gage load cell. The load cell therefore senses the weight of the hopper itself, its supporting cables, and the material in the hopper.

At the bottom of the hopper is a discharge pipe, containing a solenoid-operated dump valve. This is used to remove the weighed material from the hopper and transfer it to the next stage of the production process.

12-6-2 The electronic weighing circuitry

The electronic weighing circuit schematic is drawn in Fig. 12-9(b).

The Load Cell Bridge. The weight signal is taken from the load cell, which is a strain gage assembly. The four strain gages are mounted on a *proving ring*, as shown in Fig. 12-9(c). The proving ring and strain gages, taken together, comprise the load cell. As the proving ring is subjected to a tension load, strain gages 2 and 3 are stretched, causing their resistances to increase. Strain gages 1 and 4, being mounted on the top and bottom of the inside of the ring, are compressed as the ring is loaded. Therefore their resistances decrease. The gages are wired into the Wheatstone bridge circuit so that the two sides of the bridge tend to produce opposite changes in resistance ratio. That is, as the ratio of R_1 to R_2 gets smaller, the ratio of R_3 to R_4 gets larger. By making all four arms of the bridge respond to the load on the

Figure 12-9. (a) Mechanical layout of the screw conveyor, the hopper, and the load cell. (b) Schematic diagram of the weighing circuitry. The servo system causes the indicated weight to equal the gross weight minus the tare weight. (c) Closeup view of the load cell. It is comprised of four strain gages carefully bonded to a proving ring.

482

load cell, the available output voltage of the bridge is increased. A typical industrial load cell produces an output voltage of 30 millivolts at full load when excited by a 15-V ac supply.

Since load cell output voltages are so puny, it is very important to keep the weighing signal circuits free from electrical noise. This is the purpose of the *ac line filter* wired into the 115-V supply lines to transformer T_1 in Fig. 12-9(b). Any high-frequency noise signals appearing on the ac lines are filtered out before they reach the T_1 primary winding. In addition to this precaution, all the signal wires leading up to the op amp input terminals would be shielded. The shielding has not been shown in Fig. 12-9(b) in order to avoid cluttering the diagram.

The Tare Circuitry. Since part of the gross weight signal is due to the weight of the hopper and its support cables, provision is made to subtract this weight from the load cell signal. The weight which is subtracted from the gross weight indicated by the load cell is called the *tare weight*. The pot which produces the tare weight subtraction is called the *tare pot*. The final signal obtained after the tare weight has been subtracted from the gross weight is called the *measured net weight*. To understand the tare weight subtraction process, refer to the circuitry in Fig. 12-9(b).

The load cell bridge is excited by a stable 15-V ac supply, from secondary winding A of transformer T_1. This winding is identified as S_A in the schematic diagram. The phase relationships between the various voltages in this circuit are important, so the phases are clearly marked by phasing dots. The ac signal taken from the Wheatstone bridge is a small voltage, only a few millivolts, and it represents the gross weight supported by the load cell. This signal is labeled GROSS WEIGHT in Fig. 12-9(b). At some instant in time it will be positive on the top and negative on the bottom, as indicated. The TARE WEIGHT signal is taken from the manually adjusted tare pot, which is excited by the S_B winding. This signal is positive on the left and negative on the right at that same instant in time. This polarity relationship is established by the phase relationship between the two secondary windings S_A and S_B.

Since the GROSS WEIGHT signal and the TARE WEIGHT signal are opposite in phase, the resultant signal is the *difference* between these two voltages. In other words, the TARE WEIGHT signal has been subtracted from the GROSS WEIGHT signal. This voltage difference is labeled the MEASURED NET WEIGHT signal in Fig. 12-9(b).

Of course, someone had to adjust the manual tare pot to the proper position before any material was loaded into the hopper. This is done by simply turning the tare pot until the scale weight indicator reads zero when the hopper is empty.

The Scale Weight Indicator (a Servo System). The scale weight indicator is a servo system with the positioned object being the scale-indicating pointer. The scale-indicating pointer is a needle moving over a calibrated weight dial, just like a penny scale in a shopping arcade. As shown in Fig. 12-9(b), the position of the scale-indicating pointer is represented electronically by the wiper position of the scale indicator potentiometer. The scale indicator pot wiper is attached to the same shaft as the scale-indicating pointer. This is brought out clearly in Fig. 12-10(a). Therefore, both the scale indicator pot wiper and the scale-indicating pointer indicate the measured net weight. The scale-indicating pointer does it mechanically/visually, and the scale-indicating potentiometer does it electronically.

The INDICATED WEIGHT signal in Fig. 12-9(b) is positive on the bottom at the reference time instant. This is due to the S_C secondary winding driving the scale indicator pot. The INDICATED WEIGHT signal and the MEASURED NET WEIGHT signal are thus in phase opposition. The difference between these two signals is the error signal, labeled V_{error} in Fig. 12-9(b).

V_{error} is applied to an op amp, which furnishes the input stage of the servo amplifier. The op amp is connected as a differential amplifier, with a voltage gain of 5000 (500 K/100 Ω = 5000). Therefore V_{out} from the differential amplifier is 5000 times as large as V_{error}. This very high voltage gain is necessary because the signals which are being handled are so small.

V_{out} from the op amp differential amplifier is used to drive the control winding amplifier, which is a *complementary symmetry amplifier*. The complementary symmetry amplifier is distinguished by the use of one *npn* and one *pnp* transistor. The complementary symmetry transistors are specially matched to have identical current-voltage characteristics, except, of course, that the polarities are opposite from each other. Q_1 handles and amplifies the positive half cycles of V_{out}, and Q_2 handles the negative half cycles of V_{out}. This amplification scheme is an alternative to the push-pull amplifiers discussed in Chapter 10. It provides the same advantage, namely cooler operation of the output transistors because they dissipate no dc power.

Notice that the Q_1-Q_2 emitter tie point is grounded; the +12 and −12-V supplies are not grounded. These supplies deliver current to the servo motor control winding only, and they are completely separate from the ±15-V supplies to the op amp. There is no ground reference between these two pairs of power supplies.

V_{out}, the op amp output signal, swings positive and negative relative to ground. As it goes positive it forward-biases the base-emitter junction of Q_1, causing that transistor to conduct. Q_1 then passes current through the servo motor control winding from left to right. The source of this current is the +12-V supply.

As V_{out} from the op amp goes negative relative to ground, it forward-biases the base-emitter junction of *pnp* transistor Q_2, causing it to conduct. Q_2 then passes current through the control winding from right to left. The source of this current is the −12-V supply. The control winding current is therefore an ac current synchronized with the voltages in the measuring circuits, which are themselves synchronized with the ac power line. The servo motor fixed winding current is 90° out of phase with the ac power line. This is due to the fact that the fixed winding is excited by the T_2 secondary, through a phase-shift capacitor. Transformer T_2 is powered by the same ac power lines that drive T_1. The T_2 primary winding, although not shown in Fig. 12-9(b), would be connected to the ac lines to the left of the line filter.

The servo motor will run whenever an error voltage (V_{error}) exists. As illustrated

Tens

The Units Shaft makes
10 complete Revolutions as
the Tens (Servo) Shaft makes
1 Revolution

(a)

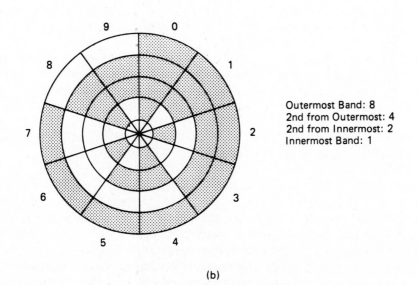

Outermost Band: 8
2nd from Outermost: 4
2nd from Innermost: 2
Innermost Band: 1

(b)

Figure 12-10. (a) Side view of the servomechanism. The servo motor shaft on the left positions the weight-indicating pointer and the scale indicator pot wiper. The servo motor shaft on the right positions the two binary-coded wheels. These wheels optically convert the shaft position into a digital weight signal. (b) Face view of one of the binary-coded wheels.

in Fig. 12-10(a), the servo motor shaft is linked to the wiper arm of the scale indicator pot, and always drives it in the proper direction to reduce V_{error} to zero. Thus the servo system continually equalizes the INDICATED WEIGHT signal and the MEASURED NET WEIGHT signal. In this way the scale-indicating pointer constantly points to the correct net weight on the calibrated weight dial.

12-6-3 The optical weight readout

A side view of the servomechanism is shown in Fig. 12-10(a). The servo motor is geared down so its output shaft spins slowly. The output shaft is referred to as the *servo shaft*, and it is double-ended. One end of the servo shaft is used to position the scale-indicating pointer and the scale indicator pot wiper. The other end is used to position two binary-coded wheels. The binary-coded wheels enable the optoelectronic circuits to read the indicated weight in a digital manner.

Here is how the coded wheels read the weight. The face view of one of the binary-coded wheels is shown in Fig. 12-10(b). Both wheels are identical, but let us temporarily concentrate on the tens wheel only. As can be seen, the wheel is divided into 10 equal sectors, each sector representing one of the decimal digits 0–9. The distance that the servo shaft rotates determines which one of these sectors will come into the topmost position, between the light sources and the photocells. If the number 5 sector, for example, comes into the top position, between the light sources and photocells, the information given out by the photocells should represent the decimal number 5. If the number 6 sector comes into the top position, the photocells should represent the decimal number 6, and so on.

The decimal numbers associated with the wheel sectors are coded in the familiar 8421 binary code by virtue of the light and dark areas in the sectors. Notice that the wheel (and each sector) is composed of four concentric *bands*, or rings. The outermost band indicates the presence or absence of the binary 8 bit. The next to the outermost band indicates the 4 bit. The next to the innermost band represents the 2 bit, and the innermost band represents the 1 bit. With this in mind let us from now on refer to the bands as the *8 band*, the *4 band*, and so on.

When a given sector moves into position between the lights and photocells it will pass the light beams in those bands where it is light, and it will block the light beams in those bands where it is dark. The presence of light at the photocell receptor stands for a binary 1, and the absence of light stands for a binary 0. In this way, a four-bit binary number can be represented.

As an example, consider the number 5 sector in Fig. 12-10(b). Its 8 band and its 2 band are both dark, but its 4 band and its 1 band are light. If the number 5 sector moves into the top position, photocell 4 and photocell 1 will be illuminated. Photocell 8 and photocell 2 will remain dark. The output of the photocells is thus 0101, reading from outermost to innermost (from 8 to 1). This is the binary code for the decimal number 5.

It is a fairly easy task to verify that the coded wheel satisfies the binary code for each of the 10 decimal digits.

The units wheel does exactly the same thing as the tens wheel, except that it rotates *10 times as far*. This is accomplished by mechanical gearing, as shown in Fig. 10-10(a). The gear on the servo shaft has a diameter 10 times as great as the gear on the units shaft. Therefore, for one revolution of the servo shaft, the units shaft goes through 10 revolutions.

To get the idea of how the entire readout mechanism works, assume that the full scale weight is 100 lb. That is, the servo shaft goes through a complete revolution when 100 lb of powder is loaded into the hopper. This being so, the servo shaft will go through one tenth of a revolution for each 10 lb of powder loaded into the hopper. This translates into one *complete* revolution of the *units shaft*.

Therefore, as 10 lb of material is slowly poured into the hopper, the units wheel goes through and indicates each one of its 10 digits in turn. When it returns to 0 after a complete revolution, the tens wheel has just completed one tenth of a revolution, and it is changing from the 0 sector to the 1 sector. This action is repeated as the weight of material goes from 10 to 20 lb, and again for every 10 lb thereafter, all the way up to 99 lb (the mechanism cannot read 100 lb with only two wheels).

As an example, suppose that 72 lb of material is loaded into the hopper. The servo shaft will turn a little more than seven tenths of a revolution, so the number 7 sector of the tens wheel will be in the top position. The units shaft will have turned seven complete revolutions plus two tenths of another revolution. Therefore the number 2 sector of the units wheel will be in the top position. The number 7 sector is dark-light-light-light, reading from the outermost ring to the innermost ring, and the number 2 sector is dark-dark-light-dark, again reading from outermost to innermost. Therefore the output of the two groups of photocells will be

$$0111 \quad 0010,$$

which represents 72 in BCD.

12-6-4 The automatic cycle logic

Figure 12-11 shows the logic circuitry for controlling the system. Notice first that each readout photocell is amplified by a transistor switch and then put into a logic inverter. The inverters then feed into BCD to 1-of-10 decoders of the same type that we saw in Chapter 3 (Fig. 3-15). The net weight of material in the hopper thus appears at the output of the two decoders. For example, if there are 72 lb of powder in the hopper, the 7 output terminal of the TENS DECODER in Fig. 12-11 will go HI and the 2 output terminal of the UNITS DECODER will go HI. All the other 18 output terminals will stay LO.

The decoders feed two pairs of selector switches. The first pair of switches, which are called the DESIRED WEIGHT SWITCHES, set the *desired* weight of material. The second pair of switches, which are called the SLOW FEED SWITCHES, set the weight at which the loading of the hopper shifts from fast speed to slow speed. As the material in the hopper approaches the desired weight, the hopper loading speed is shifted in order to prevent overshoot of the final desired weight.

In Fig. 12-11, AND2 is the gate that detects when the slow feed weight has been reached. Its inputs come from the common terminals of the SLOW FEED

Figure 12-11. The weighing cycle control circuitry. The photocell detectors on the far left are the ones shown physically in Fig. 12-10(a). When the outputs of the two decoders match the setting on the two SLOW FEED SELECTOR SWITCHES, the powder feed rate slows down. When the decoder outputs match the setting on the two DESIRED WEIGHT SELECTOR SWITCHES, the feed conveyor stops altogether.

SELECTOR SWITCHES. AND3 is the gate that detects when the full desired weight has been loaded into the hopper. Its inputs are from the common terminals of the DESIRED WEIGHT SELECTOR SWITCHES. In addition, there is a HOPPER EMPTY detection gate, AND1. Its inputs are hard-wired to the 0 output terminals of the TENS and UNITS DECODERS. The output of AND1 goes HI when the hopper is empty, and the net weight equals 0 lb.

The cycle operation is not complicated. After the scale has been manually tared by adjusting the tare pot, the operator presses the START FEED push button. The output of the switch filter applies a HI to NORI, which delivers a negative clock edge to the FAST FEED FLIP-FLOP. Both J and K are HI, so the flip flop toggles into the ON state. This enables output amplifier OA1, which energizes the FAST FEED MOTOR STARTER, MSF. The screw conveyor starts running at high speed, feeding powder into the hopper rapidly.

This continues until the optical weight detectors reach the number set on the SLOW FEED SELECTOR SWITCHES. At this instant, both inputs of AND2 go HI, and its output goes HI. This HI appears at the inputs of NOR1 and NOR2, causing their outputs to go LO. Both the FAST FEED and the SLOW FEED FLIP-FLOPS receive a negative clock edge, and they both toggle. The FAST FEED FLIP-FLOP turns OFF, deenergizing MSF, and the SLOW SPEED FLIP-FLOP turns ON, energizing MSS, the SLOW FEED MOTOR STARTER. The screw conveyor slows down and continues feeding powder into the hopper at a slower rate.

This continues until the weight detectors reach the number set on the DESIRED WEIGHT SELECTOR SWITCHES. At this instant, both inputs of AND3 go HI, and the output delivers a HI to NOR2 and I1. As the negative clock edges arrive at their CLK terminals, the SLOW FEED FLIP-FLOP turns OFF, and the STOP FEED FLIP-FLOP turns ON. The Q output of the SLOW FEED FLIP-FLOP goes LO, disabling OA2. The SLOW SPEED MOTOR STARTER deenergizes, stopping the rotation of the screw conveyor. Therefore, the flow of powder stops when the net weight in the hopper equals the desired weight.

At the same instant, the Q output of the STOP FEED FLIP-FLOP energizes the discharge solenoid through OA3. The discharge valve opens, and the weighed powder flows out of the hopper. When the hopper is empty, the HOPPER EMPTY line from AND1 goes HI. This causes both NAND1 inputs to be HI, which delivers a negative edge to the trigger terminal of the one-shot. The one-shot fires, causing a LO to appear at its $\bar{Q}$ output. This LO is applied to the clear input of the STOP FEED FLIP-FLOP, causing it to turn OFF. The discharge solenoid deenergizes, and the cycle is finished.

12-6-5 Other codes and encoding methods

The optical encoder presented in Sec. 12-6-3 uses true binary code to represent the number of each sector. This is the most obvious way to represent a number on an encoding wheel, and it has the desirable feature of being easily decoded. That is, it is simple to decode the wheel information by inserting it into a BCD to 1-of-10 decoder, as shown in Fig. 12-11. However, the true binary code has a very serious drawback when used on an encoding wheel. The problem arises when the wheel stops right on the dividing line between two sectors.

For example, consider what might happen if the wheel of Fig. 12-10(b) stopped exactly on the line between the number 2 sector and the number 3 sector. Since the 1 band changes from dark to light at the sector dividing line, photocell 1 will be half ON and half OFF and will not know what to do. Photocells 2, 4, and 8 have no ambiguity since there is no color change at the sector dividing line in those bands. Therefore, the overall photocell output might be a decimal 2 (0010) or it might be a decimal 3 (0011); there is no way to know what it will do. In this particular example the problem is not too serious because if the position is halfway between 2 and 3 it doesn't really matter what the encoder calls it; either result is very close to the truth.

However, suppose that the wheel lands on the dividing line between the number 7 sector and the number 8 sector. In this case, *all* the bands change color at the dividing line. With all four photocells in an ambiguous condition, it is conceivable that the photodetectors might *all* read 0. In this case, the error is serious. We wouldn't have minded if the encoder had called the number a 7 and we wouldn't have minded if the encoder had called the number an 8, because either of those results is quite close to the truth. But we definitely *do* mind when the encoder calls the number a 0.

The essential difference between the 2-3 situation and the 7-8 situation is that only a single band changed color at the 2-3 dividing line, while several bands (four) changed color at the 7-8 dividing line. Whenever more than one band changes color at a sector dividing line, serious encoding errors are possible. To solve this problem, other codes have been invented in which *only one bit changes at a time.* That is, as you move from one number to the next number, only one of the constituent binary bits changes state. These other codes are still *binary* codes because they involve only 0s and 1s, but we make a distinction between them and the familiar 8421 binary code, which is referred to as *true binary.*

The most famous and popular of these other binary codes is the *Gray code.* Table 12-1 shows the Gray code equivalent for each of the decimal numbers 0–15. For purposes of comparison, the true binary representation is shown alongside the Gray code.

In Table 12-1, notice that one and *only one* bit changes state in the Gray code as the coded number changes. This eliminates the possibility of serious encoding errors of the type described above. It is not worthwhile to memorize the Gray code, since it is not encountered nearly as often as true binary. Its chief application is in optical (and mechanical) position encoders. The Gray code is sufficiently popular for use with position encoders that the IC manufacturers package and sell Gray code to 1-of-10 decoders.

Another item worth mentioning is that two encoding wheels geared together as shown in Fig. 12-10(a) is not a popular arrangement. The preferred method is

TABLE 12-1. The Gray Code equivalents to the decimal numbers 0 through 15, along with the binary code equivalents.

Decimal	Gray Code	True Binary
0	0000	0000
1	0001	0001
2	0011	0010
3	0010	0011
4	0110	0100
5	0111	0101
6	0101	0110
7	0100	0111
8	1100	1000
9	1101	1001
10	1111	1010
11	1110	1011
12	1010	1100
13	1011	1101
14	1001	1110
15	1000	1111

simply to divide the encoding wheel into more sectors and encode each sector uniquely. After all, the true binary code and the Gray code don't stop at 9; they go on forever. For this example, the wheel could have been divided into 100 sectors, and each sector could have had its own unique Gray code representation. The number of bands would have to be greater, of course. It would require seven bits to encode up to the decimal number 100. Therefore it would take a seven-banded encoding wheel to encode a 100-lb scale in 1-lb graduations. The decoding circuitry would also have to be different, but such decoding circuitry is available.

12-7 A CARBON DIOXIDE CONTROLLER FOR A CARBURIZING FURNACE

12-7-1 The carburizing process

A steel part can be given a very hard exterior layer by diffusing free carbon into its surface. The metallurgical process of diffusing carbon into steel is called *carburizing*. Carburizing is normally accomplished by subjecting the steel to a fairly high temperature, about 1700°F, for several hours, in the presence of a *carburizing atmosphere*. A carburizing atmosphere is a mixture of normal combustion products with a specially manufactured gas. The manufactured gas contains heavy concentrations of carbon monoxide (CO) and carbon dioxide (CO_2). By adjusting the composition of the gases in the carburizing atmosphere, the carbon content and carbon depth of the steel can be varied to meet different requirements. Usually the depth of carbon penetration is about 0.050 in. into the surface.

The physical arrangement for controlling the carburizing atmosphere is illustrated in Fig. 12-12. The combustion of fuel and air takes place at the burner mounted on the sidewall of the furnace. The combustion products expand and occupy the interior space of the carburizing furnace. At the same time, the special carburizing gas flows through a variable valve and into the furnace chamber. If the valve is opened further to admit a greater flow of carburizing gas, the concentration of carbon compounds in contact with the steel increases. If the valve is closed somewhat, the flow of carburizing gas decreases, and the concentration of carbon compounds in the furnace decreases.

Figure 12-12. Physical layout of a carburizing furnace. The fuel and combustion air are shown entering on the left, the carburizing gas enters on the right, and an atmosphere sample is drawn from the center.

The valve position is controlled by a system which compares the actual composition of the atmosphere to the desired composition. If the control circuitry finds any discrepancy between actual and desired composition, it adjusts the valve opening accordingly.

The control system obtains a sample of the atmosphere via a small tube leading out of the furnace chamber. A blower at the other end of the tube continually sucks a fresh sample of atmosphere gas through the tube to the system's measurement device.

It has been found that the best all-around method of controlling a carburizing atmosphere is to control its *carbon dioxide concentration*. Therefore, modern atmosphere control systems measure the concentration of CO_2 in the atmosphere sample and control that concentration.

The CO_2 control system circuitry is fairly extensive, and is illustrated in Figs. 12-13, 12-14, and 12-15. Figure 12-13 shows the measurement circuitry for determining the CO_2 concentration. Figure 12-14 shows the mechanical construction of the measurement device itself. Figure 12-15 shows the error detector, controller, and final correcting device circuitry, which make up the rest of the closed-loop control system.

12-7-2 Measuring the CO_2 concentration

Let us look first at Fig. 12-13. On the far left is an oscillator. It is crystal controlled at an operating frequency of 10 MHz. Transistor Q_1 is biased in the conducting state by the $+15$- and -15-V supplies in conjunction with R_1 and R_2. The oscillator feedback path is from the center of the transformer primary winding to the R_2-R_3 junction in the emitter lead. The transistor is wired in common-base configuration, because oscillators built around a common-base amplifier are inherently more temperature stable than oscillators built around a common-emitter amplifier.

This oscillator has a frequency stability of better than 0.01% under normal operating conditions. That is, its output frequency will not stray more than 0.01% from the nominal operating frequency of 10 MHz. The 10 MHz ac signal at point A creates a 10-MHz sine-wave signal across the secondary winding of the coupling transformer. This winding is connected directly to capacitor C_v, which is a *variable* capacitance. Its exact capacitance varies in relation to the CO_2 concentration in the atmosphere sample being measured. We will see shortly how this is accomplished.

The secondary winding inductance in combination with the C_v capacitance comprise an LC tank circuit. The values of the winding inductance L and the capacitance C_v are such that this tank circuit has a resonant frequency of about 10 MHz. However, the LC circuit is thrown slightly out of resonance as the size of C_v varies. If the value of C_v is varied at a frequency of 9 Hz (which is the case, as we will see), the 10-MHz carrier signal appearing across the secondary winding will be *amplitude modulated* at a frequency of 9 Hz. This happens because the 10-MHz current circulating in the LC tank circuit changes in magnitude as the tank circuit comes in and out of resonance. The amplitude-modulated 10 MHz signal is shown at point B in the circuit.

Let us stop here in our investigation of Fig. 12-13 and transfer our attention to Fig. 12-14. This figure illustrates the construction of the apparatus that causes C_v to vary in relation to the CO_2 concentration. The apparatus has two infrared radiation sources, which radiate energy into two gas-filled cylinders, or *cells*. These two cells are called the *reference cell* and the *sample cell*, as shown in Fig. 12-14(a). The reference cell is sealed tight and contains pure oxygen. The sample cell experiences a constant flow of fresh atmosphere gas. It "samples" the carburizing atmosphere of the furnace.

Gaseous oxygen and gaseous carbon dioxide show an interesting and useful contrast, on which this whole measuring method is based. This contrast is that

Figure 12-13. Circuitry for measuring CO_2 concentration. The waveforms at various points in the circuit are drawn in *A–H*. (*Courtesy of Beckman Instruments, Inc.*)

Figure 12-14. Physical appearance of the CO_2 measurement transducer. (a) At an instant when the infrared chopper is *not* blocking the infrared sources, infrared radiation is being admitted to the reference and sample cells. (b) At an instant when the infrared chopper *is* blocking the infrared sources, no infrared radiation enters the cells. (c) Face view of the infrared chopper. (d) Face view of the LED chopper. The angular position shown corresponds to the angular position shown in part (c) for the infrared chopper. (*Courtesy of Beckman Instruments, Inc.*)

oxygen gas *will not* absorb infrared energy, but carbon dioxide gas *will* absorb infrared energy. Furthermore, the amount of infrared energy that a gas *mixture* absorbs depends on the *concentration of CO_2* in the gas. If the CO_2 concentration is light, not very much infrared radiant energy will be absorbed; if the CO_2 concentration is heavy, a lot of infrared radiant energy will be absorbed.

The two infrared sources radiate an equal amount of energy into each cell. *All* the energy radiated into the reference (oxygen-filled) cell passes through the cell without being absorbed. It emanates from the other end of the reference cell and passes into a *detector cell*. The detector cell is divided into two separate compartments, sealed off from each other by a thin metal diaphragm. Both compartments are filled with pure carbon dioxide. The CO_2 gas in the left compartment, therefore, receives and absorbs *all* the infrared energy that was radiated from the infrared source on the left of Fig. 12-14(a). This causes a tiny increase in the temperature of the CO_2 in the left side, resulting in a tiny increase in pressure on the left side of the thin metal diaphragm.

Now consider what happens on the right-hand side. The energy radiated from this infrared source does not all pass through the sample cell. A certain amount is absorbed by the CO_2 in the atmosphere sample, and a reduced amount of energy emanates from the other end. This is suggested by the fact that there are only two wavy lines leaving the sample cell, although *three* wavy lines are shown entering the sample cell in Fig. 12-14(a). The CO_2 in the right side of the detector cell absorbs this diminished infrared energy, and it too experiences an increase in temperature. The right side does not experience as much change as the left side, though, because it absorbs less energy than the left side. Therefore, the pressure on the right side does not increase as much as the pressure on the left side of the metal diaphragm. The pressure imbalance causes the flexible metal diaphragm to bulge to the right, as shown. There is a lead wire connected to the metal diaphragm, and there is a stationary metal plate next to the diaphragm, also having a wire leading out of the cell. These two metal objects (the stationary plate and the metal diaphragm) form a capacitor, whose capacitance naturally depends on the spacing between the objects. If they are close together, the capacitance C_v is high; if they are farther apart, C_v is low.

It should be clear that the capacitance of C_v depends on the CO_2 concentration in the sample gas. The greater the CO_2 concentration, the more energy is absorbed in the sample cell. The more energy absorbed in the sample cell, the less energy is delivered to the right side of the detector cell. The less energy delivered to the right side, the lower the pressure on the right side. This lower pressure results in the greater bulging of the metal diaphragm and consequently a reduction in capacitance. The overall result is that a high concentration of CO_2 causes a low value of C_v.

The infrared energy sources are not permitted to continually radiate into the reference and sample cells. A slow speed gear motor (the *chopper motor*) slowly rotates the *infrared chopper*, which is a blade shaped as shown in Fig. 10-14(c). As the chopper motor shaft rotates, the blades alternately expose and block the infrared sources. Each rotation of the shaft causes two cycles of exposure and blocking. When the infrared sources are blocked, no energy enters the reference cell, the sample cell, or the detector cell. The temperatures and pressures on the left and right sides of the detector cell have time to equalize, and the diaphragm returns to normal shape. This situation is illustrated in Fig. 12-14(b), in which the infrared chopper blade has rotated 90° and is blocking the sources.

The chopper blade rotates at 4.5 revolutions/sec (270 rpm). Each revolution produces two cycles of bulging and retraction of the metal diaphragm and therefore two cycles of capacitance variation. Thus the capacitance C_v varies at a frequency of 9 cycles/sec, and the magnitude of the capacitance variation is proportional to the concentration of carbon dioxide in the sample.

The 9-Hz variation of C_v produces the AM signal appearing at point *B* in Fig. 12-13. We will now continue with our discussion of the Fig. 12-13 circuit.

The AM signal is applied to a *demodulator* consisting of a diode and a 0.2-μF capacitor. These components *demodulate* the signal, removing the 10-MHz carrier while retaining the 9-Hz information signal. This 9-Hz signal appears at point *C* in the circuit.

The 9-Hz sine wave is applied to a 100-K pot, where a portion of it is picked off and fed into op amp 1. Op amp 1 is wired as a noninverting amplifier, having a voltage gain of about 200. The 9-Hz sine wave therefore appears in an amplified condition at point *D*. The amplified signal is routed down to op amp 2, which is wired as an inverting amplifier with a gain of 1 (a phase inverter). Therefore the only action of op amp 2 is to invert the 9-Hz signal. The signal appearing at *E* is thus 180° out of phase with the signal at *D*.

Both of the signals are delivered to the R_8, R_9, PC1, PC2 network. This network creates a full-wave rectified signal at *F*. Here is how it works. The *LED chopper* in Fig. 12-14 is mounted on the other end of the chopper motor shaft. It is shaped as shown in Fig. 12-14(d), and it is mounted so that it blocks the beam to photocell 1 (PC1) and passes the beam to photocell 2 (PC2) at the time the infrared chopper is passing the infrared radiation. This is illustrated in Fig. 12-14(a). When the shaft rotates 180°, the LED chopper blocks the beam to PC2 and passes the beam to PC1; at this instant the infrared chopper is blocking the infrared radiation. This is illustrated in Fig. 12-14(b). To visualize the synchronization between the infrared chopper and the LED chopper, look at Figs. 12-14(c) and (d) together. The positions of the infrared sources and photocells are clearly shown. Imagine these two choppers rotating in unison, and you will see the relationship between the photocell chopping and the infrared chopping.

PC1 becomes nearly a short circuit (it is illuminated) during the positive half cycle of the *D* sine wave, while PC2 is nearly an open circuit (it is dark). Therefore the positive half cycle of *D* appears at *F*. PC2 becomes nearly a short circuit during the negative half cycle of *D*, which is the positive half cycle of *E*. Therefore the positive half cycle of *E* appears at *F*. The resultant waveform at *F* is the full-wave rectified signal which is illustrated.

This full-wave rectified signal is passed into a dc filter consisting of R_{10}, R_{11}, and the three associated capacitors. The output of the filter is a small dc voltage, which falls between 0 and 250 mV. This is shown at point G in Fig. 12-13. The voltage is made to equal 250 mV when the CO_2 concentration equals some arbitrarily selected maximum value. This adjustment is made by the 100-K pot feeding op amp 1.

Op amp 3 is a noninverting amplifier with a voltage gain of 20. Therefore the output of op amp 3, labeled V_{meas}, falls between 0 and 5 V, depending on the carbon dioxide concentration. The purpose of the 1-μF feedback capacitor is to slow down the transient response of the amplifier. A sudden temporary variation in CO_2 concentration cannot cause a sudden variation in V_{meas}; a concentration change must persist for several seconds before V_{meas} will reflect it. This eliminates the effect of short-lived fluke variations in the sample concentration.

12-7-3 The error detector, the controller, and the final correcting device

The error detector is an op amp differential amplifier, shown at the left of Fig. 12-15(a). The inputs to the differential amplifier are two voltages, V_{meas} and $V_{st\ pt}$. V_{meas} represents the *measured* CO_2 concentration, and $V_{st\ pt}$ represents the *desired* CO_2 concentration. $V_{st\ pt}$ is derived from the set-point pot, which is excited by a stable 5-V supply. The voltage gain of the differential amplifier is 200 K/5 K = 40. The output, V_{error}, is given by

$$V_{error} = 40(V_{meas} - V_{st\ pt}).$$

Of course, V_{error} is subject to the saturation restriction of the op amp. For ±15-V supplies, V_{error} cannot exceed about ±13 V.

V_{error} is applied to the parallel combination of C_9 and R_{19}, but it is applied only once every 3 minutes, when the Q_2 transistor switch is pulsed ON. The turn-ON pulse has a duration of 100 msec, which is enough time to charge the 250μF cap to the full value of V_{error}. The C_9-R_{19} combination has a discharge time constant

$$= (4\ K)(250\mu F) = 1\ sec.$$

Therefore, when Q_2 turns back OFF after charging C_9 to V_{error}, the capacitor voltage (V_{C_9}) discharges back to zero with a 1-sec time constant (5 sec for full discharge).

Consider the graph in Fig. 12-15(b). These curves represent the discharge behavior of V_{C_9} for various values of V_{error}. The curves show a discharge from a *positive* voltage to zero, but keep in mind that the actual situation could be a discharge from a *negative* voltage to zero. The polarity of V_{error} depends on the direction of the deviation from set point.

The 3 V line is important in the discharge curves, and it is specially marked in Fig. 12-15(b). As long as V_{C_9} is greater than 3 V, it can operate one of the relaxation oscillators shown in Fig. 12-15(a). First let us consider what happens if V_{C_9} is more positive than $+3$ V. Then we will consider what happens when V_{C_9} is more negative than -3 V.

Figure 12-15. (a) Schematic diagram of the error detection circuitry, the controller circuitry, and the final correcting device. (b) Family of discharge curves for C_9. The greater the error signal (V_{error}), the longer it takes for C_9 to discharge to 3V. (c) Waveforms of ac supply voltage, triac main terminal voltage, and gate voltage. As long as the gate pulses occur, the triac continues to turn ON immediately after a zero crossover. When the gate pulses cease, the triac remains OFF.

(b)

(c)

Figure 12-15. (Cont.)

500

If V_{C_9} is more positive than $+3$ V, it will forward-bias diode D1 and still be able to break down the 1.5-V zener diode ZD1. It will require at least 2.1 V to accomplish this, since the voltage necessary to turn on this diode pair is given by

$$V_{\text{diodes}} = 0.6 \text{ V} + V_z = 0.6 \text{ V} + 1.5 \text{ V} = 2.1 \text{ V}.$$

The remaining 0.9 V is then available to drive the relaxation oscillator, which requires a supply voltage of about 0.9 V to operate properly. Therefore as long as V_{C_9} is greater than 3 V, the pair of diodes will be broken down, and sufficient voltage will appear at the top of relaxation oscillator 1. Relaxation oscillator 1 has an operating frequency given by Eq. (5-5):

$$f = \frac{1}{R_E C_E} = \frac{1}{(100 \times 10^3)(0.005 \times 10^{-6})} = 2 \text{ kHz}.$$

Gate pulses are thus delivered to triac 1 at a frequency of 2 kHz. This barrage of gate pulses keeps triac 1 turned completely ON during the time that the pulses continue. With gate pulses arriving at such a high frequency, the triac is bound to be fired very shortly after the supply voltage passes through 0 V. The circuit waveforms are shown in Fig. 12-15(c). The gate pulses which appear during the remainder of the half cycle have no effect on the triac, since it has already fired. However, the continuation of the rapid string of pulses ensures that the triac will fire again just as soon as the *next* half cycle begins.

The upshot is that as long as V_{C_9} exceeds 3 V, triac 1 will remain ON, and the valve motor will continue to run. The direction of rotation is such that it *closes* the control valve. This is the proper direction of rotation, since a positive V_{error} means that the measured CO_2 concentration is greater than desired. The closing of the atmosphere control valve will restrict the flow of carburizing gas and decrease the CO_2 concentration.

The length of time that the motor runs is determined by the length of time that V_{C_9} stays above 3 V. As Fig. 12-15(b) clearly shows, that length of time depends on the initial charge on C_9, that is, on V_{error}. If V_{error} is large, V_{C_9} remains above 3 V for a longer time, resulting in a greater correction to the flow of carburizing gas. This is the proper action because a large positive V_{error} means that the measured CO_2 concentration is too high by a large amount.

Now consider what happens if V_{error} is more negative than -3 V. In this case, D2 and ZD2 are turned on, and a negative voltage can appear at the bottom of relaxation oscillator 2. This relaxation oscillator delivers 2-kHz gate pulses to triac 2, which drives the valve motor in the opposite direction. With triac 2 firing, capacitor C_{12} is effectively in series with motor winding A, whereas when triac 1 was firing, C_{12} was in series with motor winding B. This reverses the direction of rotation, as explained in Chapter 10.

Again, the length of time that V_{C_9} stays more negative than -3 V depends on the initial charge on C_9. The curves of Fig. 12-15(b) still apply, even though the initial voltage is negative instead of positive. If V_{error} is a large negative voltage, the motor will run for a longer time, opening the control valve a greater amount.

This is the proper action since a large negative V_{error} means that measured CO_2 concentration is too low by a large amount.

This atmosphere control system has a very long transfer lag and transportation lag. The transportation lag is due to the necessity to suck a new atmosphere sample through the tube leading to the CO_2 detector. This tube is usually quite long for a carburizer, perhaps 50 ft or more. The transfer lag depends on how quickly a change in carburizing gas flow can make its effects felt throughout the carburizing chamber. As can be imagined, this does not happen quickly. Because of these long lags, *continuous control* of carburizing atmospheres is never attempted. Instead, a control action is followed by 3 min of no control. At the end of 3 min, another control action is initiated if needed.

12-8 CONTROL OF RELATIVE HUMIDITY IN A TEXTILE MOISTENING PROCESS

In certain textile finishing processes, the moving textile strip is passed through a moistening chamber to be moistened and softened. The relative humidity in the chamber is maintained at a high level by spraying water into a duct through which the chamber air recirculates. This system layout is illustrated in Fig. 12-16(a).

The moving textile strip enters the moistening chamber through the wall on the left. It passes over several moving rolls as it moves through the chamber to where it exits through the wall on the right. The strip is dry when it enters the chamber and wet when it leaves the chamber, so it is constantly removing water vapor from the air. This loss of water vapor must be continually replenished to maintain the relative humidity at the proper value (about 80%).

The water vapor is replenished by sucking the chamber air into the recirculating duct on the right-hand side of Fig. 12-16(a). As the air passes through the recirculating duct, it encounters a battery of mist nozzles. It picks some of the water vapor from the mist and reenters the moistening chamber on the left. The air reenters much wetter than when it left the chamber.

The amount of water vapor that the air absorbs as it flows through the recirculating duct depends on the water flow admitted to the mist nozzles. This flow is controlled by the pneumatic diaphragm control valve in the water supply line. If the valve is farther open, it allows a greater water flow to the mist nozzles, which then provide more vigorous misting.

The diaphragm control valve is the electropneumatic type shown in Fig. 10-4. The amount of valve opening is proportional to the amount of current through the electromagnet input coil which moves the balance beam. Therefore, the water flow rate and the amount of mist are determined by the amount of current delivered to the coil. The electropneumatic apparatus is designed to respond to currents in the range of 2 to 10 mA. That is, if the coil current drops to 2 mA or less, the valve shuts completely off. If the current rises to 10 mA, the valve opens wide. For

(b)

Figure 12-16. A strip moistening system. (a) Physical layout of the moistening chamber and water feed pipes. (b) Circuit for detecting and controlling the humidity.

current values between 2 and 10 mA, the valve is somewhere in the throttling range between wide open and completely closed.

The sensing coil is driven by an op amp voltage-to-current converter. The operation of a voltage-to-current converter was explained in Sec. 8-9. The electronic circuit which drives the sensing coil is shown in Fig. 12-16(b). Here is how it works.

The Wheatstone bridge is driven by a stable 10-V rms ac supply. A hygroscopic humidity transducer (hygrometer) is used for resistor R_3. A resistive hygrometer must be excited by ac voltage only. If dc current passes through it for any length of time, it will become chemically polarized, and its characteristics will change. Therefore, it cannot be used in a dc Wheatstone bridge. This transducer has the resistance characteristics shown in the graph of Fig. 11-22(b). The important data point from that graph that concerns us is

Relative Humidity	*Resistance*
80%	3.7 K

Let us assume that the humidity in the textile moistening chamber is at 80%, which is in the center of the acceptable range. The value of R_3 is then 3.7 K. The Wheatstone bridge ac output voltage, V_{bridge}, is equal to the difference between the voltage across R_2 and the voltage across R_4, or

$$V_{bridge} = V_{R_2} - V_{R_4}.$$

V_{R_2} can be calculated as

$$\frac{V_{R_2}}{10 \text{ V rms}} = \frac{R_2}{R_2 + R_1} = \frac{600 \text{ }\Omega}{600 \text{ }\Omega + 150 \text{ }\Omega},$$

$$V_{R_2} = 8.0 \text{ V rms}.$$

When the relative humidity equals 80%, V_{R_4} is given by

$$\frac{V_{R_4}}{10 \text{ V rms}} = \frac{R_4}{R_4 + R_3} = \frac{4.7 \text{ K}}{4.7 \text{ K} + 3.7 \text{ K}},$$

$$V_{R_4} = 5.6 \text{ V rms}.$$

Therefore, at 80% RH,

$$V_{bridge} = 8.0 \text{ V rms} - 5.6 \text{ V rms} = 2.4 \text{ V rms}.$$

The peak voltage supplied to the ac terminals of the bridge rectifier is given by

$$V_p = (1.41)(V_{bridge}) = 1.41(2.4 \text{ V}) = 3.4 V_p$$

The bridge rectifier introduces a total voltage drop of 0.4 V, since the applied voltage must overcome two germanium diodes at 0.2 V each. Therefore,

$$V_{in} = 3.4 V_p - 0.4 \text{ V} = 3.0 \text{ V dc},$$

which is the dc input voltage to the op amp voltage-to-current converter. The current through the load coil can be calculated from Eq. (8-8):

$$I = \frac{V_{in}}{R_6} = \frac{3.0 \text{ V}}{500 \text{ }\Omega} = 6.0 \text{ mA}.$$

Therefore, the sensing coil is carrying a current of 6.0 mA when the relative humidity in the moistening chamber is 80%. 6.0 mA is the middle of the proportioning current range of the diaphragm-actuated control valve, so the control valve should be about half open.

Now if the humidity in the moistening chamber should drop below 80% for some reason, here is what would happen. The resistance of the hygrometer will rise, causing a reduced voltage drop across R_4. This will throw the bridge further out of balance, causing V_{bridge} to become larger. The greater V_{bridge} will result in a greater dc input to the op amp circuit, causing a larger current flow through the sensing coil. This opens the water supply valve further and tends to drive the humidity back up.

Let us try to calculate how far the relative humidity would have to drop in order to open the water supply valve wide open.

To drive the valve wide open, a sensing coil current of 10 mA is required. Therefore, V_{in} is given by

$$V_{in} = (10 \text{ mA})(500 \text{ }\Omega) = 5.0 \text{ V}.$$

To have 5.0 V delivered to the 10-μF filter capacitor, the peak input voltage to the rectifier bridge must be $5.4V_p$ because 0.4 V will be dropped across the diodes. Therefore the rms output voltage from the Wheatstone bridge must be

$$V_{bridge} = \frac{V_p}{1.41} = \frac{5.4 \text{ V}}{1.41} = 3.9 \text{ V rms}.$$

For the output voltage to be 3.9 V rms, the voltage across R_4 must be

$$V_{R_4} = V_{R_2} - V_{bridge} = 8.0 \text{ V} - 3.9 \text{ V} = 4.1 \text{ V rms}.$$

If the voltage across R_4 equals 4.1 V rms, the voltage across R_3 is given by

$$V_{R_3} = 10.0 \text{ V} - 4.1 \text{ V} = 5.9 \text{ V rms}.$$

Therefore we can find the resistance of R_3 from

$$\frac{R_3}{R_4} = \frac{V_{R_3}}{V_{R_4}},$$

$$\frac{R_3}{4.7 \text{ K}} = \frac{5.9 \text{ V}}{4.1 \text{ V}},$$

$$R_3 = 6.8 \text{ K}.$$

This means that if the resistance of R_3 rises to 6.8 K, the sensing coil will cause the water valve to open wide. From the graph of Fig. 11-22(b) we can find the relative humidity which would cause the transducer resistance to be 6.8 K. It appears to be about 70% *RH*. Therefore a drop in relative humidity to 70% is necessary to call for and produce full water flow to the mist nozzles.

12-9 A WAREHOUSE HUMIDITY CONTROLLER

In some storage warehouses, it is essential that the relative humidity be maintained above a certain level. Two good examples are warehouses that store explosives and warehouses that store grain. For both of these commodities, it is dangerous to allow the relative humidity to fall below about 50%. If grain dust gets too dry, spontaneous combustion may occur. Explosives are dangerous to handle and store under very dry conditions also. Figure 12-17(a) shows a method of maintaining a warehouse atmosphere's humidity above a safe value.

This system is an On-Off control system. The recirculating air blower runs at all times, distributing air evenly to all parts of the storage area. This ensures a uniform atmosphere throughout the warehouse, preventing any dry pockets from forming. When the water solenoid valve opens, it allows water to reach the series of drip emitters located in the air duct downstream from the blower. The recirculating air blower sucks air out through the roof of the warehouse and forces it past the drip emitters. The moving air absorbs some of this dripping water, and the wetted air is then distributed via ductwork to various locations throughout the warehouse.

The signal to turn on the water originates in the wet bulb-dry bulb sampling chamber. A continuous stream of air from the storage area is sucked into the sampling chamber by a small sampling blower. The sampled air passes over both temperature bulbs, as described in Sec. 11-9-2. The air is always *pulled* into the sampling chamber rather than *pushed* in, so that no heat energy is imparted to the sampled air by the blades of the sampling blower. This might cause erroneous temperature readings. Also, the dry bulb is always located nearer the entry to the sampling chamber. This is so the dry bulb cannot be affected by moisture picked up by the air as it passes over the wet bulb.

In this system, the temperature detectors are nickel wire RTDs.

The nickel RTDs have a resistance of 20 K at 60°F. The dry bulb RTD is placed in a Wheatstone bridge in the R_4 position, and the wet bulb RTD is placed in the R_3 position as shown in Fig. 12-17(b). The resistances of the RTDs are labeled R_{dry} and R_{wet}, respectively. The left side of the Wheatstone bridge divides the 10-V dc supply evenly, since R_1 equals R_2. The right side will not divide the supply voltage evenly because R_{wet} will be less than R_{dry}, due to its lower temperature. Therefore the bridge will be unbalanced, with the bridge output voltage furnishing the input signal to op amp 1. This input voltage is identified as V_{bridge} in Fig. 12-17(b).

The relationship between the relative humidity and V_{bridge} is this: As the relative humidity goes down, the temperature difference between the bulbs becomes greater [refer to Fig. 11-23(b)]. As the temperature difference becomes greater, the difference between R_{wet} and R_{dry} also increases, throwing the bridge further out of balance. Therefore, a decrease in relative humidity causes an increase in V_{bridge}.

In this system, the temperature of the storage area is maintained near 60°F by an independent temperature control system. If the relative humidity drops below the acceptable level, 50%, the temperature difference between bulbs will reach 10°F. This is shown in Fig. 11-23(b). Find the row which indicates a dry bulb temperature of 60°F. Move along that row to the column which indicates a temperature difference of 10°F; the relative humidity value is given as 49%. Therefore, if the bulb temperature difference becomes as great as 10°F, it means the humidity is too low and must be brought back up.

Figure 12-17. (a) Physical system for maintaining warehouse humidity. (b) The psychrometer detector and water flow control circuit.

We will now calculate the value of V_{bridge} for a temperature difference of 10°F. Nickel wire has a temperature coefficient of resistance of about 0.42%/°F (0.0042/°F). This value can be obtained from Fig. 11-9(a).

Therefore, if the temperature difference is 10°F, R_{dry} and R_{wet} will differ by 4.2%, since (0.42%/°F)(10°F) = 4.2%. This means that R_{wet} will be 19.19 K for a temperature difference of 10°F, since

$$\frac{20 \text{ K} - 19.19 \text{ K}}{19.19 \text{ K}} = 0.042 = 4.2\%.$$

Under these conditions, V_{bridge} is given by

$$V_{bridge} = V_{R_{dry}} - V_{R_2} = V_{R_{dry}} - 5.00 \text{ V} \tag{12-4}$$

The voltage across R_{dry} is given by

$$\frac{V_{R_{dry}}}{10 \text{ V}} = \frac{R_{dry}}{R_{dry} + R_{wet}} = \frac{20 \text{ K}}{20 \text{ K} + 19.19 \text{ K}}.$$

From Eq. (12-4), V_{bridge} is given by

$$V_{bridge} = 5.10 \text{ V} - 5.00 \text{ V} = 0.10 \text{ V}.$$

This means that if V_{bridge} reaches a value of 0.10 V, the relative humidity has dropped too low and must be corrected.

V_{bridge} is amplified in the op amp 1 noninverting amplifier, which has a voltage gain of 80. Therefore when V_{bridge} reaches 0.10 V, V_{out_1} reaches

$$V_{out_1} = 80(V_{bridge}) = 80(0.10 \text{ V}) = 8.0 \text{ V}.$$

Op amp 2 is a voltage comparer, as discussed in Sec. 8-2. It compares the value of V_{out_1} to +8.0 V. The 8-V reference is supplied by the 7-K/8-K voltage divider in Fig. 12-17(b). If V_{out_1} is less than +8.0 V, the output of the voltage comparer is −13 V. If V_{out_1} is greater than +8.0 V, the output of the voltage comparer is +13 V. Therefore V_{out_2} switches from −13 V dc to +13 V dc when V_{out_1} reaches +8.0 V. When this happens, transistor Q_1 switches ON and energizes relay R. This relay closes a contact in the 115-V ac circuit, which energizes the water solenoid valve. This adds moisture to the recirculating air, thereby driving the relative humidity back up above 50%.

In this control system, the desired relative humidity can be changed easily, just by changing the value of the 8-K resistor in the voltage divider which feeds op amp 2.

QUESTIONS AND PROBLEMS

1. In Fig. 12-1(b), what is the purpose of the *RC* circuit in parallel with the triac?

2. In Fig. 12-1(a), what would be the best physical location for the thermistor probe?

3. In Fig. 12-1(b), if R_2 were larger, would the motor run faster or slower? Explain.

Questions 4–8 refer to Fig. 12-2.

4. Explain the purpose of the span adjust pot.

5. Explain the purpose of the set-point zero pot.

6. If V_{out_2} goes to $+12.5$ V, what would the voltage drop be across R_8?

7. Explain just why it is that for transistor Q_1, the collector voltage relative to ground is the same as the emitter voltage relative to ground (but of opposite polarity).

8. If the maximum current through the load coil were 11 mA and you wished to correct it to 10 mA, which way would you turn the maximum current adjust pot: to increase or decrease its resistance? Explain.

Questions 9 and 10 refer to Fig. 12-4.

9. Why is the dc signal from the Wheatstone bridge chopped?

10. Explain why V_{out_3} has to be inverted before being applied to R_{29}.

Questions 11 and 12 refer to Fig. 12-6.

11. If it was found that the amount of time that the draw motor was being run was not long enough, that it was taking too many corrections to get the tension back in the acceptable range, what could you do about it? What would you change in the circuit?

12. Explain carefully the operation of D5 and D6.

Questions 13 and 14 refer to Fig. 12-8.

13. If the lamp got dimmer, explain the consequences.

14. What is the purpose of R_2?

Questions 15–18 refer to Fig. 12-9.

15. What is the purpose of the tare pot?

16. What would happen if the tare winding (S_B) were hooked up backwards (if its phase were reversed)?

17. Why is it so important that the weighing circuitry itself be well shielded from noise?

18. What is the purpose of slow feed in an automatic weighing system?

Questions 19–22 refer to Figs. 12–13, 12–14, and 12–15.

19. What does the demodulator do (the diode and C_1)?

20. What does the C_5, C_6, C_7, R_{10}, R_{11} combination do?

21. Why is it important to have the reference cell filled with oxygen and tightly sealed against leaks?

22. Why does C_9 have to be a nonpolarized capacitor?

Questions 23–25 refer to Fig. 12-16.

23. Why must the Wheatstone bridge be an ac bridge instead of a dc bridge?

24. What is the purpose of the R_5-C_1 combination ahead of the voltage-to-current converter?

25. If R_2 were larger, would that tend to raise the humidity set point or lower it? Explain.

Questions 26–28 refer to Fig. 12–17.

26. What control mode is being used in Fig. 12-17?

27. What is the purpose of the diode in parallel with relay coil R?

28. If it were desired to raise the *RH* set point, would you increase the 8-K resistor or decrease it? Explain.

13

Motor Speed Control Systems

In most industrial situations, motors are operated directly from the ac or dc supply lines. That is, the motor winding terminals are connected directly to the lines which supply the electric current. In these situations, the operating behavior of the motor is determined by the nature of the mechanical load connected to the motor's shaft. In simple terms, if the load is easy to drive, the motor will tend to deliver relatively little torque, and it will run at a high speed. If the load is difficult to drive, the motor will tend to deliver a lot of torque, and it will run at a lower speed. The point is that the operating behavior of the motor is set by its *load* (for a fixed supply line voltage), and the operator has no control over motor behavior.

In modern industrial situations, there are many applications which require that the operator be able to intervene to control the motor speed. Such control is usually accomplished with thyristors. The combination of the motor, the controlling thyristor(s), and the associated electronic components is referred to as a *speed control system* or a *drive system*.

After completing this chapter, you will be able to:

1. Explain the two basic methods of adjusting the speed of a dc shunt motor
2. Discuss the relative advantages and disadvantages of speed adjustment of a dc motor from the field and from the armature
3. Explain why armature control with thyristors is superior to all other dc motor speed control methods
4. Discuss how counter-EMF feedback can be used to improve a motor's load regulation
5. Calculate a motor's load regulation, given a graph of shaft speed versus load torque (or horsepower)
6. Explain the operation of the single-phase and three-phase thyristor drive systems which are presented
7. Explain the operation of a switchgear-controlled reversible drive system
8. Describe the basic operating principles of variable frequency ac drive systems using inverters or converters

13-1 DC MOTORS—OPERATION AND CHARACTERISTICS

Dc motors are important in industrial control because they are more adaptable than rotating field ac motors to adjustable speed systems.

Figure 13-1 shows the schematic symbol of a dc *shunt motor*. The motor's field winding is drawn as a coil. Physically, the field winding is composed of many turns of thin (high-resistance) wire wrapped around the *field poles*. The field poles are ferromagnetic metal cores, which are attached to the stator of the machine. The high resistance of the field winding limits the field current to a fairly small value, allowing the field winding to be connected directly across the dc supply lines. However, the relatively small field current (I_F) is compensated for by the field winding's large number of turns, enabling the winding to create a strong magnetic field.

The field winding is unaffected by changing conditions in the armature. That is, as the armature current varies to meet varying load conditions, the field winding current stays essentially constant, and the strength of the resulting magnetic field stays constant. The field current can be found easily from Ohm's law as

$$I_F = \frac{V_s}{R_F},\qquad(13\text{-}1)$$

where V_s is the source voltage applied to the field winding [Fig. 13-1(a)] and R_F is the winding's dc resistance.

The armature winding is shown in Fig. 13-1(a) as a circle contacted by two

Figure 13-1. (a) Schematic representation of a dc shunt motor. (b) A rheostat in series with the field winding for controlling the motor's speed. (c) A rheostat in series with the armature for controlling the motor's speed.

small squares. It is drawn like this because the armature winding is constructed on the cylindrical rotor of the machine and current is carried to and from the armature winding via carbon brushes contacting commutator segments.

The armature winding of a dc motor is constructed of relatively few turns of thicker wire, so it has a low dc resistance. The armature winding resistance of a medium- or large-sized dc motor is usually less than 1 ohm.

When power is first applied to the armature winding, only the dc ohmic resistance of the winding is available to limit the current, so the inrush current surge is quite great. However, as the motor begins to accelerate, it begins to induce (generate) a *counter-EMF* by the usual generator action. This counter-EMF opposes the applied source voltage and limits the armature current to a reasonable value.

When a dc motor has reached normal operating speed, its counter-EMF is about 90% as great as the applied armature voltage [V_s in Fig. 13-1(a)]. The *IR* voltage drop across the armature winding resistance accounts for the other 10% of the applied voltage, neglecting any voltage drop across the carbon brushes.

The exact size of the counter-EMF generated by the armature winding depends on two things:

a. The strength of the magnetic field. The stronger the magnetic field, the greater the counter-EMF tends to be.

b. The speed of rotation. The greater the speed, the greater the counter-EMF tends to be.

Equation (13-2) expresses the dependence of counter-EMF on field strength and rotational speed:

$$E_C = kB(RPM). \tag{13-2}$$

In Eq. (13-2), E_C stands for the counter-EMF created by the spinning armature winding, B stands for the strength of the magnetic field created by the field winding, and *RPM* is the rotational speed in revolutions per minute. The proportionality constant k depends on the construction details of the armature (the number of winding turns, the length of the conductors, etc.).

Kirchhoff's voltage law for the armature loop is expressed in Eq. (13-3), which simply states that the applied armature voltage is equal to the sum of the voltage drops in the armature. The sum of the voltage drops in the armature winding equals the counter-EMF added to the *IR* resistive voltage drop, again neglecting the minor effect of the brush drop.

$$V_s = E_C + I_A R_A. \tag{13-3}$$

In Eq. (13-3), R_A stands for the dc resistance of the armature winding, and of course I_A is the armature current.

13-1-1 Varying the speed of a dc shunt motor

Basically, there are two ways of varying the running speed of a dc shunt motor:

a. Adjusting the voltage (and current) applied to the field winding. As the field voltage is increased, the motor *slows down*. This method is suggested by Fig. 13-1(b).

b. Adjusting the voltage (and current) applied to the armature. As the armature voltage is increased, the motor speeds up. This method is suggested by Fig. 13-1(c).

Field Control. Here is how method **a**, adjusting the field voltage, works. As the field voltage is increased, by reducing R_v in Fig. 13-1(b), for example, the field current is increased. This results in a stronger magnetic field, which induces a greater counter-EMF in the armature winding. The greater counter-EMF tends to oppose the applied dc voltage and thus reduces the armature current, I_A. Therefore, an increased field current causes the motor to slow down until the induced counter-EMF has returned to its normal value (approximately).

Going in the other direction, if the field current is reduced, the magnetic field gets weaker. This causes a reduction in counter-EMF created by the rotating armature winding. The armature current increases, forcing the motor to spin faster, until the counter-EMF is once again approximately equal to what it was before.

The reduction in magnetic field strength is "compensated" for by an increase in armature speed.

This method of speed control has certain good features. It can be accomplished by a small, inexpensive rheostat, since the current in the field winding is fairly low due to the large R_F. Also, because of the low value of I_F, the rheostat R_v does not dissipate very much energy. Therefore, this method is energy-efficient.

However, there is one major drawback to speed control from the field winding: To increase the speed, you must reduce I_F and weaken the magnetic field, thereby lessening the motor's torque-producing ability. The ability of a motor to create torque depends on two things: the current in the armature conductors and the strength of the magnetic field. As I_F is reduced, the magnetic field is weakened, and the motor's torque-producing ability declines. Unfortunately, it is just now that the motor needs all the torque-producing ability it can get, since it probably requires greater torque to drive the load at a faster speed.

Therefore, there is a fundamental conflict involved with field control. To make the motor spin faster, which requires it to deliver more torque, you must do something which tends to rob the motor of its ability to produce torque.

Armature Control. From the torque-producing point of view, method **b**, armature control, is much better. As the armature voltage and current are increased [by reducing R_v in Fig. 13-1(c)], the motor starts running faster, which normally requires more torque. The reason for the rise in speed is that the increased armature voltage demands an increased counter-EMF to limit the increase in armature current to a reasonable amount. The only way the counter-EMF can increase is for the armature winding to spin faster, since the magnetic field strength is fixed. In this instance, the ingredients are all present for increased torque production, since the magnetic field strength is maintained constant and I_A is increased.

The problem with the armature control method of Fig. 13-1(c) is that R_v, the rheostat, must handle the armature current, which is relatively large. Therefore the rheostat must be physically large and expensive, and it will waste a considerable amount of energy.

Of the two methods illustrated in Figs. 13-1(b) and (c), the field control method is usually preferred.

13-2 THYRISTOR CONTROL OF ARMATURE VOLTAGE AND CURRENT

As we saw in Chapter 4, an SCR can perform most of the duties of a rheostat in controlling the average current to a load. Furthermore, an SCR, or any power thyristor, does not have the shortcomings of high-power rheostats. SCRs are small, inexpensive, and energy-efficient. It is therefore natural to match the dc shunt motor and the SCR to provide armature control of motor speed. The general layout of an SCR speed control system is illustrated in Fig. 13-2.

Figure 13-2. SCR in series with the armature to control motor speed.

In Fig. 13-2, the ac power is rectified to produce dc power for the field winding. The SCR then provides *half-wave rectification and control* to the armature winding. By firing the SCR early, the average armature voltage and current are increased, and the motor can run faster. By firing the SCR later (increasing the firing delay angle), the average armature voltage and current are reduced, and the motor runs slower. The gate trigger control circuit can be either an open-loop circuit or a closed-loop, automatically correcting circuit.

Of course, Fig. 13-2 is not the only arrangement of an armature with an SCR that is acceptable. Either of the circuits shown in Figs. 4-11 or 4-13(b) will also work, with the motor armature being the circuit's load. The circuits of Figs. 4-11 and 4-13(b) may even be preferable to the circuit in Fig. 13-2 because they provide *full-wave* power control instead of half-wave control.

13-3 A SINGLE-PHASE HALF-WAVE
SPEED CONTROL SYSTEM
FOR A DC SHUNT MOTOR

Figure 13-3 shows a simple half-wave speed control circuit for a dc motor. The motor speed is adjusted by the 25-K speed adjust pot. As that pot is turned up (wiper moved away from ground) the motor speed increases. This happens because the gate voltage relative to ground becomes a greater portion of the ac line voltage, thus allowing the gate-to-cathode voltage to reach the firing voltage of the SCR early in the cycle.

As the speed adjust pot is moved downward, the gate-to-ground voltage becomes a smaller portion of the line voltage, so it takes longer for V_{GK} to reach the value necessary to fire the SCR.

The relationship between speed and firing delay angle for this system is graphed in Fig. 13-3(b). As can be seen, it is impossible for the motor to attain 100% of its full rated speed because the system can deliver only half-wave power to the armature.

This system has a desirable feature which tends to stabilize the motor speed

Figure 13-3. (a) Schematic diagram of a half-wave SCR drive circuit. (b) Graph of shaft speed vs. firing delay angle for the circuit of part (a). (c) Graph of shaft speed vs. torque for a fixed setting of the speed adjust pot in part (a).

even in the face of load changes. The feature is called *counter-EMF feedback*. Here is how it works.

Suppose the speed adjust pot is positioned to provide a shaft speed of 1500 rpm. If the torque load on the motor now increases, there is a natural tendency for the motor to slow down. It does this so that the counter-EMF can decrease slightly, thus allowing an increased armature current to flow. The increased armature current provides the boost in torque needed to drive the heavier load. This is the natural reaction of all motors.

In the system of Fig. 13-3, though, when the counter-EMF decreases, the cathode-to-ground voltage (V_K) decreases, since V_K depends in large part on the counter-EMF generated by the armature winding. If V_K decreases, the firing of the SCR takes place earlier because V_G does not have to climb as high as before to make V_{GK} large enough to fire the SCR. Therefore, an increase in torque load *automatically* produces a reduction in firing delay angle and a consequent increase in average armature voltage and current. This action holds the motor speed almost constant, even in the face of varying torque load. The graph of motor speed versus load torque is presented in Fig. 13-3(c), assuming an initial no-load shaft speed of 1500 rpm.

The ability of a speed control system to maintain a fairly constant motor speed in the face of varying load is called its *load regulation*. In formula form, the load regulation is given as

$$\text{load reg.} = \frac{RPM_{NL} - RPM_{FL}}{RPM_{FL}}, \tag{13-4}$$

where RPM_{NL} stands for the rotational speed at *no load*. The phrase *no load* means that the load countertorque tending to slow down the motor shaft equals zero. RPM_{FL} stands for the rotational speed at *full load*, meaning that the load countertorque tending to slow down the motor shaft is at maximum. It can be seen from Eq. (13-2) that the smaller the change in speed from the no-load condition to the full-load condition, the smaller the load regulation. Therefore, the smaller the load regulation figure, the better the control system.

The drive system of Fig. 13-3 provides good load regulation. This is another advantage over the speed control methods described in Sec. 13-1.

As a specific example of calculating load regulation, refer to Fig. 13-3(c). We can see that the no-load speed is 1500 rpm and that the full-load speed is 1475 rpm. Therefore, the load regulation is given by

$$\text{load reg.} = \frac{1500 \text{ rpm} - 1475 \text{ rpm}}{1475 \text{ rpm}} = 0.017 \quad \text{or} \quad 1.7\%.$$

For many industrial applications, a load regulation of 1.7% is quite adequate.

13-4 ANOTHER SINGLE-PHASE SPEED CONTROL SYSTEM

Figure 13-4 shows another speed control circuit. Here is how it operates.

The incoming ac power is rectified in a full-wave bridge, whose pulsating dc output voltage is applied to the armature control circuit. Capacitor C is charged by current flowing down through the low-resistance armature winding, through D_2

Figure 13-4. Another SCR drive circuit. The load regulation of this circuit will be superior to the load regulation in Fig. 13-3. The field circuit is not shown in this diagram.

and the speed adjust pot, and on to the top plate of the capacitor. The capacitor charges until it reaches the breakover voltage of the four-layer diode. At that instant, the four-layer diode allows part of the capacitor charge to be dumped into the gate of the SCR, firing the SCR. The firing delay angle is determined by the resistance of the speed adjust pot, which sets the charging rate of C.

Diode D_3 suppresses any inductive kickback which is produced by the inductive armature winding at the completion of each half cycle. When the SCR turns OFF at the end of a half cycle, current continues circulating in the armature-D_3 loop for a short while. This dissipates the energy stored in the armature inductance.

The purpose of the R_1-D_1 combination is to provide a discharge path for capacitor C. Recall that a four-layer diode does not break back all the way to 0 V when it fires. Therefore the capacitor is not able to dump *all* its charge through the gate-cathode circuit of the SCR. Some of the charge remains on the top plate of C, even after the SCR has fired. As the dc supply pulsations approach 0 V, the remaining charge on C is discharged through R_1 and D_1 into the field winding. Therefore the capacitor starts with a clean slate on the next pulsation from the bridge.

This system also provides counter-EMF feedback, and it therefore has good load regulation. Here is how the counter-EMF feedback works.

Suppose the speed adjust pot is set to give a shaft speed of 2000 rpm at a certain torque load. If the load should increase for some reason, the first thing the motor wants to do is slow down a little to admit more armature current. When this happens, the armature's counter-EMF decreases a bit.

As the counter-EMF decreases, the voltage available for charging capacitor C *increases*. This happens because the voltage available for charging C is the difference between the bridge pulsation voltage and the counter-EMF created by the armature. This can be understood by referring to the counter-EMF polarity markings in Fig. 13-4.

With more voltage available to charge C, it is natural that C will charge to the firing voltage sooner, thus increasing the average voltage supplied to the armature. This corrects the tendency of the motor to slow down, and brings it back up to virtually the same speed as before.

13-5 REVERSIBLE SPEED CONTROL

Some industrial speed control applications require that the rotation of a motor be *reversible*. That is, the motor must be able to spin either clockwise or counter-clockwise, besides having adjustable speed. Reversal of the direction of rotation can be accomplished in either of two ways:

a. Reversing the direction of the field current, leaving the armature current the same.
b. Reversing the direction of the armature current, leaving the field current the same.

The circuits of Fig. 10-27 show how armature current direction can be reversed in a half-wave control system. Figure 13-5 shows how armature current can be reversed in a full-wave speed control system. The most straightforward method of reversing armature or field current is by the use of two separate motor starter contactors. The *forward* contactor causes current to go through the armature in one direction, while the other contactor, the *reverse* contactor, causes current to flow in the opposite direction.

In Fig. 13-5(a), the FOR contactor is energized by pressing the FORWARD START push button. As long as the REV contactor is dropped out at that time, the FOR contactor will energize and seal itself in around the N.O. push-button switch. The operator can then release the FORWARD START button, and the contactor will remain energized until the STOP push button is pressed.

In Fig. 13-5(b), it can be seen that when the FOR contacts are closed, the current flows through the armature from bottom to top, thereby causing rotation in a certain direction (assume clockwise). When the REV contacts are closed, armature current flows from top to bottom, thus causing rotation in the counter-clockwise direction. As usual, the speed of rotation is controlled by the firing delay of the SCRs.

Reversible full-wave control can be accomplished without the use of switchgear (contactors, push buttons, etc.) with the circuit of Fig. 13-6. In Fig. 13-6, the direction of rotation is determined by which trigger circuit is enabled. If the forward trigger circuit is enabled, the top two SCRs will fire on alternate half cycles of the ac line, and they will pass current through the armature from right to left. If the reverse triggering circuit is enabled, the bottom two SCRs will fire on alternate half cycles of the ac line, and they will pass current through the armature from left to right, as indicated. The method of enabling one trigger circuit while disabling the other has not been shown in Fig. 13-6.

(a)

(b)

Figure 13-5. A reversible full-wave SCR drive system. (a) Motor starter control circuit. (b) Armature circuit. The SCRs fire on alternate half cycles, causing the armature voltage to have the polarity shown. The armature current direction depends on whether the FOR or REV contacts are closed.

13-6 THREE-PHASE DRIVE SYSTEMS FOR DC MOTORS

For dc motors larger than about 10 hp, a three-phase drive system is superior to a single-phase system. This is because a three-phase system provides more pulsations of armature voltage per cycle of the ac line and thus gives greater average armature current flow.

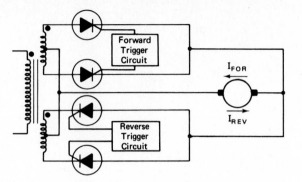

Figure 13-6. An all-solid-state reversible full-wave drive system.

The simplest possible three-phase drive system is illustrated in Fig. 13-7(a). Although this system gives only half-wave control, it is capable of keeping current flowing through the armature continually. It can do this because when any one phase goes negative, at least one of the other phases is bound to be positive. If a certain phase is driving the armature, at the instant it reverses polarity, one of the other two phases is ready to take over. Thus it is possible to keep armature current flowing continually.

If no neutral fourth wire is available, three-phase half-wave control can be accomplished by the addition of three rectifier diodes, as shown in Fig. 13-7(b). During the time that line voltage AB is driving the armature, the flow path is down line A, through SCRA, through the armature, and through D_B into line B. When line voltage BC is driving the armature, the armature current passes through SCRB and D_C. At the instant that line voltage CA is driving the armature, current passes through SCRC, through the armature, and back into line A through diode D_A.

13-7 AN EXAMPLE OF A THREE-PHASE DRIVE SYSTEM

Figure 13-8 shows the complete schematic diagram of a three-phase drive system. The 230-V, three-phase power is brought in at the upper left in this figure. *Thyrectors* are installed across each of the three phases to protect the solid-state drive circuits from high-voltage transient surges which may appear on the power lines. A thyrector acts like two zener diodes connected back to back. If any momentary voltage surge appears which exceeds the breakdown voltage of the thyrector, the thyrector shorts out the excess. That is, the thyrector acts like a zener diode in that it will allow only a certain amount of voltage to appear across its terminals. If a greater voltage tries to appear, the thyrector becomes a short circuit to any amount of voltage in excess of its rating. In this example, with 230 V rms between the ac power lines, the peak voltage is about 324 V. The thyrectors would be chosen to have a breakdown voltage rating a little larger than the peak line voltage. In this case they would have a rating of 350 or 400 V.

(a)

(b)

Figure 13-7. (a) Three-phase drive system containing a neutral fourth wire. (b) Three-phase drive system with no neutral wire. The rectifier diodes complete the circuit of the armature loop.

The 230 V ac appearing between power lines B and C (V_{BC}) is half-wave rectified by diode D_1 and applied to the motor field winding. Diode D_2, called a *free-wheeling diode*, allows the field winding current to continue flowing during the negative half cycle of the V_{BC} voltage. When V_{BC} enters its negative half cycle, the field winding induces a voltage which is positive on the right and negative on the left. This induced voltage forward-biases D_2, causing current to flow through the loop comprised of the field winding and D_2 in the same direction as before.

Figure 13-8. Complete schematic diagram of a three-phase drive system. Trigger circuits *A*, *B*, and *C* are all identical, so only trigger circuit *A* is drawn. Pulse transformer T_3 in the trigger circuit is wired to the gate and cathode of the corresponding SCR. This drive system would have very good load regulation due to the counter-EMF feedback to Q_1.

Relay RFF is a *field-failure relay*. Its job is to constantly monitor the current in the field winding. Since it must sense *current*, rather than *voltage*, its coil is comprised of just a few turns of relatively heavy wire. As long as the field current is flowing, RFF will remain energized. If the field current is interrupted for any reason, RFF drops out. If it does drop out, it causes all power to be removed from the armature winding, as described below. This is necessary because a dc motor can be destroyed if its armature remains powered up when there is no magnetic field present. With no magnetic field present, the armature winding is incapable of generating enough counter-EMF to limit the armature current to a safe value. Under this condition, the armature current would quickly surge up to a destructive level, thereby overheating and ruining the armature conductors and/or insulation. Even if the armature winding could withstand the electrical strain, the rotor speed would increase drastically, in a vain attempt to induce sufficient counter-EMF. This drastic increase in rotational speed can cause mechanical destruction due to overheated bearings or armature windings thrown out of their slots by centrifugal force.

Transformer T_1 steps the 230-V line voltage down to 115 V for use in the START-STOP circuit. This is done to provide safety to the operators who use the START and STOP push buttons. When the START button is depressed, motor starter M will be energized, as long as RFF is energized. The auxiliary M contact then seals in around the START button, allowing the START push button to be released. The main M contacts are in series with the motor armature itself; when those contacts close, armature current can flow, and the motor can begin running.

The motor will stop whenever the M coil deenergizes. This will happen if the N.C. STOP push button is temporarily depressed or if the RFF contact opens.

In the speed control circuit proper, SCRA controls armature current when V_{AB} is in its positive half cycle (past the 60° point). Diode D_B then carries the armature current back to line B. During the interval when V_{BC} drives the armature, the armature current flows through SCRB and D_C. When V_{CA} drives the armature, current flows through SCRC and D_A. SCRA is triggered by TRIGGER CIRCUIT A, SCRB is triggered by TRIGGER CIRCUIT B, and SCRC is triggered by TRIGGER CIRCUIT C. All three SCRs are fired at approximately the same firing delay angle within their respective conduction cycles. This is accomplished by *ganging* the three control potentiometers together.

TRIGGER CIRCUITS A, B, and C are all identical. The schematic diagram of TRIGGER CIRCUIT A is the only one of the three which is drawn schematically in Fig. 13-8.

The Trigger Circuit. In the upper left of the trigger circuit, transformer T_2 steps V_{AB} down to 20 V ac. The T_2 output voltage is rectified by diode D_3 to provide a 28-V dc supply for the triggering circuitry. To understand the operation of the rest of the trigger circuit, assume for a moment that the N.C. M contact has been open for a while and that C_2 is fully charged. We will return later to see what these components do.

The base current of *pnp* transistor Q_1 is set by the speed adjust pot. The base current flow path is down from the +28-V dc supply, through the emitter-base junction of Q_1, through R_4 and R_2, and into the 10-K pot. As the pot wiper is adjusted downwards, the voltage available for driving the Q_1 base increases, and the base current increases. As the pot wiper is moved up, Q_1 base current decreases. The Q_1 base current determines the Q_1 collector current, which charges capacitor C_E through R_5. As we saw in Chapter 5, the faster C_E charges, the earlier V_P of the UJT is reached, and the earlier the SCR fires.

This trigger circuit has built-in speed regulation (load regulation) also, just like the circuits of Figs. 13-3 and 13-4. The regulation is provided by 220-K resistor R_7. Here is how it works.

Suppose that the speed adjust pot has been adjusted to provide 2000 rpm at a certain load torque. Recognize that the counter-EMF voltage developed by the armature tends to *reverse-bias* the emitter-base junction of Q_1, through R_7. The reverse-bias tendency due to this circuit is overcome by the R_4-R_2-speed adjust pot combination which keeps Q_1 turned on and conducting. If the torque load on the motor shaft increases, the motor slows down a little bit, and the counter-EMF decreases. This reduces the reverse-bias tendency through R_7. Therefore the Q_1 base current increases a little bit, and the transistor is able to charge C_E faster. Because of this, the UJT fires earlier, and the SCR fires earlier. This raises the average voltage that the SCR applies to the armature, tending to correct the motor speed.

Let us return now to consider the circuit comprised of C_2, R_3, and the N.C. contact of M. The purpose of this circuit is to accelerate the motor up to speed slowly when it is first started. It does this by limiting the armature current for a certain amount of time after the starter is energized. Prior to starter M being energized, the N.C. M contact is closed and is holding capacitor C_2 discharged. There is thus a short circuit across the path consisting of R_4 and the Q_1 base-emitter junction. Q_1 is held cut-OFF at this time.

When the N.C. M contact opens as the motor starter energizes, C_2 begins charging through resistor R_2 and the speed adjust pot. As C_2 charges, it begins acting more and more like an open circuit. As this takes place, the Q_1 base current slowly builds up to the steady-state value determined by the pot setting. Until that steady-state value is reached, the firing of the UJT and SCR is delayed past the normal firing instant. In this manner the armature current is temporarily retarded for a while after the starter energizes. Therefore, the motor accelerates slowly to its set speed, and the great initial inrush of armature current is avoided.

13-8 SPEED CONTROL
OF AC INDUCTION MOTORS

As we saw in Chapter 10, ac induction motors spin at a speed which is a little bit slower than the synchronous speed of the rotating field. The synchronous speed of the rotating field is determined by the number of poles of the stator winding and the frequency of the applied ac voltage. In formula form,

$$\text{synchronous speed in rpm} = \frac{120\,f}{P},$$

where f stands for the frequency of the applied stator voltage in hertz and P is the number of poles of the stator winding.

In electronic speed control of ac motors, the frequency of the voltage applied to the stator is varied to change the synchronous speed. The change in synchronous speed then causes the rotor shaft speed to change.

There are two approaches to obtaining a variable frequency ac voltage for an ac motor drive system:

a. Changing dc power into variable frequency ac. A circuit which can do this is called an *inverter*.

b. Changing 60-Hz ac power into variable frequency ac. A circuit which can do this is called a *converter*.

Both inverters and converters are built with SCRs. The idea of an inverter is illustrated in Fig. 13-9. The three SCRs are fired in sequence, one immediately

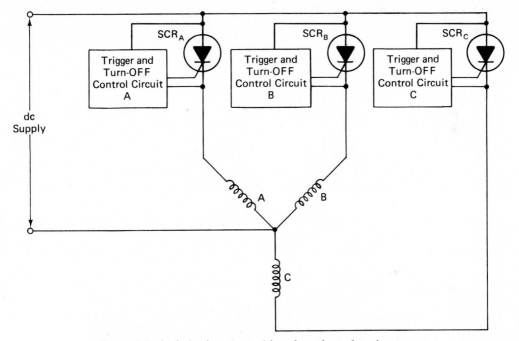

Figure 13-9. Ac induction motor driven by a three-phase inverter. The effective frequency delivered to the stator windings depends on how long the SCRs remain ON.

following the other. Each SCR is fired and later turned OFF by its *trigger and turn-OFF control circuit*. Thus, SCR*A* is fired by its control circuit and a certain time later is turned OFF by that circuit. At the instant SCR*A* turns OFF, SCR*B* is fired by its control circuit. SCR*B* remains ON for the same amount of time that SCR*A* was ON; then it is turned OFF. At the instant SCR*B* turns OFF, SCR*C* is fired by its control circuit. After an equal amount of time, SCR*C* is turned OFF, and SCR*A* is fired again to repeat the cycle. It can be seen that the effective frequency of the voltage applied to the stator windings can be varied by varying the amount of time that an SCR is allowed to remain ON. If the SCRs remain ON for a long time, the effective stator frequency is low, the synchronous speed of the rotating field is low, and the motor shaft turns slowly. If the SCRs remain ON for only a short time, the stator frequency increases, the synchronous speed rises, and the motor shaft speeds up.

The operation of a variable frequency converter is basically the same as that of an inverter. The frequency of the applied stator voltage is determined by the length of time that the SCRs are allowed to remain ON. A schematic representation of a half-wave converter is shown in Fig. 13-10. Notice that it requires no less than 18 SCRs to construct a variable frequency converter. Full-wave converters are even more elaborate, requiring 36 SCRs (and 36 trigger and turn-OFF control circuits).

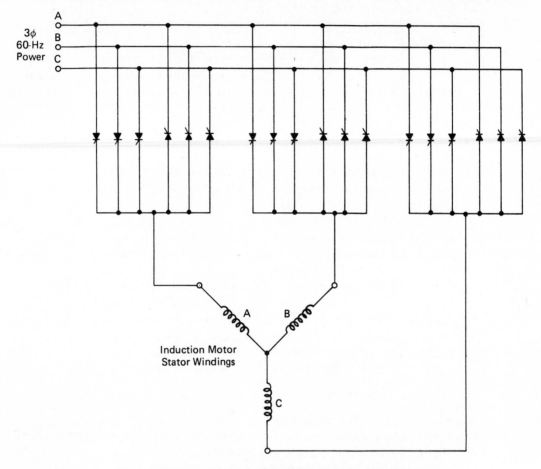

Figure 13-10. Layout of a half-wave three-phase converter. The effective stator frequency can be varied by varying the rate at which SCR triplets are sequentially fired.

There are other methods used to adjust the speed of ac motors besides varying the stator voltage frequency. Two or more discrete fixed speeds can be obtained by using magnetic switchgear to change the number of poles of the stator winding. A specially designed stator winding, with all winding leads brought out of the machine, is then necessary.

Another method is to use a *wound rotor induction motor* which has conventional wire armature windings instead of the squirrel-cage construction illustrated in Fig. 10-15. The armature winding phase leads are brought out of the machine via brushes and slip rings. The armature winding leads are terminated in a three-phase rheostat, which is then adjusted to control the rotor speed. Increasing the resistance of the three-phase rheostat causes the rotor to slow down, while decreasing the resistance causes the rotor to speed up.

QUESTIONS AND PROBLEMS

1. What is the advantage of dc motors over ac motors in industrial variable speed systems?
2. Which is greater in a dc shunt motor, armature current or field current? Would this be true of a dc series motor?
3. Explain why *decreasing* the field current in a dc shunt motor *increases* its rotational speed.
4. What is the main drawback of field control of a dc shunt motor?
5. Explain why increasing the average armature voltage to a dc shunt motor causes it to speed up.
6. What is the main drawback of armature control using a series rheostat?
7. Describe the sequence of events as a shunt motor is started with an "across-the-line" starter. Describe how the following three variables change: armature current, counter-EMF, and shaft speed.
8. Why is armature control by thyristor better than armature control by series rheostat?
9. In Fig. 13-3(b), why can't the motor reach 100% of its full rated speed?
10. Does Fig. 13-3(c) represent speed variation for different pot settings or for a fixed pot setting? Explain.
11. If the drive system of Fig. 13-3 could provide load regulation of 0%, what would the graph of Fig. 13-3(c) look like?

Questions 12–15 refer to Fig. 13-4.

12. What is the purpose of R_1 and D_1?
13. In which direction should the speed adjust pot wiper be moved to speed up the motor? Should it be moved to the right or to the left?
14. What is the purpose of D_3?
15. Do the bridge rectifier diodes have to be heavy current diodes, or can they be relatively light current diodes? Why?
16. In a dc shunt motor, will the rotation be reversed if *both* the field current and armature current are reversed? Explain.
17. In Fig. 13-5, what is the purpose of the N.C. REV and FOR contacts?
18. Explain the distinction between *switchgear* control and *electronic* control.
19. Generally speaking, when is a three-phase drive system used instead of a single-phase drive system?

Questions 20–27 refer to the three-phase drive system of Fig. 13-8.

20. Why do the manufacturers of drive systems install thyrectors across the incoming power lines?

21. Explain the purpose and operation of the field-failure relay, RFF.

22. What is the maximum number of degrees per half cycle for which any SCR is allowed to conduct? Why can't the SCRs be allowed to conduct for 180°?

23. Why is step-down transformer T_1 used? Why don't we simply design the motor starter coil to operate on 230 V ac?

24. Give a step-by-step explanation of why the motor slows down as the speed adjust pot wiper is moved up.

25. If the firing delay angle of *one* SCR is changed, do the other two SCRs also change, or are they all independent? Explain.

26. Explain how R_7 provides counter-EMF feedback to the trigger control circuit to yield improved load regulation.

27. What is the purpose of the circuit consisting of C_2, R_2, R_3, and the N.C. *M* contact? Explain how it works.

28. Explain why changing the *frequency* of the voltage applied to an ac induction motor stator winding causes its shaft speed to vary.

29. What does an inverter do?

30. What does a converter do?

Index